The
Cold War
and
Soviet
Insecurity

The Cold War and Soviet Insecurity

The Stalin Years

VOJTECH MASTNY

New York   Oxford

Oxford University Press

1996

Oxford University Press

Oxford   New York
Athens   Auckland   Bangkok   Bombay
Calcutta   Cape Town   Dar es Salaam   Delhi
Florence   Hong Kong   Istanbul   Karachi
Kuala Lumpur   Madras   Madrid   Melbourne
Mexico City   Nairobi   Paris   Singapore
Taipei   Tokyo   Toronto

and associated companies in
Berlin   Ibadan

Copyright © 1996 by Vojtech Mastny

Published by Oxford University Press, Inc.
198 Madison Avenue, New York, New York 10016
Oxford is a registered trademark of Oxford University Press

Library of Congress Cataloging-in-Publication Data
Mastny, Vojtech, 1936–
The Cold War and Soviet insecurity : the Stalin years / Vojtech Mastny.
p. cm.   Includes bibliographical references and index.
ISBN 0-19-510616-4
1. Soviet Union—Politics and government—1925–1953.
2. National security—Soviet Union.
I. Title.
DK267.M3567   1996
947.084'2—dc20   95-49341

1 3 5 7 9 8 6 4 2

Printed in the United States of America
on acid-free paper

For Kathryn

Preface

This book has been in the making for a long time. It has given its author the extraordinary experience of having lived through the events it describes and then being able to study them from the inside evidence as Soviet and other communist archives opened their doors. The price of that experience has been much writing and rewriting, sometimes with no end in sight.

I first conceived the study of the last Stalin years as a sequel to my *Russia's Road to the Cold War* during the Reagan years of the "second Cold War." Although a Soviet-certified "enemy of socialism and détente dressed in a scholarly garb,"* I found it difficult to believe that the adversary could possibly be as formidable as it appeared at a distance to so many American conservatives, including some of my colleagues at Boston University, where I was teaching at that time. A desire to find out more about the sources of Soviet insecurity gave the initial impetus to my research in U.S. libraries and archives, which the university's Center for International Relations generously supported.

In 1988–90 my leave of absence in Western Europe provided an exposure to different, though not necessarily more accurate, views of the Soviet Union, informed by its greater proximity. The Federal Institute for East European and International Studies in Cologne and the Netherlands Institute for Advanced Study in the Humanities and Social Sciences in Wassenaar, where I was successively a visiting fellow, offered a stimulating intellectual environment and excellent working conditions, thus allowing the project to move forward. Its progress would have been much faster if I had not been distracted by too many other worthwhile projects as the Cold War was coming to an end.

The slowdown turned out to be a benefit in disguise once the end of the confrontation as well as of the Soviet empire brought both new

---

* "Podstrekateli" [The Firebrands], *Krasnaia Zvezda*, February 12, 1981.

perspectives and new documents. At that memorable time, a Fulbright professorship at the University of Bonn, followed by an invitation to join the faculty of the Bologna Center of the Johns Hopkins University, enabled me to witness firsthand such historic events as the coming down of the Berlin Wall and the coming up of the Russian flag over the Kremlin, where it replaced the old Soviet one. Such impressions could not fail to leave an imprint on the book.

For my research in the Moscow archives, I received support from the Diplomatic Academy of the Russian Foreign Ministry, the American Association of Professional Schools of International Affairs, the Cold War International History Project, and the Bertelsmann publishing house. Aleksandr A. Gordievskii of the Russian Diplomatic Academy, Sergei Mironenko, now director of the Archives of the Russian Federation, and Leonid Ia. Gibianskii of the Institute of Slavic and Balkan Studies of the Russian Academy of Sciences, as well as Aleksandr Chubarian and Aleksei Filitov of its Institute of Universal History, assisted my undertaking in a variety of indispensable ways.

In Prague, my research proceeded especially thanks to Oldřich Sládek, director of the archival administration of the Czech Republic responsible for the records of the communist party central committee, and Marta Kapalínová, director of the archives and documentation division of the Foreign Ministry. In Berlin, the staff of the former archives of the East German communist party proved consistently helpful.

At the Johns Hopkins School of Advanced International Studies, Bologna Center, I benefited from a succession of dedicated and efficient research assistants from among my students: Klaus Wiegrefe, P. J. Simmons, Roger Malone, Jancsi Strohmayer, and Marc Mezey. Gail Martin was always able to get the books I needed on interlibrary loan, and Lori Cohen did a superior job editing my manuscript.

When the manuscript was nearing completion, a fellowship at the Norwegian Nobel Institute in Oslo enabled me to test my findings in discussions with its staff and guests while putting the finishing touches on my work during the spectacular Nordic summer of 1994. My thanks for that experience go especially to Geir Lundestad, Odd Arne Westad, and Anne C. Kjelling.

In the fall of 1995, I was able to do the last-minute updating of the already finished manuscript thanks to the excellent facilities of the Slavic Research Center at the University of Hokkaido, where I was spending a semester as a research fellow.

At Oxford University Press, Susan Ecklund proved a splendid copy editor, Joellyn M. Ausanka incorporated my extensive revisions with cheerful competence, and Nancy Lane managed the whole publication process with an admirable expertise.

Sapporo, Japan Vojtech Mastny
December 1995

Contents

Introduction: The Cold War as History   3

1. Stalin's Quest for Soviet Security   **11**

    Defense of the Despotism   11
    The Making of an Empire   17
    The Unwanted Cold War   23

2. The Specter of Communism
(September 1947–June 1948)   **30**

    The Birth of the Cominform   30
    The Balkan Trap   35
    The Stakes Increased   40

3. A Harvest of Blunders
(June 1948–April 1949)   **47**

    The Berlin Fiasco   47
    The Dictator's Delusions   53
    Militarization of the Cold War   58

4. Retreat and Consolidation
(April 1949–November 1949)   **63**

    Cutting the Losses   63
    The Trials of Past Errors   67
    No End to the Cold War   74

5. Resuming Advance
(November 1949–June 1950)   80

Real and Imaginary Subversion   80
The Second Front in Asia   85
The Design for Aggression   91

6. The Test of Strength
(June 1950–January 1951)   98

An Adventure Gone Wrong   98
War Expanded   104
At the Brink   110

7. On the Defensive
(January 1951–March 1952)   116

Eastern Europe under Siege   116
The Inauspicious Negotiating   121
"I Trust No One"   128

8. Fits and Starts
(March 1952–November 1952)   134

Stalin's German Illusion   134
The Despot Loses His Grip   140
Win the War at Home   145

9. The Dead End
(November 1952–March 1953)   153

The Nadir of Stalinism   153
Stalin's Last Plot   158
The Unopened Window of Opportunity   164

10. Coping with the Stalin Legacy
(March–July 1953)   171

The Prevailing Consensus   171
Beriia and the German Crisis   178
The Saving of the Soviet System   185

Conclusion: The Soviet Threat in Retrospect   191

Abbreviations Used in the Notes   199

Notes   201

Archival Sources   263

Index   267

The Cold War and Soviet Insecurity

Introduction:
The Cold War as History

Rarely has the present receded into the past more quickly than in the years that followed the end of the Cold War. Normally, people sense a continuity; even after great wars that break it do they feel an urge to come to grips with their recent experience by relating it to their new condition. No such urge has grown out of the sudden and unexpected denouement of the East-West conflict, breeding instead the bizarre notion that history itself may have ended.[1] Although a proliferation of crises soon exposed the fallacy of such a notion, the forty-year rivalry has continued to fade from memory. That a contest of such intensity and magnitude could safely be relegated to oblivion would seem too good to be true; even if it were so, the reasons why would all the more cry for an explanation.

This book would long have been finished if the disappearance of its protagonist—the Soviet Union—had not intervened while the research was in progress, thus changing the perspective, though not necessarily the substance, of the argument. What used to be an open-ended development became closed, thus allowing us to see not only how the Cold War started but also how it ended. At the same time, our ability to tell what really happened and why has vastly increased. The partial opening of Russian archives and the eagerness of Soviet witnesses to testify have added the inside dimension that was previously missing. It has become possible not only to conceive of the Cold War as history[2] but also to study it as such.

The change of perspective has altered both the questions to be asked and the answers that can be given. With the Soviet adversary out of the way, the questions are less "critical," but the answers can be more definite. Gone is the lucrative business of "Soviet estimate"; rather than to estimate, however, we may be able to actually tell what Moscow's intentions and capabilities were. Similarly, now the fight is over, the old inquiry about who started the Cold War is less interesting than the new one about why it ended the way it did. Less constrained to apportion the blame, we can be more relaxed in narrating a fascinating story.

There is, of course, more to the story than merely its fascination. The new questions are not only interesting but also important. They can affect nothing less than the permanence of the vaunted new world order, which would be flawed without taking into account the strengths and weaknesses of the one it is to replace. But was the "long peace"[3] that was the Cold War really an order, or rather disorder in disguise? It made few contemporaries comfortable, and its demise was greeted with all but universal relief. Only more recently, as the shape of the substitute order seems nowhere to be seen, could be detected a nostalgia for the one irretrievably lost.

The Western perception of a Soviet threat was at the essence of the Cold War.[4] Could it be that the threat was an empty one and its costly containment an aberration? Or was it because of this very exertion that a real threat was thwarted? Whatever the answer, are there lessons to be drawn from the experience—or those to be avoided? Perhaps there are none, and the Cold War—neither a tragedy nor a farce—is best forgotten. Yet it has been said that those who forget history are fated to relive it—not a prospect to be relished.

In looking at the Cold War as history, much of the distinction between the "orthodox" and the "revisionist" schools, with their implied prescriptions about Washington's proper conduct in its ongoing contest with Moscow, has become blurred. Nor is the "postrevisionist" attempt at taking the best of both schools topical after the change of the perspective. The very distinction between Right and Left, not to speak of that between the "hawks" and the "doves," is no longer the same in a postcommunist world where no serious military threats appear to be looming. Hence also the thrust of this book differs from that of its 1979 predecessor, *Russia's Road to the Cold War*.[5] It considers not only the Western perception and misperception of the Soviet threat but also the putative Western threat to the Soviet Union.

Of the new questions about the Cold War, none is more troubling than the one of why its end came so unexpectedly. The failure of Western scholarship to envisage what actually took place as even theoretically possible has shaken theories of international relations devised in response to the conflict as a guidance for its managers.[6] The Soviet Union's collapse without attempting to defend what were supposed to be its vital interests casts doubt upon both our ability to design credible models of political behavior and the very desirability of such an endeavor.

The course of events would seem to have vindicated the greater relevance of the historical approach, with its focus on human beings in their infinite variety and unpredictability as well as its aim to understand and explain rather than to posit and "prove." Nevertheless, although historians were better equipped by their trade to expect the unexpected, they actually did not do so better than anyone else. Nor can they plead that the future is not their business and still claim for history the mantle of a *magistra vitae.*

By taking a long-term view, at least it was possible to conceive that the "enduring balance," which ostensibly guaranteed East-West stability for the foreseable future,[7] could not in fact be more enduring than any work of man. Still, in assessing Soviet prospects in 1984, the present author only went so far as concluding that "both a revitalization of the sick system ... and its collapse because of the rulers' inability to rule, ... while neither ... highly probable, are nevertheless much more possible than they have ever been."[8]

The Cold War's unpredicted finale has by no means consigned the whole body of Western scholarship it generated to the proverbial rubbish heap of history. Yet some of the conceptual tools developed while the conflict was still in progress have since proved more durable than others. One of them has been the concept of totalitarianism, prematurely written off as a relic of primitive anticommunism, another the notion of security, which the transformation of its meaning has not made any less timely. They both figure prominently in this book.

The concept of totalitarianism, with its accent upon the incompatibility of both fascism and communism with freedom, was pronounced by Western pundits as incapable of explaining the Soviet system's evolution toward more freedom—only to be resurrected after communism's downfall by those who had personally experienced it as well suited to account for the limits of that evolution.[9] Certainly, it is against Stalin's version of the totalitarian model that its later modifications must be measured. Even in his time, however, how much did the system's performance actually conform to the model?

Although the Soviet regime's collapse shattered any totalitarian pretensions of its omnipotence, the outcome does not necessarily detract from the significance of its ideological underpinnings as long as it lasted. Unless evidence is produced to the contrary, at least its professed principles should be taken seriously rather than as mere window dressing or camouflage for something else. If the Cold War and its ending demonstrated anything, they showed that beliefs—including especially the "false consciousness" whose potency recognized none lesser than Karl Marx—can be as powerful as realities and illusions more compelling than interests. But was this the case also under Stalin or only in the later years of the regime's decadence?

Nor does hindsight detract from the key position of the leader in a system with a totalitarian bent and the crucial influence of his personality on policy. The passing—for better or for worse—of "the age of giants" in world politics has if anything made a monster like Stalin loom larger in retrospect than he did while his lesser successors were still able to convey the wrong impression of their capacity to manage and overcome his legacy. But did he manage the system he bequeathed to them any better, or was he its true gravedigger? In particular, did Stalin's failure to avert the Cold War spell the Soviet Union's ultimate undoing?

The more Moscow's power continued to unravel, the more it became evident that the Cold War would not end until the Soviet subjugation of Eastern Europe that had started it had been reversed. By then the empire Stalin had originally acquired as an indispensable safeguard of Soviet security as he understood it had already proved the fatal source of Soviet insecurity. Thus security was at the heart of Moscow's predicament because of the very changes the notion underwent during and as a result of the Cold War.

It was the reexamination and redefinition by each side of what security meant that made the end of their rivalry possible. Thus an idea was used to first explain and then change the world—just as Marx exhorted philosophers to do. To no negligible degree, the feat was due to the subtle workings of the "Helsinki process" of the Conference on Security and Cooperation in Europe, which expanded security to include the "human rights"—a shorthand for those of its mainly nonmilitary attributes that make governments and their subjects feel more secure or insecure internally.[10] This has always been a greater problem for regimes aspiring at total control than for pluralistic ones.

The notion of security is what links Soviet domestic and foreign policy without necessarily ascribing primacy to either. Their relationship expressed not only the military preoccupations that kept the Cold War going but also the nonmilitary ones that ultimately brought it to an end. Since the system Stalin founded could never be sufficiently secure either internally or externally, how deeply did its guardians feel the Western menace that they constantly invoked? The intricate interplay of mutual threat perceptions is a recurrent theme in the analysis that follows.

The consequences of the Soviet Union's domestic system for its international conduct were not unequivocal. To be sure, Stalin was not —to paraphrase A. J. P. Taylor's view of Hitler—merely a traditional though no doubt wicked Russian statesman.[11] Nor did he preside over a system "not distinguished from non-Communist ones by any particular characteristics."[12] But did the peculiarities of his regime, like Hitler's, inevitably make it pose a threat to the outside world?

Depending upon the circumstances, not only an aggressive but also an accommodating foreign posture can be a product of internal insecurity, just as domestic repression—as well as its relaxation—can result from perceived external threat. In deciding which way the Soviet policy would turn, the relative importance at any given time of political and military considerations was crucial when the East-West rivalry stimulated an uprecedented growth of armaments while also generating unaccustomed inhibitions against their being put to use. How well did the West understand Moscow's political and military priorities?

The end of the Stalin era coincided with the beginning of the nuclear age, whose policy implications were generally regarded to be great but otherwise not very clear. The qualitative distinction between the nuclear and conventional weaponry was not yet so sharp at a time when the awareness of potentially catastrophic consequences of any East-West military clash was only gradually developing. Could it be said, then, that it was the fear of nuclear weapons, rather than other considerations, that kept the Cold War cold?

Since Stalin's last years have so far been the most obscure in Soviet history, the availability of new sources ought to provide better answers to these and other questions. But is the opening of the archives conducive to altering drastically the picture that could be drawn even without them? The example of Nazi Germany, whose aims and policies were grasped accurately enough from publicly available evidence before its secret files became known, is sobering. Although secret documents may not be indispensable to getting the right answers, however, once the archives have been opened the answers cannot be convincing without taking into account what can be seen there.

This is what the present study aims at doing as best as it could. The Russian archives are still far from open but are merely opening—and sometimes again closing—while their future remains uncertain.[13] The resulting lack of complete evidence is not necessarily an insuperable obstacle; making the best of the evidence at hand matters more than chasing after all the missing pieces that might never be found or may not have existed in the first place. Thus the records of top policy-making still kept hidden in the Kremlin "Presidential" archives, presumably because of their abiding operational significance, are unlikely to be as illuminating as those of more normal governments, especially for the Stalin period. Not only did the despot have little taste for paperwork, dispensing with a formal process of proposing, discussing, and hammering out policies, but his successors' complicity in his crimes has also made many an embarrassing folder disappear a long time ago.

Much of Stalin's policy will probably forever have to be extrapolated from other sources, such as the extensively available files of the party central committee or the foreign ministry. The former are important to the extent that ideology was important, although they may for that very reason be less informative than the latter. As the Soviet Union rose to a superpower after World War II, so did the relative weight of its governmental institutions, including the foreign ministry under Stalin's utterly devoted aide, Viacheslav M. Molotov. In view of the sheer volume of business that had to be transacted, have subordinates at this second level, where policies were prepared and their implementation supervised, been able to influence the decisions that the aging dictator could not or would not effectively handle within his short working day?

Regrettably for the study of the militaristic police state, the massive files of the Soviet armed forces and secret services remain almost entirely out of reach or, worse, susceptible to being rifled in an effort to pull out bits and pieces for profit and other sordid reasons.[14] Such disconnected documents might give an air of exaggerated importance to something that is merely sensational but leave unanswered the truly important questions: Did spying and counterspying, or the dictates of the military machine, change the course of policy at any time? How much were those responsible for it able to find out about the secrets of their adversaries, and what, if any, conclusions did they draw from their knowledge and their ignorance? Did the Soviet estimates mirror the American ones, or were they very different?

The pertinent files not only in Moscow but also in Washington are still not sufficiently accessible to allow for conclusive answers, the prodigious U.S. declassification efforts notwithstanding. Meanwhile, at least partial answers may be found by substituting data from archives in Eastern Europe, especially of those countries whose governments used to toe the Soviet line with particular zeal, such as the former Czechoslovakia or East Germany, as well as from the records of Western communist parties responsive to Moscow's direction. Yet the stereotype of the foreign communists' dependence upon it may be deceptive. Even under Stalin's watchful eye, were they able—provided that they were willing—to pursue their own agendas by design or by default and consequently shape Soviet policies in unexpected ways, thus making "the tail wag the dog"?

Diverse sources are also required to complement and verify the many a dubious allegation arising from the flood of Soviet memoirs—by those who have something important to say and those who have not. In a country lacking the British or French tradition of the genre, their testimonies are to be treated with caution, coming as they often do from witnesses whom a lifetime exposure to deception and prevarication may have rendered incapable of distinguishing the truth from the untruth even if they wanted to.[15] Still, their accounts have the value of immediacy and a proximity to their subject that cannot be ignored.

All things considered, the new inside evidence can make enough difference, though not by itself. The documents cannot be left to "speak by themselves," for they seldom do so with any clear voice or a single one. They can speak with many voices—those of their interpreters, if asked the right questions. Yet the questions a historian asks are not always those an archivist can answer or even understand, nor are those that interest a Western scholar usually the same that preoccupy most Russians. Moreover, in the piles of seemingly worthless paper, getting at what really matters can be an arduous and time-consuming task.

The experience, however, is rewarding by providing the authentic flavor of Soviet thinking that permeates the once confidential files. Perhaps the greatest surprise so far to have come out of the Russian archives is that there was no surprise: the thinking of the insiders conformed substantially to what Moscow was publicly saying. Some of the most secret documents could have been published in *Pravda* without anybody's noticing. There was no double bookkeeping; it was the single Marxist-Leninist one whose defects spelled the bankruptcy of the Soviet enterprise in the long run.

A growing amount of impressive work by non-Russian and Russian authors keeps rolling off the presses. Typically, in rejecting the old dogma the latter have been reverting to what used to be the state of historical art before the imposed Soviet orthodoxy—nineteenth-century German positivism, with its worship of the facts for their own sake. This often tends to obscure the forest for the trees, but at least there is now enough timber from which the building of an edifice may be started.

It is not too early to attempt a synthesis regardless of the additional information that will continue coming in. Its accumulation is already sufficient to pause and reflect how the pieces could be related to a whole, thus explaining the workings of Soviet insecurity under Stalin and their consequences for the Cold War. More evidence is apt to bring diminishing results. "Don't look too deep," Emerson warned historians, "the real reasons are usually quite simple." Yet this makes them all the more interesting. Why, if they were indeed so simple, did the Soviet threat seem so inscrutable? This is no idle question to be asked about the superpowers that were capable of plunging humankind into a terminal war. Even if the scare never materialized we can do worse than try to get the answers right.

1

Stalin's Quest
for Soviet Security

Created by a minority coup rather than by exertion of popular will, the Soviet state was always intrinsically insecure. It was the product of momentary confusion during an avoidable war rather than of any insoluble internal crisis of the tsarist regime it replaced. Its founders did not consider it secure unless the revolution they promoted to keep themselves in power would triumph abroad as well. Although they tried their best to spread it, they did not succeed; disunity among their enemies nevertheless allowed their state to survive and grow. All the same, as taught by Marxist doctrine, they continued to insist that the outside world remained implacably hostile. Whether this was true or not, their constant perception of a threat prevented Soviet leaders from ever feeling sufficiently secure. This made them different from other leaders, and Stalin more different than most.

Defense of the Despotism

The possible pathological bent of Stalin's mind is an intriguing but not particularly illuminating proposition;[1] more pertinent to policy are the experiences, ideas, and events that shaped his extraordinary behavior. A Russianized Georgian, he was but a secondary figure in the revolution. By the time he began his ascendancy after Lenin's death in 1924, any serious military threat to the Soviet state had already passed, if it had ever existed in the first place. Rather than by defending the revolution against foreign enemies, Stalin rose to power by excelling in the deadly intrigue endemic in the Bolshevik party.

The party's expert on nationalism, Stalin lacked the internationalist outlook of the Bolshevik Old Guard and had little firsthand familiarity with foreign countries. In 1920 he played an undistinguished role in the Red Army's abortive revolutionary crusade against Poland, whose workers confounded Bolshevik calculations by defending their country alongside their class enemies. He reversed the reasoning of his archrival Leon Trotskii by positing that the revolution's global victory presupposed its consolidation in the Soviet Union first. Stalin insisted that all internal enemies must be destroyed before the foreign ones could be defeated. In giving priority to internal rather than external security, he identified the security of the state with that of his personal power.

Marxism in its Leninist variety was congenial to Stalin. Not only did it complement Marx's dismal notion of class struggle with Lenin's sinister prescription that the ends justify the means; the manipulative Marxian dialectic, with its blurring of the difference between truth and untruth, offered a convenient method to identify the enemy at will. The doctrine that pretended to explain everything made it possible to justify arbitrary acts by alleged historical necessity while giving them a pseudoscientific guarantee of success.

Despite Stalin's ideological dedication, revolution was for him a means to power rather than a goal in itself. This distinction was at the heart of his dispute with Trotskii, the champion of a "permanent revolution," whom he bested not so much by the force of his arguments as by his ability to slander, isolate, exile, and finally murder the man whom he absolutely detested. The fuzzy heresy of "Trotskyism" could henceforth apply in the Stalinist parlance to any actual or intended revolutionary activity that the despot did not specifically authorize. The Comintern, originally founded as the general staff of the world revolution, degenerated under his guidance into a secondary instrument of his foreign policy.[2]

Although Stalin was not averse to using revolutionary movements abroad for whatever they might be worth, they were of little value to him unless he could control them. He abhorred spontaneity, and his suspiciousness bordered on paranoia. He demanded from his foreign disciples unconditional loyalty and slavish obedience even if they were willing to follow him voluntarily. Habitually skeptical about their ability to muster enough popular support, he was prone to underestimate their chances of succeeding on their own.

In 1927 Stalin directed the Chinese communists to ally themselves with their nationalist rivals against their purported common capitalist enemies, only to see the nationalists turn against the communists and crush them.[3] Although the experience did not destroy the Chinese communists' loyalty to Moscow, it made them more reliant on their own devices. This did not make Stalin any more comfortable with the Asian and other non-European revolutionaries who derived their support from the peasantry, a class he despised. Although he was in many ways the archetypal "Oriental despot,"[4] his horizon was essentially European.

Moscow's ability to complement conventional diplomacy with manipulation of fanatically dedicated communist followers impressed contemporaries more than it deserved. For all its aggressive rhetoric, the Soviet Union remained throughout the interwar period more a nuisance and a disgrace than a threat to the rest of the world. Crippled by its devastating revolution and self-imposed isolation, it remained a second-rate power, handicapped by the discrepancy between its ambitious goals and limited means.

Stalin, a latecomer to foreign policy, did not take full charge of it until after destroying all his rivals in the early 1930s.[5] But his lack of experience in international affairs was less problematic than his unwillingness to admit mistakes and learn from them. He typically blamed his own miscalculations on scapegoats while raising the alarm about the country's security. His Chinese fiasco, compounded by the Comintern's additional setbacks in Europe, coincided with his anti-Trotskii campaign and his cry that war was imminent.

While the war scare Stalin raised was deliberately exaggerated, it was not totally disingenuous.[6] He was inordinately concerned about war. After all, Lenin had taught that as long as the capitalist system existed, its inherent contradictions made wars among its rival states inevitable, as well as desirable to help precipitate their revolutionary overthrow.[7] What this meant in terms of a possible threat to Soviet security, however, was not entirely clear.[8] Stalin undertook the nation's greatest peacetime military buildup in 1928–35, when it was least threatened by foreign enemies.[9]

Acutely sensitive to the dynamics of power and weakness, the dictator gauged his country's relative standing in the world by pondering trends in the "correlation of forces"—the sum total of military and other attributes of power determining at any given time the relationship between the hostile forces of socialism and capitalism. The construct had the advantage of sensitivity to conflicts of interest and of putting momentary developments into long-term perspective; its disadvantage was in overestimating the severity of the conflicts and distorting the perspective by reducing international politics to a zero-sum game. It did not allow Stalin to perceive accurately the danger of fascism and the rising German threat.

While central to the Soviet view of Europe, Germany was the cause of some of Stalin's most egregious misjudgments.[10] It had been the graveyard of Soviet hopes ever since the Bolsheviks had expected the world revolution to start there—in the country of Marx and Engels, where the contradictions of capitalism and the class consciousness of the proletariat were presumably at their highest—only to see the German workers let them down. They had better experiences with the German capitalists and militarists, with whom, after World War I, they had maintained a mutually satisfactory collaboration for a decade on the grounds of their common rejection of the hated Versailles system.[11]

When the system was undermined by the Great Depression, Stalin proved susceptible to the proposition that the fascist ascendancy in Germany was the beginning of capitalism's end[12]—only to witness the emergence of a regime that made any previous capitalist threat pale by comparison. The unexpected turn of events did not diminish Stalin's exaggerated respect for things German, but rather compounded it by his sneaking admiration for Hitler, who utilized congenial methods so effectively. For Soviet, though not for German policy, the two dictators' affinity was more consequential than their ideological hostility.[13]

Hitler's Blood Purge of the Nazi radicals in 1934 provided a model for Stalin's own ensuing Great Purge of the Bolshevik Old Guard. The creation in the process of the self-perpetuating machinery of terror that became the hallmark of Stalinism ensured a "permanent purge" in a neverending search for internal security.[14] Nor did the purge, which may have originated more in the dynamics of the system the despot tried to manage than in his design,[15] add to the country's external security. In fact, it shattered the nation, including its military establishment, at the very time they needed strengthening to better withstand the impending German assault.

This consideration may not have been crucial to Stalin if he believed that he could spare his country the test of war. His turn toward "collective" security policy in 1934, which coincided with the onset of the purge, implied such a belief, aiming to deflect the German threat by diplomatic and other political means. Far from being truly collective, the policy sought to shift onto the Western powers—which, unlike the Soviet Union, were close to the probable scene of conflict—the main risk of resisting the aggressor.[16] By simultaneously directing communists to form "popular fronts" with other left-wing opponents of fascism, Moscow sought to prod the reluctant French and British governments into action.

Marking Stalin's final repudiation of the bankrupt formula of security by revolution, the turnabout entailed an unaccustomed risk on his part. He could capitalize on the respect and even admiration that the debacle of old-fashioned capitalism inspired in parts of Western society for the ostensible vitality of his brand of socialism. But bringing into his calculations this mixed assortment of idealists, adventurers, and fools of mainly bourgeois origin and individualistic bent presupposed a subtlety he had not demonstrated before. He was able to recruit from among this group some of his most valuable spies,[17] but otherwise the results were mostly disappointing.

The Soviet intervention in the Spanish civil war of 1936–39, in which Stalin engaged sympathetic Western leftists together with communists in the Comintern-run "international brigades," tested the efficacy of the collective security and popular front formulas, and found them wanting.[18] A quixotic attempt to encourage the West's resistance to fascism while providing his Spanish clients enough aid to fight but not enough to win, it was doomed to failure anyway.[19] Nor were its prospects for success enhanced by Stalin's applying his homegrown methods to suppress real or imaginary "Trotskyites" in a distant country of its own revolutionary traditions.

Even before the Spanish fiasco, Stalin's quest for collective security, reluctantly undertaken for want of other options, had run its course. Always suspicious of the democracies, whose values he despised, he never ruled out accommodation with the congenial Nazi dictatorship. His dissolution in 1938 of Poland's communist party signaled his disposition to sacrifice the country to rapprochement with Germany. But Stalin could not make a deal as long as Hitler was able to gain more from continued Western appeasement; only its reversal in the spring of 1939 made Berlin amenable to Moscow's overtures. The dismissal in May of Soviet Foreign Commissar Maksim Litvinov, prominently identified with the abortive collective security alternative, and his replacement with Stalin's right-hand man Viacheslav M. Molotov, led in another three months to the Nazi-Soviet pact.[20]

A landmark of the century, the pact of August 25, 1939, was a Soviet rather than a German accomplishment. It fell short of Hitler's expectations by failing to dissuade the Western powers from responding to his invasion of Poland a week later by a declaration of war; it gratified Stalin by precipitating their embroilment with Germany while keeping his own country out of the fray. Here was Lenin's classic war between imperialists from which the world revolution ought to benefit. By this time, however, the Soviet Union had not only postponed its pursuit of revolution indefinitely but had itself become an imperialist power.

The Soviet state had not initially been imperialist in the same sense as its tsarist predecessor. Denouncing imperialism as a deplorable product of capitalism, after World War I the Bolsheviks had tried to export their revolution by invoking proletarian internationalism. Condemning nationalism, especially "Great Russian chauvinism," they had expected the neighboring peoples to follow their example, and then reunite with Russia voluntarily in a supranational commonwealth. Even after this failed to materialize, leaving the Soviet Union smaller than tsarist Russia, the recovery of the lost lands had not commanded immediate operational priority.

Only in response to the rising German threat did Stalin develop the concept of an empire as an operational goal related to security—his principal contribution to the theory and practice of Soviet foreign policy. During his duplicitous negotiations with the British and the French in the summer of 1939, which paralleled his progressing rapprochement with Germany,[21] he demanded the right to counter not only direct but also "indirect" aggression, which meant intervention in neighboring states whenever, in his opinion, their safety might be endangered. Rebuffed by the West, Stalin got what he wanted from Hitler.

Rather than a temporary expedient adopted unwillingly to buy time, the pact with the Nazi chief was for Stalin the foundation of an empire intended to give the Soviet state lasting security. More important than the signatories' public nonaggression pledges was their secret agreement providing for Moscow's outright annexation of eastern Poland and an additional Soviet sphere of influence in Eastern Europe. In 1940, this agreement allowed for the absorption by the Soviet Union of the Baltic States and other territories as well, with the extension there of the abominations of Stalinism. The unsavory deal, whose very existence the Soviet regime tried to deny almost till its end,[22] epitomized the evil of the empire that lasted until it collapsed under the weight of its sins half a century later.

Stalin cultivated the liaison with Nazi Germany by seeking from it assurances of goodwill and offering his own in return. In November 1940 he sent Molotov to Berlin to request concessions that would allow still further expansion of the Soviet sphere of influence, particularly towards the north and south.[23] At the same time, he had the Comintern steer European communists, most of whom retained their loyalty to him despite his deal with their sworn enemy, away from contesting German primacy in the parts of the continent Hitler regarded as his own. In the shady dealings between Soviet secret services and their Nazi counterparts, survivors of the Comintern's international army from Spain interned in camps in France played ambivalent roles.

The Soviet leader did not behave as if he thought his rapprochement with Germany was bound to fail. Although he had defensive installations constructed in the strategic territories he had acquired, he did not do his best to prepare for the possibility of their having to be used. He tried all the more vigorously to subdue the local populations by means of mass deportations and other instruments of terror, thus giving further proof of how much he regarded security an internal rather than an external matter.

The unexpected Nazi invasion in July 1941 exposed another of Stalin's miscalculations about Germany—the most disastrous but not the last. It showed that the quest for security by empire, far from providing real security, bred a false feeling of it, thus making the blow that came even worse than it need have been. Yet the dictator proved notably unwilling to discard the concept to which he had become so deeply committed; instead, the catastrophe generated in the Soviet mind obsession with surprise attack and worship of military might. In outlining his vision of the future to visiting British foreign secretary Anthony Eden in December 1941, Stalin again envisaged a Europe divided into spheres of influence—only this time with Great Britain as the partner.[24] Yet, contrary to hindsight appearances, there was no straight road from that vision to the Continent's eventual division during the Cold War.

The Making of an Empire

Having failed in 1942 to obtain from the British an advance commitment to the division of Europe because of American opposition, Stalin
never tried again. As long as the price of victory remained uncertain, it
was better for him not to tie his hands lest he ask for too much, or else
too little. His minimum aims, without which the war would have
presumably not been worth winning, included the preservation of the
territorial gains made as a result of his collusion with Hitler, the
prevention of any regional combinations in Eastern Europe capable of
resisting Soviet wishes, and the establishment throughout the Continent of communist centers of power to help implement in different
countries whatever policies Stalin would eventually decide to pursue
there. Any further aims were flexible rather than fixed.

Once the Comintern had reactivated its foreign disciples after the
German attack, its directives did not conceptually differentiate between
Eastern and Western Europe. They urged communists everywhere to
form "national fronts," distinguished from the previous "popular" ones
by positing collaboration with not only left-wing but any enemies of
fascism.[25] The collaboration did not rule out, but neither did it postu-
late, eventual seizure of power by the communists; as Stalin explained
to the French party leader, Maurice Thorez, their most important task
was to expand their base of support by finding allies.[26] Other things
being equal, they should have best been able to maintain the balance of
power in their countries, which could then be manipulated from
Moscow.

Not only were things not equal, however, but the wartime conditions often made it difficult to ascertain at a distance just how unequal
they were. Contrary to the impression informed by later developments,
the numbers of Soviet intelligence personnel in those days were,
according to one of its veterans, "simply laughable."[27] In some places,
such as occupied Czechoslovakia, the police managed to infiltrate the
communist underground to the very top.[28] In Poland Marceli Nowotko,
one of the emissaries parachuted there from Moscow to restore the
party that Stalin had gratuitously disbanded before, was murdered by
another member of the group; his successor, Paweł Finder, in turn died
by German hand after having been arrested under suspicious circumstances.[29] Being left largely on their own did not necessarily make the
communists disloyal; the Yugoslav ones, in particular, were eager to
oblige Stalin when able to find out what his wishes were.[30] Yet their
inability to do so made them suspect in his eyes.

Once Soviet postwar planning began in earnest after the Red Army's
decisive victory at Stalingrad in February 1943, Moscow's relations with
foreign communists were first on the agenda. The dissolution three
months later of the Comintern came as a surprise and a shock to them;

other people either wanted to believe that Stalin had finally given up communism, or suspected a ruse. Yet the official announcement, which suggested that the wartime developments had accentuated differences in various countries, thus necessitating more flexible communist policies, offered a fair indication of what really happened.[31]

As Stalin explained to the Comintern's top officials, its liquidation was needed to make room for launching special committees to organize and indoctrinate prisoners of war from enemy countries, particularly Germany, Italy, Hungary, and Romania.[32] Once established, the committees would proceed to signal to the potential opposition back home that the Soviet Union might be amenable to terminating the war by a negotiated peace short of total victory.[33] While keeping this option open pending clarification of the further course of the war and its exact outcome, the creation—for the first time—of an international department in the Soviet party central committee allowed for more flexible long-term management of foreign communists. Under the expert supervision of the Bulgarian former general secretary of the Comintern Georgii Dimitrov, who was thoroughly subservient to Stalin, the new organization could better assist particular Soviet goals in different countries regardless of whether a country qualified as Hitler's victim or his accomplice.

In anticipation of the great Allied conferences scheduled to meet in late 1943 to clarify the respective war aims, three commissions on postwar planning were established in Moscow in the fall of that year. The first, presided over by former Soviet ambassador to London Ivan M. Maiskii, was to elaborate proposals for the treatment of Germany and its allies, especially for extracting reparations from them. Another, headed by former foreign commissar Litvinov, generated ideas about the overall postwar order and the peace treaties that would underwrite it. The third commission, led by politburo member Marshal Klimentii Ye. Voroshilov, tackled military matters, including the conditions of the armistice.[34]

By the time the commissions started their work in early 1944, the Western representatives at the Allied conferences in Moscow and especially Teheran had given Stalin the impression that they were more disposed to condone his political aspirations than he had expected but less ready to assist Soviet military advance by opening the promised "second front" in Europe.[35] While its launching six months after Teheran removed this particular uncertainty from his mind, reducing the prisoner-of-war committees from potential instruments of a negotiated peace into mere propaganda tools, enough other uncertainties remained to make Moscow's quest for future European order more tentative than generally believed.

Stalin's underlings, including the reputedly moderate and pro-Western Maiskii and Litvinov, sometimes produced more radical and confrontational proposals than he subsequently allowed to be pursued. In northern Europe, in particular, military and foreign commissariat officials wanted to go farther than he considered prudent. In early 1944 they envisaged not only Eastern Europe and most of Germany but also half of Norway as the Red Army's future area of occupation—an idea Stalin ruled out after the Western Allies had promised him more of German territory as a Soviet occupation zone than he previously had reason to expect.[36]

Subordinates also authored unfulfilled Soviet plans for forcing Norway to yield control of the Svalbard (Spitsbergen) archipelago and Bear Island,[37] as well as for establishing bases on the Danish Bornholm Island to control access to the Baltic Sea.[38] Even though these projects did not materialize, they showed that in the open-ended and increasingly complex wartime situation Soviet policy was being shaped not solely by Stalin's vision but also by an interplay of often contradictory impulses from different officials and agencies, evolving haphazardly rather than by design.

Nowhere was the resulting incoherence more evident than in the Soviet plans for Germany. Having been burned before, Stalin seemed peculiarly reluctant to commit himself to any definite course in dealing with so unpredictable a country lest it again misfire. Trying to find out from his Western partners what they proposed to do, he at first favored but then became wary of its partition or dismemberment,[39] which he feared might breed a lasting desire for revenge. While he preferred weakening Germany by truncating its territory and restoring an independent Austria, he kept delaying a decision.

While Stalin provided no authoritative guidance, Maiskii recommended taking "out of Germany whatever can be taken out except for the 'starvation minimum'" and ensuring that for its people "work be tantamount to forced labor."[40] More realistically, A. A. Smirnov and Vladimir Semenov, of the foreign commissariat's German division, envisaged collaboration with the other occupation powers as a necessary precondition for access to the country's resources.[41] Concerning its political future, the German communist exiles in Moscow looked forward to a "militant democracy"—their term for a one-party state; before the war ended, however, their project to that effect had been discarded without a substitute.[42]

Poland, rather than Germany, was for Stalin the key to Soviet security. He sought a postwar settlement that would depend less on the solution of the tricky German question, which required Western cooperation, than on the seemingly more straightforward Polish one, which he could hope to be better able to achieve on his own. He wanted a subservient, though not necessarily communist, Poland—or, as Moscow's 1943 directive to party secretary Finder called it, "democratic" rather than "socialist"[43]—but had difficulty finding "any Poles one could talk to."[44] To ensure their subservience, he insisted on territorial demands which he knew no Polish government could accept without appearing to its people as a Soviet puppet.

Finder's successor, Władysław Gomułka, knew better than Stalin that communists were unacceptable as partners to any but marginal groups within the country's overwhelmingly anti-Soviet political spectrum. In the spring of 1944 he nevertheless responded to Moscow's pressure by trying to woo potential noncommunist sympathizers from among those loyal to the London government-in-exile, but predictably failed.[45] Afterward Stalin put the communists and their fellow travelers in power on his army's bayonets, still hoping that more respectable individuals would jump on the bandwagon to boost the regime's international credibility. Although hardly any did, he remained confident that "the alliance will not break up over Poland."[46]

Elsewhere in Europe, Moscow's stakes were not so high nor were they necessarily fixed in advance; to a large extent they depended on the vicissitudes of the war. In the Balkans—unlike for tsarist Russia, not an area of primary strategic importance for the Soviet Union—the communists' ascendancy in the Yugoslav and Greek resistance movements embarrassed Stalin because of their revolutionary proclivities. After British Prime Minister Winston Churchill in October 1944 proposed to him to divide influence in the peninsula by percentages, signaling a willingness to accept more Soviet influence than Moscow had so far been aiming at, Stalin concurred.[47]

Although the exact meaning of the deal was never clarified, the Soviet leader showed by his behavior that he regarded Greece, though not Yugoslavia, as London's bailiwick. As long as the British appeared to be accommodating, he encouraged Yugoslav party chief Josip Broz Tito to proceed with his pet idea of a communist-led confederation of Balkan States. But when London, in early 1945, raised objections, Stalin ordered the project shelved.[48] Opportunistic rather than reckless, he was not impervious to pressure—his main difference from Hitler.

Even before the Red Army's advance into Eastern Europe was an accomplished fact, its other feats had predisposed the West to accept a future Soviet sphere of influence there informally, if not formally. No agreement was ever reached about the limits of the sphere or the nature of the influence, thus allowing each side to interpret them according to its preferences. Moscow did not consider the spheres as coequal but presumed that in the eastern part of the Continent its influence would be paramount while the western one would be open to political competition.

Rather than a divided Europe, Stalin could confidently envisage it undivided but so weak and fragmented that none of its states would be capable of resisting his will. He left no doubt that he considered France's demise as a great power to be irreversible and sought to thwart British efforts at supranational integration of its smaller neighbors. The political collapse of Europe brought about by Germany's aggression opened up for the Soviet Union the enticing prospect of its being able to act after the war as the arbiter of the continent.

In Stalin's scheme of things, military seizure of territory for political gain was less crucial than it has usually been assumed. In their readiness to sanction in advance conquests he had not yet made, the Western powers mistook his ability to use force for a determination to use it. This may not have made a difference to the fate of Poland but could have to that of the less strategically important Hungary, where the Red Army moved in only when the British and Americans, for their own military reasons, chose not to act on Stalin's proddings to land in the upper Adriatic and advance toward Hungary from there.[49] Like them, he preferred to use his armies for bringing the war closer to an end in ways he considered the most effective, rather than delaying it on political grounds. In 1944, Finland possibly escaped occupation by Soviet troops because they were needed to press on toward Berlin.[50]

Stalin made the attainment of his preferred postwar order dependent less on the vagaries of war than on the emergence after its conclusion of a congenial international environment. He tried to accomplish what he wanted with rather than against his powerful Western allies, whose support, or at least acquiescence, he deemed indispensable for achieving the kind of security he craved. "It was to our advantage to preserve the alliance," Molotov later reminisced,[51] and his underlings drafted their policy papers accordingly. Ambassador to Washington Andrei A. Gromyko predicted that since the United States "would be interested in economic and political cooperation with the Soviet Union," their aims would be largely compatible.[52] Nowhere beyond what Moscow considered the Soviet borders did its policies foresee the establishment of communist regimes.

In occupied Hungary the Soviet authorities introduced four political parties of their choice; they included the communist party as their agency but directed its leaders to respect the setup without any time limitations.[53] Even in Bulgaria—the first Eastern European country where already in 1944 the communists had taken control of the government—Stalin mused that he would not have minded fascists as an opposition in the parliament, presumably because they were particularly corruptible.[54] Not opposed in principle to Tito's Balkan unification scheme, he at the same time condoned the resistance by Bulgaria's chief negotiator, Traicho Kostov, against the variant that would have amounted to his country's absorption into a greater Yugoslavia.[55] All things considered, the Soviet ruler preferred Eastern Europe divided and pliable rather than communist.

Of all the countries in the region, Czechoslovakia best suited Stalin's purposes as the potential "main conduit of our influence in central and southeastern Europe."[56] Historically Russophile, even before the war it had a substantial communist electorate, and now in London, a government-in-exile not only capable of getting along with the communists but also intent on a special relationship with Moscow, meaning subordination in foreign policy. In 1943 the Czechoslovak president, Edvard Beneš, took it upon himself to conclude with the Soviet Union the first of the "friendship" treaties that would eventually tie to its fold all the Eastern European states comprising its empire.[57] Moscow dropped hints that the model might be suitable for France and Italy as well.[58]

In shaping the postwar order, the February 1945 Yalta conference was not as crucial as its later notoriety suggested. No deal about the division of Europe into spheres of influence was struck there; more important was its effect on the Far East, where Stalin got an advance on his promise to help in the war against Japan. At U.S. initiative and much to his gratification, the Kurile Islands were given to the Soviet Union.[59] As far as Europe is concerned, what went wrong at Yalta was the way how the participants badly misjudged each other's intentions.

The Western representatives could not believe that Stalin would find it in his interest to impose unrepresentative governments in Poland and elsewhere in Eastern Europe, whereas he could not imagine how they could possibly expect him to do anything else. Directing Molotov to sign the American-drafted "Declaration of Liberated Europe," which affirmed its peoples' right to self-determination, he did not "mind signing it. We can fulfill it in our own way. What matters is the correlation of forces."[60]

Despite their growing apprehension about Soviet intentions, the Western Allies did little to discourage Stalin from thinking that he could take their acquiescence for granted. The outcome of the August 1945 summit conference at Potsdam was seen in Moscow as amounting to their acknowledgment of having "lost eastern Europe and the Balkans."[61] By the end of the year they accepted, however unhappily, the Soviet faits accomplis in the region after changes had been made in the local governments that attached "some fig leaves of democratic procedure to hide the nakedness of Stalinist dictatorship."[62] Stalin's quest for security by empire could have hardly been more successful.

Potsdam confirmed the principle of joint responsibility for the administration of occupied Germany—for Moscow the necessary prerequisite for a satisfactory solution of the German question. A peace conference that would render the defeated enemy permanently harmless was generally expected to follow soon. The Council of Foreign Ministers, inaugurated in London in September 1945, was cherished by the Soviet Union as a great-power directorate which, by enabling it to share in all important international decisions, would help secure its "foreign policy gains made during the War."[63] Yet the failure of the other ministers to entertain Molotov's expanding desiderata should have been taken in Moscow as a warning; instead it was received there with surprise and indignation as evidence of unwarranted Western hostility.

The Unwanted Cold War

The end of a major war tends to breed the mirage of absolute security,[64] susceptibility to which Stalin had already shown in the excesses he had committed in trying to make himself more secure internally. The victory in World War II promised his country more security than it had ever had, yet not enough for him. His insatiable craving for it was the root cause of the growing East-West tension, regardless of his and his Western partners' desire for manageable, if not necessarily cordial, relations. The forthcoming Cold War was both unintended and unexpected; it was predetermined all the same.

None other than Litvinov, the Soviet planner of the postwar order, expressed through his indiscretions to visiting Westerners the opinion that his government's drive for security without clearly defined limits was the primary cause of trouble, while the West's failure to resist it early and firmly enough was an important secondary cause.[65] But Molotov, Stalin's chief diplomat, saw nothing wrong. He retrospectively explained the origins and nature of the quarrel as follows: "All this simply happened because we were advancing. They [the Western powers], of course, hardened against us, and we had to firm up what we had conquered.... Everywhere it was necessary to make order, suppress the capitalist ways. That's what the 'Cold War' was about."[66] Molotov agreed with Litvinov in his judgment that the conflict had arisen from Soviet actions, but insisted that Stalin "knew the limits."

This may have been true in the sense of Stalin's knowing that there were limits somewhere but not exactly where. When in 1945 Marshal Semen Budennyi, the dubious hero of the Red Army's unsuccessful invasion of Poland after World War I, reportedly egged him on that he should have let Soviet troops keep marching on into western Europe, he demurred, asking the pertinent question of who would then feed all its people.[67] Nor did Stalin tolerate the attempts by his Yugoslav disciples to grab parts of Italy and Austria, for which he would have been held accountable by the West. For the same reason, he discouraged their Greek comrades from resuming the civil war against the pro-British Athens government.[68]

Yet in other places the postvictory euphoria made the Soviet ruler less resistant to the imperialistic temptations he had earlier not allowed his subordinates to translate into action. Perhaps goaded by his sycophantic security chief and fellow Georgian Lavrentii P. Beriia, he attempted to intimidate Turkey to cede to the Soviet Union territory in the Caucasus and military bases in the Black Sea straits.[69] Delaying the evacuation of Soviet troops from Iran, he sought to establish in its part of Azerbaijan a client regime to gain dominant influence in this other historic area of Russian expansion.[70] And he cast his eyes beyond that area by pursuing Litvinov's original idea of claiming a share in the administration of Italy's former African colonies.[71]

Rebuffed and then abandoned, such probes of the limits of Western tolerance bred more tension—as had similar conduct by Stalin's tsarist predecessors—but not necessarily an irreconcilable conflict. The roots of that conflict were domestic and ideological. Rather than sharing with his people relief at the end of their wartime suffering, Stalin saw a threat to his tyranny in their expectations of a better life. He needed to justify it by convincing them that they remained surrounded by enemies. The propaganda campaign launched to that effect in 1946 ensured that eventually they would be. It gave the anonymous Soviet citizen quoted in the confidential survey of the popular mood reason to wonder whether "peace can be preserved under communism."[72]

Germany was not the cause of the mounting East-West confrontation. Having rather been the cause of the Soviet Union's unprecedented wartime collaboration with the West, it nevertheless became their main battlefield once their relations turned sour for other reasons. The conflict in Germany developed not so much because of what the Western powers did or failed to do as because of the Soviet difficulty in implementing the contradictory policies that Stalin let his different agents pursue there.[73] His professed desire to "gain sympathy in Germany," thus turning it into "a Soviet ally,"[74] squared poorly with the atrocities by his soldiery against its populace and the pillaging of its patrimony by his reparation gangs.

Although Stalin as early as 1945 feared that Germany might be divided,[75] he proceeded on the assumption that its division could be prevented. In trying to secure the decisive influence not only in the Soviet but also in the Western zones, he had launched in Berlin—much like in Hungary the year before—a limited number of political parties of his choice, as well as rudimentary organs of central administration, all designed to gradually extend their operation throughout the whole country.[76] He expressed to the German communists his wish to make it "democratic," namely, ruled by a government that they would not exclusively control, yet responsive to Soviet will rather than the will of its people. He expected such a government to enforce fundamental changes by undemocratic means: "the purge of the state administration, public ownership of enterprises, ... expropriation of big landowners."[77]

In the fall of 1946, any hopes for a German government to Stalin's liking faded as a result of the huge anticommunist protest vote in the nation's local elections, especially significant in the Soviet-controlled sector of Berlin. Viewed from Moscow, this marked a more important turning point than the suspension by Washington earlier that year of the reparations deliveries from the U.S. occupation zone to the Soviet Union.[78] Yet Stalin's German failure did not make him shed the illusion that, despite persisting in his self-defeating policies, he could still get the kind of Germany he wanted.

In January 1947 Stalin assured the visiting delegation of top East German communists that he wanted the country to rise again and that, if his reparations requirements were an obstacle, they ought to be abolished.[79] Yet he turned a deaf ear to the visitors' pleas to terminate the dismantling of industries in the Soviet zone; instead, he berated the Americans for giving lip service to the economic unity of Germany while obstructing the establishment there of a centralized government. He resisted their alternative of a federal system of government—ostensibly because of its high cost and low efficiency but in reality because of its being conducive to self-government, incompatible with his vision of a Germany that could be manipulated from Moscow.

In this respect, Stalin's vision of Germany did not differ in its substance from that of other countries within the Soviet sphere of influence. The East German communists, like their comrades elsewhere, followed under Moscow's guidance a "national road to socialism."[80] Calculated to make the best use of the political assets available locally, the formula did not preclude, but neither did it guarantee, free elections or genuine coalition governments. It did not entail a timetable for the attainment of "socialism" or even its precise definition, but presupposed a favorable, preferably improving, "correlation of forces."[81] In the event, however, the opposite had been happening—not only in Germany but also in Austria and Hungary, where the communists had lost in free elections, though not in Czechoslovakia, the only country where they won.

It was because of this unexpected difficulty of achieving in the region the kind of security Stalin wanted—rather in response to any particular Western moves—that he proceeded, with the help of local communists, to curtail still further even the limited pluralism he had previously found expedient to foster. In doing so, he was not implementing a design but reluctantly discarding the one that had not been working. But the gradual abandonment—in practice if not in theory—of the "national front" strategy, which obliged communists to collaborate with others in genuine or sham coalitions, did not yet require them to foresake their own "national roads to socialism."

Nor did Stalin relinquish easily the hope of being able to steer to his advantage the postwar left-wing surge in Western Europe. He professed to believe that "today socialism is possible even under English monarchy."[82] He courted the visiting pro-Soviet ideologue of the British Labor party, Harold Laski, propositioning a liaison.[83] Yet in the struggle for the minds and hearts of the Western democratic Left, supported by Washington in a "quiet revolution"[84] to counter the one that was being imposed in Eastern Europe, Moscow was losing, too.

As Stalin's relations with his erstwhile Western partners were turning from bad to worse, he had less incentive than before to defer to their sensitivities. He turned his attention to the Balkans, where no vital Soviet interests were at stake but the most aggressively anti-

Western communist regime held power in Belgrade. In May 1946 he mentioned to its chief, Tito, the desirability of restoring the Comintern in another form.[85] Fretting about getting old, the Soviet dictator treated the Yugoslav one as something of an heir apparent, who would "remain for Europe" once he himself would be gone.[86] By 1947 he no longer discouraged the Yugoslavs from proceeding with their plans for a regional confederation of communist states and their Greek comrades from resuming armed struggle against the Athens government.[87]

The Balkan situation prompted the United States to undertake a radical revision of its posture toward the Soviet Union even before Stalin felt compelled to do the same in regard to Washington. Heralding the policy of containment, which would remain at the essence of America's conduct in the Cold War for the rest of its duration, the March 1947 Truman Doctrine resulted in the delivery of military assistance to not only the Greek government (which was immediately threatened) but also the Turkish government (which was not). At Moscow's address, the president proclaimed American determination to uphold by whatever means necessary the integrity of states endangered by communist subversion.

Despite the ringing rhetoric of the proclamation, the modest military aid to countries peripheral to both American and Russian interests did not pose a security threat to Moscow. Its internal assessments showed no sense of alarm,[88] and the lack of any response on its part other than verbal condemnation showed that Stalin correctly understood the presidential doctrine as not applying to the part of Europe he already controlled.[89] In its effects on Soviet policy, it was not the turning point it was later made out to be.[90]

The Soviet government placed exaggerated hopes in the April 1947 Moscow conference of foreign ministers. Although the Americans and British tentatively approved there the formation in Germany of the central administrative agencies that were crucial for Stalin's design for its future, the French vetoed them.[91] In trying to salvage at least something of the design, Molotov tried to be more accommodating about Soviet reparations demands, but not enough about the key issues of the management of Germany's economy and the country's constitutional setup. The failure of the conference still did not prompt a radical revision of Soviet policy; the impetus for its change eventually came from Western Europe, and originated in internal rather than international politics.

By then, the presence of communist ministers in several Western European governments had become as anomalous as the presence of noncommunist ones in Eastern European governments. Under growing pressure from their local enemies, the French communists panicked, provoking a showdown in April 1947 over the trivial issue of a wage freeze, which resulted in the ejection of their ministers from the cabinet.[92] The same happened, for different domestic reasons, to their counterparts in Belgium and Italy,[93] giving Stalin's aide responsible for relations with foreign communists, Andrei A. Zhdanov, a good reason to upbraid his Paris charges for having brought humiliation on themselves without having consulted with Moscow in advance.[94]

The communist setbacks raised doubts about the viability in Western Europe of the coalition strategy whose futility had already been exposed in Eastern Europe. If the methods Stalin employed to assert his control in the eastern part of the Continent had made the Cold War inevitable, his unfulfilled expectations in the western part made it irreversible. Having made Western Europe's weakness and fragmentation prerequisites of Soviet security as he understood it, he was bound to see any signs of a reversal as threatening. And the recovery was coming even before Washington lent it a powerful helping hand by launching the Marshall Plan.[95] Yet the ambivalent Soviet response to the new situation suggested that Stalin initially did not grasp—or did not want to grasp—its seriousness.

Unlike the general and loosely worded Truman Doctrine, the specific and precise proposal enunciated by Secretary of State George C. Marshall on June 5 no longer entailed merely shoring up two shaky governments on the outskirts of Europe but nothing less than a long-term American commitment to the stabilization of the economic, and hence also political, social, and moral, order in its heartlands. Intended to safeguard U.S. security in a global system of international pluralism, which best suited the interests and ideals of the world's most powerful nation, the plan required the recipient states to take the initiative in jointly calculating their needs and to cooperate in the most effective distribution of the available resources.[96]

As such, the project was deeply subversive of Stalin's hegemonial concept of international order, aimed at ensuring his country's security at the expense of all others'. By offering assistance to any European nation in the expectation that the Soviet Union would refuse it, the United States shifted onto Stalin the burden of deciding whether he would allow his Eastern European clients to accept the American aid. He had the unenviable choice of either risking the intrusion of Western influences within his sphere of power, or else insulating the sphere, thus precipitating against his will the division of the Continent into hostile blocs.[97] Although the mounting ideological conflict predetermined the outcome, the decision was not as certain in the minds of Stalin and his underlings as Washington had calculated or Moscow later wanted the world to believe.[98]

Much though their "class instincts" rebelled against it, the members of the Soviet foreign ministry team entrusted to study the project favored adherence.[99] In the same vein, on June 22 Molotov encouraged the East Europeans to make "their contribution to the elaboration of economic measures and make their claims in view of the fact that some European countries ... have already stated such desires."[100] Later

that month, he went to the Paris meeting of the prospective participants at the head of a large delegation instructed to explore how the
U.S. proposal could be deprived of any strings attached, thus making it
possible to have the American cake and eat it, too.[101]

The Soviet belief that this could be done stemmed from the miscon-
ception that the Marshall Plan was for the United States a matter of
necessity rather than of choice. In the expert opinion of Stalin's court
economist Evgenii Varga, prepared for the foreign ministry, the leading
capitalist nation was on the verge of a deep economic crisis, which
required it to get rid of its surpluses at any price, yet wanted "to draw
maximum political advantages" from what it had to do anyway.[102]

Once in Paris, Molotov acted as if Washington could be induced to
provide assistance on Soviet terms. He asked the unanswerable questions of how much credit the United States would grant and whether
Congress would approve it.[103] Testing the depth of the Western
Europeans' persisting fear of Germany, he proposed to discriminate
against former enemy states in the distribution of the aid. He was
ready to support the French idea of a steering committee of its recipi-
ents, but only if their resources were not scrutinized and the former
enemies were merely admitted as observers.[104] He wanted each country
to simply announce its needs, whereupon the Americans would be
expected to deliver.[105]

Unable to make progress, at the June 30 session of the conference
Molotov was reiterating his position when an aide handed to him what
some of those present thought was a cable from Moscow directing him
to break off talks.[106] But the message rather consisted of a Soviet intelligence report describing the recent secret Anglo-American conversations
in London, where the Marshall Plan was presented as a defense of
Western Europe against a Soviet threat.[107] Molotov suspended negotiations, and two days later, he angrily left Paris. Since his gesture had not
been premeditated, his government was left unprepared as to what to
do next.[108] It let its Eastern European allies to initially respond on their
own to the invitation for a second conference, scheduled to meet in the
French capital in ten days' time to seal the plan.

The Czechoslovaks and Poles wanted to attend; the Yugoslavs,
always wary of the American project as an insidious capitalist ploy, did
not. On July 5 Moscow sent word that it wanted all of them to go, but
only "to prevent a unanimous acceptance of the plan, and then leave
the conference, taking along as many delegates from other countries as
possible."[109] But already the next day it reversed itself, citing alleged
new information and urging them to boycott the meeting.[110] The information, said to have revealed the Western nations' unwillingness to
compromise and their intention to organize themselves against the
Soviet Union, was not really new. What was new was Moscow's
reassessment of the likely consequences of the Marshall Plan for its
relations with its allies.

In trying to reinforce their dependency, Stalin chose to teach the East Europeans a lesson in obedience. Although the Yugoslavs never wanted the American aid in the first place, now they had to reject it on Soviet orders. The Czechoslovak government, though not yet controlled by the communists, had always made its acceptance contingent on Stalin's approval,[111] and once he voiced his disapproval, merely wanted him to mitigate its embarrassment at having already publicly agreed to participate in the Paris meeting; this he conspicuously refused to do.[112] In contrast, the communist president of Poland, Bolesław Bierut, dared to take exception to the insolent Soviet announcement of his government's rejection of the invitation before the decision was actually made;[113] he had to go along all the same.

Stalin's clients then complied with his wishes in their own ways. While both the Yugoslavs and the Czechoslovaks turned down the Marshall Plan peremptorily, the former did so as good communists, the latter to curry Stalin's favor.[114] The Poles declined it more politely, with thanks and expressions of friendship.[115] Yet these chosen variations in style did not detract from the imposed uniformity in substance. U.S. ambassador to Moscow Walter B. Smith described the outcome as "nothing less than a declaration of war by the Soviet Union on the immediate issue of the control of Europe."[116] Considering Soviet weakness, this was an overstatement. But the ambassador was right in concluding that now "the lines are drawn."

After thirty years of communism, Stalin's rule brought neither to him nor to his people the security they hoped for. Its initial pursuit by revolution had degenerated into the defense of his tyranny. His tinkering with the alternative of collective security had failed to save the country from a devastating war. Although his quest for security by imperial expansion in the end triumphed, making the Soviet Union finally a great power, barely two years after its spectacular victory in that war he again had reasons to see his security imperiled. That the perceived threat was more in the eyes of the beholder and was in any case very much of his own making did not make it any less of a threat. Nor could the absence of an immediate military danger provide enough reassurance against the overwhelming material superiority of his capitalist adversaries and Washington's growing determination to make the best of its supremacy. In trying to reverse the deteriorating correlation of forces, Stalin once again called his battalions of international communism into the field.

2

The Specter of Communism
September 1947–June 1948

Never since Karl Marx had invoked the specter of communism in 1848 were its devotees so numerous, self-confident, and responsive to a single command than one hundred years later—in 1947. The Soviet victory in World War II and the resulting rise of the "fatherland of socialism" to a great power had bolstered the worldwide appeal of the idea, convincing even some of its adversaries that it might be the wave of the future. No less crucial had been the growth of Stalin's personal prestige, making him more capable than ever before or after of advancing the communist cause—if he wanted to—by imposing his leadership upon willing followers. It was a unique opportunity to use or to lose.

The Birth of the Cominform

Once Moscow had damned the Marshall Plan and made its Eastern European clients follow suit, its trusted man on the French party politburo, Jacques Duclos, expected its new directives to come soon.[1] By the end of July, however, they still had not arrived; what came instead was an invitation to a conspiratorial meeting of representatives of Europe's main communist parties at the Polish resort of Szklarska Poręba on September 22. When Duclos finally found out what the new line was, he had every reason to be astonished. Not only was the policy his party had been faithfully following at Soviet behest since World War II branded as misguided, but he personally was also taken to task for helping to implement it.

At issue at the conference, which was the Soviet response to the Truman Doctrine and the Marshall Plan, was a revision of the failed policy. The preparatory papers noted the obsolescence of the notion of diversity of conditions in different countries, which underlay the dissolution of the Comintern in 1943 and the subsequent pursuit of national roads to socialism that justified communist participation in coalition governments.[2] Instead, what all communists had in common was again to be emphasized, namely, the necessity of uniting, independently of others, in waging an anti-American struggle by all the means at their disposal.[3]

The idea of creating a substitute for the Comintern had been in the air ever since the onset of the Cold War. After Dimitrov had left his job as the principal supervisor of foreign communists at the Soviet party's department of "international information" to become the head of the government of Bulgaria, the department was reorganized to one of "foreign policy" in December 1945.[4] The change implied its more active role in Moscow's international operations, though not at the expense of the foreign ministry, whose relative weight after World War II—unlike after World War I—had increased. In managing the Soviet Union's newly won great power status, diplomacy took precedence.

Dimitrov's successor, Aleksandr S. Paniushkin, a secret police apparatchik, was not someone of comparable stature. By 1947 he had been overshadowed by Zhdanov—the party boss of Leningrad, the Soviet Union's "Western" as well as revolutionary city—to whom Stalin entrusted the main responsibility for international communism. Zhdanov had also been in charge of the Soviet policy toward Finland, where his subtle treatment of the defeated enemy showed him as an accomplished politician rather than a mere ideologue and party hack.[5] To be sure, the man notorious for steamrolling the Soviet intelligentsia into conformity was no exemplar of a liberal.

Zhdanov's secret keynote address at Szklarska Poręba, the key portions of which were published a month later,[6] proclaimed the doctrine of the two implacably antagonistic camps of "socialism" and "capitalism." This formulation, approved by Stalin, had been missing from the original text,[7] thus indicating his last-minute decision to give the statement a more aggressive thrust. In a *tour d'horizon* of the world divided into hostile parts, Zhdanov blurred the division, though not the hostility, by describing such countries as Indonesia and Indochina as being "associated" with the Soviet side because of the war they had been waging against their Western colonial masters and by suggesting that India, Egypt, and Syria were at least "sympathetic" to it. Insisting that peace could only be maintained by fighting Western "imperialism" more vigorously, he saw "the chief danger to the working class" in its "underrating its own strength and overrating the strength of the enemy."[8]

The policy implications followed less from what Zhdanov declared in his published speech than from what was said and done during the confidential meetings afterward. He and Stalin's other current favorite, Georgii M. Malenkov, explained to the assembled communist dignitaries that the purpose of their gathering was not merely the exchange of information and experiences mentioned in the original invitations but also the establishment of the Cominform as a coordinating center to ensure that they would fight the capitalist enemy together rather than separately.[9]

Whether deliberately or, more likely, because of the last-minute Soviet planning, the participants had not been alerted in advance of the intent to create the organization. Hence, at first they did not endorse the idea unanimously—as they would have done had it been properly orchestrated beforehand. Polish party secretary Gomułka questioned the Cominform's compatibility with the "national roads to socialism"; the Yugoslavs wondered why the Greek communists, waging their struggle with the force of arms, were not to be included.[10] But all those present were disciplined enough not to press their questions. They fell quickly into the Soviet line, which envisaged seven members from Eastern Europe—excluding Albania and East Germany—and from Western Europe the key French and Italian parties.

The reason for including the two Western parties was revealed once the Yugoslav and Romanian delegates delivered speeches commissioned by the Soviet organizers. Milovan Đilas, the second-in-command in Belgrade, stunned Duclos by the ferocity of his attack on the "parliamentary illusions" of the French and Italian comrades. After the Frenchman had asked for time to prepare a reply, Romania's Ana Pauker took her turn hammering down on the same theme. The next day Duclos performed the ritual of self-criticism, apologizing for his party's having been so foolish as to collaborate with a "bourgeois" government that later booted out its communist ministers.

Knowing that he must not question the Soviet wisdom of having previously abetted this very policy, Duclos at least inquired why Đilas, if he had been aware of the French error before, had waited until now to say so.[11] Since there was no good answer to that question Đilas ignored it, leaving it to Edvard Kardelj, Belgrade's chief ideologist, to harp on the "fetishism" of coalition politics. Kardelj lectured the Western European communists that they must strive to capture the "commanding heights" of power not by seeking votes but by compelling the people to follow their lead.[12]

Here was the essence of the new Soviet strategy for Western Europe. Supporting it but taking exception to its haughty presentation by the Yugoslavs, Italian representative Luigi Longo grasped correctly that it had been the recent hardening of East-West lines which made the communists' continued opposition to the division of Europe obsolete; what needed to be opposed now was the consolidation of its Western part under U.S. hegemony. Without having to be told, he understood that the communists should resort to extraparliamentary means, such as crippling strikes and "direct action" in the streets by the organizations of wartime resistance veterans that they controlled.[13]

Because of the fundamental consensus prevailing at the conference, Moscow did not have to spell out any specific directives there. Their absence left some of its disciples wondering whether the new course entailed a "serious qualitative change in policy and tactics"[14] or whether, as Stalin told the visiting left-wing British Labor M.P. Konni Zilliacus a month later, there was none.[15] Interpreting Soviet thinking to the Italian party directorate, its general secretary Palmiro Togliatti answered the pertinent question of whether preparing an insurrection was appropriate if there were no chances for its success by saying that the question was the wrong one although "a communist cannot avoid it forever."[16] In any case, Stalin's wishes had to be heeded regardless of the cost.

Duclos understood the assignment as requiring the communists to do their best to "destroy the capitalist economy"[17] and topple the pro-American governments in their countries, but without trying to seize power in the process. Accordingly, the French party provoked massive strikes and public disturbances, leading to clashes with the security forces and casualties. The confrontation climaxed on November 14, when the police discovered an arms cache at a repatriation camp of former Soviet prisoners of war—an incident which the Paris government tried to play down but Moscow chose to play up by crying provocation.[18] In a country where one-fourth of the electorate voted communist, the threat of destabilization was real.

For eastern Europe, the mission of the Cominform was neither so urgent nor so consequential. Its preparatories mentioned problems in the region, such as the communists' diminishing popularity in Czechoslovakia or indications that in Yugoslavia their appeal applied not so much to the party as to its individual leaders.[19] Nevertheless, the situation reports presented at Szklarska Poręba by the Eastern European representatives exuded self-satisfaction and confidence in the future. Nor did their Soviet supervisors cast doubt on the viability of their "national roads to socialism,"[20] which still entailed vestiges of power-sharing. Yet the tenor of the conference, particularly Zhdanov's ominous thesis that "the correlation of forces within each country depends on the disposition of socio-political forces at the international level,"[21] tolled the bell for any remnants of pluralism in the "people's democracies."

Like their comrades in the West, the Eastern European communists did not have to be told what to do. Gomułka explained to the gathering that he understood his party's task as cracking down more vigorously to eliminate any organized opposition in Poland, particularly that of the Catholic Church.[22] The Prague politburo, informed about the outcome of the deliberations by its general secretary Rudolf Slánský, concluded that the time had come to prepare for a showdown during a crisis that would enable the communists to seize power, perhaps during the national elections in May 1948.[23] And Tito, even while the secret Cominform conference was in progress, already proclaimed publicly that at least in Yugoslavia the "people's democracy" meant nothing less that the superior form of democracy exemplified by the Soviet one.[24]

During the months that followed, the sovietization of Bulgaria, Romania, and Hungary was driven ahead by the local communists without Moscow's direct guidance.[25] In his keynote address at the Cominform's founding meeting, Zhdanov had also suggestively described Finland as marching on the same "path of democratic development"[26] as these countries, although its being left out of the organization rather indicated that its future still remained undecided in Stalin's mind. In tightening his grip on the part of Europe that was at his mercy, he was still groping for the best way to organize it.

Stalin envisaged a key international role for the Yugoslavs. Not only did he choose them as his proxies to spearhead the assault on the French and Italian "errors," but he also accepted their offer to host the Cominform's headquarters in Belgrade.[27] Their later insinuations that such displays of benevolence were a sly way to blunt their vigilance in preparation for his intended blow against them cannot be substantiated by contemporary evidence. In internal Soviet documents, too, Yugoslavia was praised, especially for its militant anti-American stance.[28] Its promotion suited Stalin's purposes at that particular stage of the Cold War.

As long as avoidance of conflict with the West had been Stalin's priority, the Yugoslavs' radical proclivities had been liabilities; with the lines of conflict drawn, they became potential assets. In the summer of 1947, when they notified him of their intention to formalize closer cooperation with Bulgaria as a preparatory step for the controversial confederation of Balkan communist states, he did not interfere.[29] Their respective leaders, Tito and Dimitrov, then proceeded to conclude in July a preliminary agreement at Bled on mutual coordination of policies and eventual union.[30] This was followed by Belgrade's accord with Greek communist chief Markos Vafiades, providing for increased military assistance to his guerrillas.[31]

Once the Bled agreement was signed, however, Stalin had second thoughts. Disingenuously accusing his followers of not having consulted with him, he warned that their deed might give "reactionary Anglo-American elements ... a pretext to speed up their military intervention in Greek and Turkish affairs,"[32] with dire consequences for their own countries. Unswervingly loyal, Tito and Dimitrov promptly complied with his wish to postpone the implementation of their pact until after the general peace treaty with Bulgaria came into effect in September. The easy resolution of the dispute, which showed that Stalin objected merely to the timing rather than to the substance of the project, gave little inkling of a strain.

Unhindered by Moscow, before the year's end Tito also concluded treaties of mutual assistance with Hungary and Romania.[33] He made them directed—unlike similar pacts that Eastern European governments had entered into under Soviet auspices before—no longer against Germany but against any enemy, meaning particularly the United States. In the West, the aggressive shift lent support to the widespread contemporary view of Tito as Europe's premier troublemaker.[34] It nourished speculation about his pursuing a Soviet design to bring also Czechoslovakia and Poland into the Yugoslav-organized grouping.[35] These and other variations were reportedly being considered in Moscow, though never advanced beyond the drawing board.[36]

Tito's pompous touring of the neighboring capitals has been retrospectively credited with having alarmed Stalin about Belgrade's allegedly independent foreign policy. Yet its directors had shared the texts of the intended pacts with Moscow beforehand, dutifully incorporating the minor amendments it asked them to make.[37] Rather than double-crossing Stalin, the Yugoslav communists did not differ at that time from all the others in their trying to act on what they perceived to be his wishes. They thought of themselves as his most faithful disciples, and his own propagandists reciprocated with what struck veteran British correspondent in Moscow Alexander Werth as an "amazing and exclusive flattery of 'heroic Yugoslavia'" in the Soviet media.[38] Toward the end of 1947 the mutual relations were, if anything, getting closer than they had ever been.

What changed this unproblematic situation was the proclamation on Christmas Eve by the Greek communists of a provisional government in the part of the country which they controlled. Such a daring challenge to the American protectors of the Athens government could have hardly been undertaken against Moscow's wishes, and contrary to later speculations, it indeed was not. Several weeks earlier, the Soviet party central committee had approved not only the establishment of the rebel government but also the communist plan, code-named "Limnes," aimed at conquering enough territory to install it in Salonika, Greece's second-largest city, after its expected capture.[39] Whether Stalin had actually authorized the timing of the proclamation or, more likely, had carelessly left it up to his underlings and their Greek charges, he soon had reasons to regret it.

The Balkan Trap

The American response to the challenge was swift and effective. Serving public warnings to any country that might contemplate recognition of the self-styled government, Washington backed them with private warnings to Albania, Yugoslavia, and Bulgaria. It further considered, although it eventually ruled out, the option of sending U.S. combat troops to Greece.[40] Neither the Soviet nor any other government extended diplomatic recognition to the insurgents.

Developments in the area compelled Stalin to focus on, of all places, Albania—the country he disparaged as the epitome of insignificance. Having become in all but name a Yugoslav province, which Tito treated much like he did his own dependencies, it was a snakepit of Balkan intrigue. While currying Belgrade's favor, party boss Enver Hoxha and security chief Koçi Xoxe spun webs of conspiracy against each other and their respective retainers. In November 1947 the plotting drove politburo member Nako Spiru into suicide, which the authorities tried to camouflage as a shooting accident in his office.[41]

The episode need not have bothered Stalin had Tito not chosen to have his ambassador to Moscow, Vladimir Popović, intervene with Zhdanov about Spiru's trying to contact the Soviet embassy in Tirana before his death in a vain attempt to obtain its support against charges of harboring an "anti-Yugoslav attitude." It was in response to this intervention that the Soviet leader on December 23 invited a high-level Yugoslav delegation to come in for discussions. He asked Belgrade to make its wishes clear and assured it of his willingness to consider them favorably.[42]

The Greek communists' fait accompli, which occurred on the next day, enhanced the significance of the visit because of Albania's importance as the main staging area for the assistance they had been receiving from Tito. Having already deployed there an air force regiment, he was now planning to send in ground forces as well—ostensibly to protect the country against a possible attack by Greek government troops, but in reality to prepare for its annexation as Yugoslavia's seventh constituent republic.[43]

While the situation in Albania provided the setting for Yugoslav-Soviet conflict, the impetus for its escalation did not come from there but from the Soviet embassy in Belgrade. Before Tito's representatives reached Moscow, ambassador Anatolii I. Lavrentev had poisoned the minds of his superiors by his increasingly censorious dispatches about the Yugoslavs, which contrasted sharply with the public praise accorded them by his government. Besides accusing them of ignoring him, he faulted Tito for not giving the Soviet Union proper credit for his country's liberation and even dared to deprecate the former Comintern hand as a poor Marxist. On January 8 the envoy dispatched a whole litany of further complaints, and two days later the Soviet military attaché in Belgrade, Gen. Georgii S. Sidorovich, ominously recommended that the Comintern should look into the Yugoslavs' "errors."[44]

These were astounding allegations by midranking Soviet officials, making them seem as if they were being manipulated by someone higher up. Yet in the absence of any evidence to that effect, their behavior can be plausibly explained by their overwhelming urge to anticipate whatever might conceivably go wrong, lest they be punished later for not predicting it. And there was enough that could go wrong to make them especially vigilant. As a result, in trying to foresee Stalin's suspicions, they could plant them in his receptive mind even before he would hatch them himself, thus surreptitiously molding his policy.

Aside from the subordinates' self-fulfilling prophecies, Stalin's rift with Tito had ironically been preconditioned by the very affinity of their regimes. Not only had both arisen from bloody civil wars, but the Yugoslav party was also the most Stalinist of all, complete with its adulation of the supremo. Even Belgrade's bent for twisting the truth and its aggressive persecution complex were congenial to Moscow's habitual mendacity and expansionistic fear of encirclement, although they were more Serbian than communist in origin. Regardless of the Yugoslav communists' genuine devotion to the Soviet fatherland of socialism, their Stalinist disposition could then hardly avoid becoming intolerable to Stalin. In this incompatibility of affinities, rather than in any unbridgeable policy differences, were the roots of the confrontation that he had not originally planned any more than had Tito.

In explaining the genesis of the conflict, the sequence of events during the second half of January was crucial. When on January 17 Stalin met the Belgrade delegation, he did not object to the Yugoslavs' "swallowing" Albania any time they wished. He expressed his preference for their favorite Xoxe over the "petty bourgeois" Hoxha and promised them more economic aid.[45] Interpreting such goodwill as evidence of identical views, Tito lost no time in asking for, and receiving, Hoxha's permission for stationing his troops in the Albanian coastal town of Korçë.[46] He was not the only one who misread Stalin's intentions.

On the same day Stalin entertained the Yugoslavs, their Bulgarian ally Dimitrov was returning from Romania, having signed there another pact that alluded to the United States as the enemy. In an interview he gave on his train, he not only eulogized the idea of a communist Balkan confederation but also mentioned Greece as its future member.[47] Even Tito found this needlessly provocative, instructing Ðilas in Moscow to intercede with the Soviet comrades that they curb the Bulgarian's candor, which had meanwhile stirred an international furor.[48] He did not succeed. On January 23, *Pravda* reported the interview with implied approval.

Both Tito and Dimitrov have retrospectively been regarded as having overstepped the limits of Stalin's tolerance. Yet it was nothing in their behavior, which he had been tolerating, but rather his own reassessment of its likely consequences that accounted for the subsequent reversal of Soviet policy—though not before a warning from the West added an incentive. This was supplied by British foreign secretary Ernest Bevin in a major speech on the day before *Pravda* printed the Dimitrov statement but had already prepared it for publication.

Dwelling on the communist threat to Greece, Bevin pleaded for the unity of Western Europe to counter the advancing sovietization of Eastern Europe.[49] He gave his plea a military dimension by urging the Low Countries to join the Dunkirk treaty between Great Britain and France, concluded the year before against a possible German threat but since then made more topical by the greater need to deter a Soviet one. To reinforce the deterrent, London also took the lead in trying to persuade the United States to commit itself to the defense of Europe.[50]

Only after the cost of communist aggressiveness had been so plainly stated did Stalin act to curb his Balkan clients. When on January 26 Tito's general M. Kupresanin appeared in Tirana to implement the agreement on the stationing of troops—which Lavrentev ruefully reported to Moscow had not been coordinated with Soviet advisers[51]— the crafty Hoxha checked with Stalin before submitting it for approval to the Albanian politburo. He received the prompt answer that the troops were not needed since no danger was imminent.[52]

The truth was rather that the planned deployment risked provoking the very danger it was supposedly meant to avert—an expansion of hostilities in the region with possible U.S. involvement. "The most important thing," Stalin explained to Hoxha, "is not to provoke our former allies."[53] In a message to Dimitrov, he demanded from him a retraction of the offensive interview, pointedly enclosing the Yugoslav critique of it that he had himself received from Ðilas confidentially.[54] The demand created a panic in Sofia. Adding to Dimitrov's embarrassment, on January 28 *Pravda* distanced itself critically from his statement that it had previously endorsed.[55] This left him little choice but to publicly recant, which he did in style—by profusely expressing his gratitude to Soviet critics for allowing him to see his error while lashing at "Western agents" for having supposedly distorted his words.[56]

Although Tito similarly obliged Moscow by canceling the Albanian operation at its request, he nevertheless received from Molotov a stern message charging "serious disagreements between our governments concerning the mutual relations between our countries."[57] Protesting that there were none, he asked for guidance, to which entreaty Stalin responded by inviting both Yugoslav and Bulgarian representatives for urgent consultations.[58] The urgency resulted not so much from concern about their conduct as about Western policies—as evident from the maneuver the Soviet government hastily performed during the week before the date it had set for the talks.

In a flurry of diplomatic activity, which according to Molotov's confidential explanation was necessary to offset the unfortunate effects of Dimitrov's interview,[59] Moscow summoned a government delegation from Romania and made it sign a mutual defense treaty which again specified Germany, rather than the United States, as the enemy.[60] In this, the treaty differed from those that Bucharest had recently concluded with both Bulgaria and Hungary, as well as from those that Tito had signed with all these countries earlier, but was reminiscent of Moscow's wartime pacts with the Western powers—a pathetic appeal to their bygone alliance. Similar treaties with Bulgaria and Hungary followed in quick succession.

When Stalin received his Balkan visitors on February 10, he had already reversed his course. He made it plain to the Yugoslavs that Albania was not to be swallowed and that the Greek communist rebellion must be ended, for the Americans would never allow it to succeed.

He scolded the Bulgarians for concluding the treaty with Romania, although they had cleared it with Molotov beforehand. When they dared ask "where Comrade Dimitrov's fault lies," Stalin conceded that "we, too, commit stupidities."[61] As if to substantiate this, he then proceeded to suggest that he was not actually opposed to the Balkan confederation and would favor one between Poland and Czechoslovakia as well, besides another between Romania and Hungary. He told the Yugoslavs that they should annex Albania after all and conclude the union with Bulgaria to boot, even "tomorrow."[62]

Stalin obviously did not know what he wanted, although he knew well enough how to go about getting whatever he might want in the future. Before he let his guests go, he made them promise Molotov in writing that they would henceforth consult with Moscow on all important international matters. This is what they had been doing all along on their own initiative, but now they had to do so on Soviet orders. They complied in their own ways: the Bulgarians unconditionally, the Yugoslavs by exploring loopholes left by the ambiguity of Stalin's pronouncements.

At its meeting on February 19, the Belgrade politburo did not rule out, but neither did it act upon, the union with Bulgaria.[63] Yugoslavia terminated public support of the Greek communists, which it had enunciated after they had formed their government, but secretly promised their emissaries continued help. It reasserted control in Albania by forcing Hoxha's submission to Xoxe and even revived the idea of stationing the troops, but finally dropped the plan after Moscow voiced its displeasure in no uncertain terms.[64]

Although the Yugoslavs had been sloppy in meeting Stalin's demands, they were not trying to defy him, and neither did he give an indication that he thought they were. More important than their actions and inactions was their indulgence in loose talk. At the politburo's meeting on March 1, its members spoke their minds, unaware that the Soviet mole among them, Sreten Žujović, would pass their ruminations through Lavrentev to Moscow.[65] Besides wisecracking about Bulgaria as a Soviet "Trojan horse," they scorned Stalin's policy as not revolutionary enough, described the Cominform as an instrument of coercion, and grumbled that Moscow was bent on conquering their country economically.[66] Afterward they barred their Soviet advisers free access to data about its economy.

Reporting to Molotov on March 9, Lavrentev accused Belgrade of fussing that a Soviet diplomat in Tirana had been insulting to Tito, that the Soviet government was being slow in preparing a new trade agreement, that its technical assistance was meager, and that the ruble-dinar exchange rate it imposed was unfair. On March 13 Molotov dismissed all these relatively minor complaints in a prickly letter to Tito, but ominously ignored the more important matter of the advisers.[67] Only at this time did Stalin begin to move toward a confrontation, thus unwittingly springing the Balkan trap that he had previously managed to avoid.

Without awaiting Tito's reply, five days later Molotov raised the issue of the advisers in another message, announcing their summary withdrawal.[68] Since no new Yugoslav insolence warranted such drastic action, other reasons must have accounted for the decision to force a confrontation. These had grown out of the recent precipitous deterioration of East-West relations, particularly a turn for the worse in the German question, to which the Stalin-Tito rift became linked in a mutually reinforcing correlation.

The Stakes Increased

The deterioration was the result of an accelerating development set in motion by the Marshall Plan. When the German communists voiced their concern about the favorable impression that its progress would make on the "working masses," their Soviet mentors dismissed any worries by confidently predicting its early failure.[69] Its successful start, despite all communist efforts to sabotage it, did not augur well for the attainment of the Soviet objectives in Germany at the London conference of foreign ministers in November 1947. Commenting on the international climate to Czechoslovak ambassador Jiří Horák, Molotov did not think that a war was coming, but neither would he bet on the outcome of the London meeting.[70] Soviet officials suspected that the Western powers would merely try there to demonstrate the uselessness of Moscow's favorite Council of Foreign Ministers, after they had already decided to create a separate German government in their occupation zones.[71]

This was, on the whole, a correct estimate. While insisting that Germany's economic unity must precede its political unity, Secretary of State Marshall ruled out any common economic policy on Soviet terms, particularly any more reparations deliveries from the U.S. zone.[72] He resisted Molotov's "consistently, almost desperately, endeavoring to reach agreement which really would be [an] embarrassment to us."[73] In another desperate move, Stalin's aide also tried to foist upon the conference a delegation of the Soviet-made "German People's Congress"—a sham assembly handpicked from among communists and their sympathizers, mostly from the eastern zone.[74] Yet despite Molotov's courting the British by proposing to use their rather than the American document on Germany as the basis for discussion,[75] it was they who blocked the delivery of a message to the conference by his German clients by denying them visas.

Bevin was if anything even more difficult than Marshall. He reportedly warned the Russians that "you are putting your neck out too far, and one day you will have it chopped off. We know much more about you than you imagine. We know that you cannot stand a war. But you are behaving in such a way that one day there will be a showdown." To Bevin's question "What do you want?" Molotov is said to have replied almost plaintively: "I want a unified Germany."[76] Comparing the foreign secretary—a genuine proletarian by origin—with his aristocratic predecessor, the Bolshevik sighed that "Eden was a real gentleman; Bevin is not."[77]

Even if apocryphal, the exchange gave a fair indication of the impasse that had been reached. Thoroughly frustrated, Molotov on the last day of the conference reviewed for the record his government's quest for German unity in a lengthy postmortem.[78] Evidently undecided about what to do next, Moscow subsequently vacillated between moderation and assertiveness. At the end of December, by which time the failure of the communist offensive in Western Europe had become obvious, Zhdanov cautioned the Italian party leadership to reduce the pressure "from below," lest it provoke insurrection and defeat.[79] In East Germany, officials from the Soviet party central committee vainly tried to get rid of the hard-line Col. Sergei I. Tiulpanov, the éminence grise of the occupation administration.[80] In January 1948 Stalin turned toward a confrontation with the West in Germany just as he was steering away from it in the Balkans.

From January 21 onward, Allied members of the Berlin *Kommandatura*, the city's quadripartite military executive, noticed a change in the behavior of their Soviet counterparts. These no longer tried hard to press their desiderata, instead obstructing the deliberations by filibuster and in other ways. Robert Murphy, the chief political adviser to the U.S. military government, suspected a design to "irritate and tire the Western representatives to prod them into unwise decisions."[81]

At that time the Allies were putting the final touches on the planned economic merger of their zones, expected to be approved in London at a conference on Germany, to which the Soviet Union had not been invited. Moscow tried to block the gathering by serving notice that it would consider invalid any decisions reached there.[82] The day before the conference opened on February 19, it convened its client foreign ministers of Poland, Yugoslavia, and Czechoslovakia in Prague, to add their protest against any intended political separation of West Germany.[83] Yet the London conferees, besides approving the economic unification of the Western zones, proceeded to bring such a separation closer by insisting upon the introduction in the whole of Germany of the federal system of government that Stalin abhorred.[84]

As the stakes kept growing in Germany, the Soviet Union cast a long though ambivalent shadow over Czechoslovakia and Finland, where crises erupted almost simultaneously, but with very different results. As early as November 1947, George F. Kennan, the perceptive architect of Washington's containment policy, had predicted that Moscow, in trying to offset its setbacks in Western Europe by further consolidating its hold on Eastern Europe, would probably "clamp down completely on Czechoslovakia."[85] Indeed, ever since the founding of the Com-inform, the local communists had been looking forward to the opportu-nity to seize power, yet it was their opponents' loss of nerve that eventually created the opportunity. When on February 20, 1948, non-communist ministers in the Prague cabinet, exasperated by the manner in which the communist minister of the interior, Václav Nosek, abused his powers over the police, submitted their resignations, they miscalcu-lated in assuming that the president, instead of accepting them, would be encouraged to face down the communists.

While President Beneš, clinging to the illusion of national unity, vac-illated, Molotov's deputy Valerian Zorin sent word through a round-about channel that the Soviet government was surprised at his hesitation and expected him to let the ministers go.[86] At the same time, Zorin offered help to the communist party chairman, Klement Gottwald, whom he informed that Soviet troops were ready to march in from Hungary. According to Western intelligence, they were not.[87] Gottwald, confident that the communists could manage by themselves, did not act on the offer. To intimidate the largely passive populace, he had them stage noisy demonstrations in the streets while keeping the party militias on hand for possible action. None was needed, for unlike the communists, their opponents had no plans, and the president in the end caved in. On February 25, the national radio announced that he had accepted the ministers' resignations even before he actually did.[88] Gottwald later reminisced that he "couldn't believe it would be so easy."[89]

In Finland, the Soviet Union at first interfered more heavy-handedly. Having tried in vain to cajole the Finns into taking the initiative them-selves for signing with Moscow the same kind of "friendship" treaty that bound with it its already subdued Eastern European dependencies, Stalin on February 23 bluntly demanded that President Juho Paasikivi send a delegation to conclude the pact.[90] To a contemporary British observer, the situation looked very much as if the "unhappy govern-ment of Finland" were to be "embraced with the kiss of death" by "being invited to visit the Bluebeard's mansion."[91] Making matters worse, the Norwegian foreign minister as well received warnings that a Soviet request for a similar kind of treaty with his country might soon be forthcoming.[92]

To his Yugoslav visitors the month before, Stalin had expressed regrets that he had not occupied Finland after the war because of "too much regard for the Americans," who "wouldn't have lifted a finger."[93] Now he seemed intent to rectify what he had neglected, for he allowed the prominent Finnish communist, Hertta Kuusinen, to enunciate publicly that Czechoslovakia's road "must also be our road."[94] Yet it in fact was not, for the Finns, unlike the Czechs, did not falter in their commitment to democracy, while their president, unwilling to consort with the communists, fired Kuusinen's husband Yrjö Leino from his crucial job of minister of the interior and sent troops to protect government buildings.

Impressive though it was, however, such conduct might not have saved Finland from Stalin's tender embrace if the timing had been different. His subjugation of Czechoslovakia, despite all its efforts to oblige him, was plausibly, if wrongly, regarded abroad as the first act in the incipient westward expansion of his empire rather than the last act of its sovietization. Bevin saw Moscow as "actively preparing to extend its hold over the remaining part of continental Europe and subsequently, over the Middle East and no doubt the Balkans and Far East as well."[95] U.S. military governor in Germany Gen. Lucius D. Clay warned Washington that war "may come with dramatic suddenness," and French premier Georges Bidault inquired about the possibility of a U.S. military guarantee for his country.[96]

In the aftermath of the Prague coup, Stalin had reasons to ponder the larger cost of his dubious gain in Czechoslovakia. The perception that the Soviet Union was on the move galvanized Western resistance to its expansionism, giving a boost to the plans for military cooperation, about the progress of which Moscow was accurately informed by its British spy Donald Maclean.[97] He was present at the secret Washington talks that led to the conclusion on March 17 of the Brussels pact between the United Kingdom, France, and the Low Countries.[98] Unlike the Dunkirk treaty the year before, this was no longer aimed at Germany but at any enemy, meaning especially the Soviet Union—a shift of thrust in the opposite direction from that of the recently concluded Soviet alliances in Eastern Europe, whose primary purpose was controlling rather than defending the allies.

Instead of preparing to fight a German threat in the future, Moscow tried to preventively fend it off by advancing its scheme for a dependent Germany. Two days after the conclusion of the Brussels pact, it reconvened in Berlin the People's Congress, whose unelected delegates it had vainly tried to impose on the conference of foreign ministers before, and created a People's Council as a sort of shadow government. On the next day, March 20, Marshal Vasilii D. Sokolovskii demonstratively left the meeting of the Allied Control Council—the joint organ of the occupation powers responsible for implementing their common policy in Germany. Although he professed a readiness to return, Moscow's simultaneous hints about the supposed need to regulate the Western powers' access to their part of Berlin made them suspect that the condition for his return was a common policy to Soviet liking.[99] And, indeed, the idea of achieving this goal by blocking the access had already been outlined on March 12 in a secret memorandum for Molotov by his assistant, Andrei A. Smirnov—another instance of an important policy initiative prepared by a subordinate.[100]

It was on the eighteenth of this eventful mid-March week that Moscow, unbeknownst to the outside world, also fired the opening shot in its battle against Tito. Molotov's confrontational letter was dispatched to Belgrade on that day[101]—shortly after the conclusion of the Brussels pact had provided Stalin with an added incentive to discipline his vassals in preparation for the showdown he was fomenting in Germany. In Albania, his emissaries swiftly asserted Soviet control, enabling Hoxha to turn the tables on Xoxe and finally laying to rest the project for its annexation by Yugoslavia.[102] Having put an end to Belgrade's expansionism, Moscow distanced itself from the Greek communist insurgency, extending reassuring feelers to the Athens government.[103]

If Stalin hoped to liquidate explosive Balkan issues while more important matters were at stake in Germany, he was to be disappointed. Replying in kind to Molotov's peremptory message, Tito offered the Soviet government the unsolicited advice that it should be more careful in choosing its informants in Belgrade.[104] The tone of his letter was totally incompatible with his continued self-perception of loyalty to Moscow and with his desire to reassure rather than to provoke Stalin. Such incomprehensible behavior by a disciple who ought to know better infuriated the despot—the only plausible explanation of why Stalin on March 27 fired back a reply of otherwise inexplicable ferocity, thus foreclosing the possibility of reconciliation.[105]

The complication with Tito made it advisable for Stalin to avoid further complication with the Finns; in this sense, "the communist dictator in Belgrade helped save Finnish democracy."[106] More pertinent, however, was the heavy damage that the coup in Czechoslovakia had inflicted on Soviet interests in Western Europe, making any attempt at a repetition foolhardy. Moscow may have been particularly impressed by the proposal, secretly extended by Canadian Under-Secretary of State Escott Reid, to include Finland in the prospective Western military alliance—a bit of information that Maclean was in a position to relay to his Soviet paymasters.[107]

The developments since Prague were also bound to make Moscow more receptive to the intelligence it had been receiving from its Helsinki informants passing to it secret documents of the Finnish cabinet. These left no doubt that the proposed friendship treaty in its original form would have been rejected, thus straining the relations between the two countries at a most inopportune time.[108] Rarely did the theft of secret papers prove more beneficial to the people from whom they had been pilfered.

Having exhausted all means of procrastination, the Finnish representatives reached Moscow full of trepidation, but they were relieved to find Stalin contented with a far more benign document than the one he had previously proposed. The treaty, signed on April 6, merely obliged the Finns to consult with the Soviet government in case of a foreign threat to their country and defend its integrity—something they were only too happy to promise.[109] The outcome was the result of an extraordinary combination of good luck and their ability to mold Stalin's security interests while these were still malleable by offering him the right mix of resistance and reassurance.

Although in the long run Stalin's restraint made Finland the Soviet Union's best neighbor, in the short run it came too late to spare him further disappointments elsewhere. The perception of an undiminished Soviet threat, effectively dramatized by the United States, sealed the defeat of the communists in the crucial Italian elections of April 1948, which turned the tide of their advance in Western Europe.[110] The ebb of international communism could not fail to leave an impact on Stalin's row with Yugoslavia, which he had originally chosen as its vanguard.

Stalin's acrimonious correspondence with Tito, which pointedly ignored their recent foreign policy disagreements, was all the more replete with charges calculated to ruin the Yugoslavs' good name as communists—from their allegedly flabby revolutionary spirit to their supposed susceptibility to "Trotskyism." At issue in these absurd accusations was the mishandling of international communism as a tool of policy—which, to be sure, was Stalin's own fault. "Terribly surprised," Tito wondered "for whom such slander ... was necessary."[111] It was necessary for Stalin to divert attention from his blunders while teaching the Yugoslavs and other East Europeans an object lesson in submission.

It was the irony of the Stalinist logic that the sovietization of Eastern Europe, in which there was no longer any room for partners but only for stooges, made Tito liable to be singled out for exemplary humiliation precisely because of his readiness to follow the Soviet lead out of conviction rather than fear. Concurrently with the campaign against him, Moscow was preparing charges of heresy against leading Polish, Czechoslovak, and Hungarian communists as well.[112] Although the charges about to be raised against them were not as serious as those against the Yugoslavs, they would have been embarrassing enough and possibly fatal to individual leaders. In the event, it was the unexpected difficulty Stalin encountered in trying to subdue Tito that saved the other prospective victims—if only temporarily—by making it expedient to shelve the accusations.

While Stalin had secretly been keeping other Eastern European parties informed about the stern messages he was sending to Belgrade, he concealed from them the increasingly defiant replies he was getting back; these were only later published by the Yugoslavs. When, on May 4, he set the stage for a showdown by disclosing to Tito that the dispute had "already become the property" of those other parties, he included a long quote, purportedly unearthed in the Moscow diplomatic archives, which insinuated that three years earlier Tito's trusted aide Kardelj had made irreverent remarks about him behind his back to the Soviet ambassador.[113] Following the recent exposure of the traitor

Žujović by the Belgrade politburo, the attempt to retaliate by sowing discord within it marked the point of no return. It signaled Stalin's intention to not merely humiliate and discipline Tito but overthrow him.

Yugoslavia was not the only place where Moscow tried to drive a wedge in the ranks of its adversaries; it used the same ploy against the West. On May 11, after ambassador Smith had approached Molotov in an attempt to slow down the alarming deterioration of U.S.-Soviet relations, the Kremlin on its own accord released to the public a doctored account of their secret talks, making them appear as if the envoy had been courting the Soviet official behind the backs of America's European allies, and had been rebuffed.[114] The mischievous publication, imitating in its own way the "Ems dispatch" that Prussian chancellor Otto von Bismarck had composed in 1870 to prod France into war, could not fail to make the bad relations even worse.

After Truman had in the aftermath of the Czechoslovak coup requested from Congress restoration of the draft and increased defense spending, on June 4 the Vandenberg resolution by the Senate cleared the way to the United States' participation in a military organization for the defense of Western Europe. This was four days after the three Western allies had proclaimed at another London conference their determination to establish a separate state in Western Germany, calling for the convocation of a constitutional convention that would make it definite.

The nine months since the launching of the Cominform in an effort to shore up Moscow's deteriorating positions in Europe had witnessed their still further deterioration. Having tried and failed to reverse American ascendancy in the pivotal western part of the Continent, Stalin had been more successful in compelling his Balkan followers to desist from challenging the United States in this strategically secondary area. Yet in doing so, he imprudently embarked on an avoidable confrontation with the unimportant Yugoslavia just as he was precipitating another one in the crucial Germany. In tightening his grip on Eastern Europe to consolidate his positions in the mounting East-West conflict, he was not as reckless as he seemed; he proved ready to retreat in Finland after having overstepped in Czechoslovakia. But he was still reckless enough to provoke the West to proceed with greater determination toward creating both an anti-Soviet German state and an anti-Soviet military bloc. In deploying his special weapon of international communism, Stalin did try to calculate his costs and benefits, but he miscalculated badly.

3

A Harvest of Blunders
June 1948–April 1949

While in fomenting the German crisis Stalin had largely been reacting, he was in full control of the Yugoslav one, which was of his own creation. Planning showdowns in two different places simultaneously showed how badly he needed a success. In Berlin, the prize was inflicting a diplomatic defeat upon the Americans that would shatter their credibility as protectors of Western Europe; in Belgrade, Stalin's credibility as the undisputed master of Eastern Europe was at stake. If successful, he would be able to achieve in two bold strokes Soviet supremacy in both parts of the Continent—something that had eluded him after World War II. If he were to fail, however, the result would be bolstering American ascendancy in the Cold War even more while breaking the unity of international communism, with potentially disastrous consequences for the integrity of his own empire.

The Berlin Fiasco

The Berlin blockade started surreptitiously; from contemporary documents, it even is difficult to tell exactly when it did. Ever since May 1948, Soviet authorities had been harassing Western traffic in and out of the city; the lack of a firm response did not discourage them from trying more. On June 17, the Soviet walkout from the Berlin *Kommandatura* signaled a desire for confrontation, though not necessarily a definite plan of action.[1] The paucity of internal Soviet documents about the preparation and implementation of the blockade[2] suggests its being largely improvised on short notice rather than meticulously designed in advance. It was very much Stalin's personal undertaking, managed more casually than it deserved.

The setting for the confrontation, though not its cause, was provided by the Allies' intention to introduce a separate currency in their zones

—an indispensable measure because of the unchecked Soviet printing of the country's common currency, which fueled inflation. Moscow anticipated the separation by getting its own money ready to be introduced in its zone. When, on June 18, the Western powers officially announced their expected monetary reform, they unexpectedly exempted from it their Berlin sectors, pending further negotiations with the Soviet authorities.[3] But these chose to ignore the conciliatory gesture, instead precipitating a confrontation by preventively prohibiting the Allies from circulating their new currency in their own sectors on the spurious grounds that the entire city "comes within the Soviet zone of occupation and is economically part of the Soviet zone."[4]

Moscow did not ignore the Allies' hesitation to proceed with a complete monetary break. After two days its local representative, Marshal Sokolovskii, acknowledged "with satisfaction" their statement that the reform was not being applied in Berlin, and challenged them to acquiesce in his government's claim of supremacy in the city as a whole.[5] They did not, but they agreed to discuss the implications of their action with his economic experts.[6] In advance of the discussion, however, the Soviet Union prejudged it by announcing the introduction of its own German currency, demanding its acceptance as legal tender in West Berlin as well.[7]

The gist of the situation was that each successive Soviet step aggravated it while the opportunities to mitigate it were neglected. At the experts' meeting on June 22, Western representatives put on display their disagreements.[8] While the session was in progress, the Soviet occupation government, without awaiting its outcome, unilaterally decreed the circulation of its new money in the Western part of Berlin.[9] Only this intolerable move finally prompted the Allies to confront the challenge by introducing their own currency there.

After the lines had thus been drawn, the time came for Moscow to make its terms known. Yet this was not done. Over champagne, Sokolovskii's aide Colonel Zyrianov hinted to U.S. army officers that a redrawing of the zonal boundaries might help terminate the blockade and perhaps even prevent war.[10] Rather than the boundaries, however, the issue was the recent decision of the London conference to create a state in Western Germany. The day before imposing the blockade, the Soviet Union had reconvened in Warsaw its vassal foreign ministers to solemnly declare the decision null and void.[11] "If we were to lose in Germany," Molotov told them, "we would have lost the [last] war."[12] British intelligence found out that Moscow alerted the East Europeans of its intention to assume full control of Berlin on July 7[13]—a timetable indicative of a belief in prevailing there quickly.

Stalin did not make the blockade as tight as he could have done. He chose not to obstruct Western air access to the city, as originally envisaged in the plan prepared in his foreign ministry.[14] The loophole did not necessarily make a difference, for the survival of the besieged sectors was most unlikely to be ensured solely by the airlift inaugurated by the Allies on June 26. Indeed, the probable failure of this daring attempt to supply a population of more than two million from the air promised to humiliate them, forcing them to accept Moscow's offers of supplies and its political terms as well.

By keeping the air corridors open, the Soviet Union tried to avoid a possible military clash there—which it would have itself had to initiate if it had wanted to use its fighters to make the blockade fully effective.[15] On the ground, however, it was up to the Allies to decide whether they should attempt to force the blockade, thus taking the risk of starting hostilities under unpropitious circumstances. Although this was an unattractive option, it was not entirely out of the question.

Two of the most influential Americans in Germany, military governor Gen. Lucius Clay and his political adviser Robert Murphy, favored calling what they suspected was a Soviet bluff.[16] They may well have been right, considering the absence of any evidence of Soviet preparations for a military emergency.[17] Stalin was spared the test, however, by the U.S. Joint Chiefs of Staff, who resisted an attempt at forcing the blockade on the plausible grounds that it would fail and finally convinced the president not to authorize it.[18]

What Truman did authorize on June 28 was the dispatch to Great Britain of sixty bombers capable of carrying atomic bombs.[19] But the planes had not been configured to actually carry them. The well-publicized move went down in history as Washington's "single deliberate attempt to exploit its atomic monopoly to deter Soviet military action."[20] Since there was no such intended action to be deterred, however, the symbolic gesture influenced at most Soviet political action —though not in the desired direction.

On June 29—too soon to qualify as a reaction to the presidential announcement—Sokolovskii in a letter to his British counterpart, Gen. Brian Robertson, hinted at the possibility of lifting the blockade, but mentioned no terms. In any case, the next day the faint prospect vanished after the text of his message, having found its way to the Western press, was publicized simultaneously with General Marshall's tough statement warning that the United States was determined to remain in Berlin.[21] Although the statement was not a reaction to the Sokolovskii letter either, the coincidence made it look as if it were. Taking offense, Moscow responded by terminating also formally its participation in the *Kommandatura*. Refusing "to answer any question on the resumption of traffic unless the results of the London conference were also discussed," Sokolovskii now explicitly linked the resolution of the Berlin crisis to the overall German settlement.[22]

Until then, the Soviet terms had not been clearly stated. Only after Marshall's statement, made to sound hollow by the airlift that was unlikely to work and the dispatch of the bombers without bombs, did Stalin divulge what he hoped to accomplish and how.[23] On July 14,

Moscow's elaborate reply to the U.S. protest against the blockade point-edly singled out the West's rejection of the earlier Soviet demand for access to the industrial resources of the Ruhr as its original sin. Expounding on Stalin's vision of an undivided Germany that would be at his disposal, the document berated the progressing separation of the Western zones as the main obstacle to the preservation of its unity. Alleging that the Allies' rights in Berlin depended on their "obligatory execution . . . of the . . . agreements . . . in regard to Germany as a whole," the Soviet government proposed negotiations without precon-ditions, that is, while the blockade was in effect.

U.S. ambassador Smith saw in the dearth of Moscow's incentives evidence of its belief that it would prevail in the confrontation.[24] Such a belief was also implicit in the drafting under its auspices of a constitu-tion that was to be applicable for the whole of Germany. The docu-ment, prepared in East Berlin, envisaged Soviet-style command economy and suppression of political opposition by means of a ban on any activi-ties elastically defined as directed against the "equality of citizens."[25] Soviet planners proceeded on the correct assumption that Washington had decided not to aggravate the situation.[26]

Influential Western voices favored appeasement. They included not only the French but also the otherwise tough General Robertson, who thought that giving Moscow a share in the control of the Ruhr would be a fair price to pay for its allowing free elections and the establish-ment of a representative government in Germany, after which all foreign troops would leave.[27] But such wishful thinking did not carry weight in London, which tended to be more confrontational than Washington, albeit for doubtful reasons.[28]

Ironically, it was the the lingering belief that a united Germany might be pro-Soviet which made the British the most adamant propo-nents of a separate German state.[29] The fallacy, though not the idea it fostered, was only laid to rest by the massive upsurge among Germans of pro-Western sentiments, prompted by the blockade and the airlift. In the end the certitude of their commitment to Western democracy, much more than the uncertain Western diplomacy, was what made the Berlin airlift such a memorable success.

The longer the shuttle went on, keeping the Allies' positions from collapsing and their governments from negotiating from weakness, the more was it probable that Germany's progressing partition would extend to its still formally undivided capital as well. This threatened to make the situation that Stalin wanted to reverse worse than it had been before, thus raising doubts about the expediency of continuing the blockade. He had to win his gamble quickly if he were not to lose it.

When Stalin received the three Western ambassadors on August 2, he was conciliatory rather than aggressive, protesting any intention to oust the Allies from the city.[30] He offered to lift the blockade in return for the withdrawal from Berlin of the Western currency and the mere suspension—rather than immediate cancellation—of the London decisions. He objected not so much to the economic unification as to the political separation of the Western zones. Anxious to find out whether the envoys were authorized to negotiate about this, he was disappointed to hear they were not.[31] Instead they made the awkward offer to allow the Soviet-issued currency to circulate in West Berlin—a sensitive but secondary matter—thus generating the wrong impression that their governments were prepared to compromise on the primary issue of West German statehood.

In quest for a compromise, Stalin, assisted by Molotov, began to reduce his demands. He was no longer asking for the London accords to be deferred but merely that this should be recorded as his wish. The U.S. ambassador described the pair as "literally dripping with sweet reasonableness and desire not to embarrass"[32]—but not for long. Four days later, Molotov came back and tried to "sell the same horse twice,"[33] again demanding the deferment. He tabled a draft agreement to that effect, which the Western representatives predictably rejected. They soon reaped the benefits.

Molotov kept playing hide-and-seek for another week, complaining about the increasingly painful Western counterblockade of the Soviet zone but not making any attractive proposal. He no longer mentioned the London decisions one way or another until Stalin produced such a proposal.[34] On August 24, he offered to end the blockade, even on the next day, for nothing more than the withdrawal from Berlin of the Western currency, leaving everything else for discussion "in the near future."[35] At last Smith was satisfied—but his Washington superiors were not. They were not in a hurry after the currency reform in the Western zones had proved a spectacular success and time had begun working in their favor in other ways as well.

Reminding Stalin that a clock was ticking, the coming into effect of the Brussels treaty on August 25 brought the dreaded Western military pact yet another step closer. After several vain attempts at drafting a Berlin document that would include a proviso altering what the Allies had decided in London, two days later Molotov stopped trying for a quick settlement of the pending issues. He proposed to refer them "for discussion and agreement among Military Governors in Berlin."[36] This was the essence of the deal concluded in Moscow on August 30, though not before he once more attempted to slip into the text at least a little reference to those London decisions.

In Berlin, the talks could more readily be influenced—for better or worse—by the pressure that the Soviet Union was capable of generating locally. Its trying, though not succeeding, to foist its own appointees on the West Berlin municipal administration could not fail to influence the situation for the worse. So did the convocation on September 1 of the committee entrusted to prepare West Germany's constitution, even though the event had been scheduled some time earlier. By now the blockade had already split the city in countless ways, thus making it in any case an unpromising venue for further negotiations.[37] No sooner did they resume than they were bogged down amid recriminations.

Smith wondered whether Stalin may not have misperceived a Western willingness to acquiesce in Soviet control of the city in return for a mere face-saving appearance of its four-power administration.[38] It is more likely that he simply did not know what to do next. Not only did he avoid further conversations with the Western ambassadors, but he also postponed for as much as ten weeks his planned meeting with top East German party officials while going on a long vacation.[39] During his absence from Moscow, his bureaucracy was supplying him with selective upbeat reports, suggesting that the airlift was in trouble and would be unable to carry the West Berliners through the coming winter.[40]

In reality, while the boss was away his underlings made trouble for him. Both Walter Ulbricht, the East German party chief, and his Soviet supervisor Colonel Tiulpanov saw the blockade as a golden opportunity to advance their power as well as the cause of communism.[41] In trying to topple elected officials in the Western sectors, they were instrumental in dispatching gangs to provoke riots there even while the four-power talks were in progress. This did not help Sokolovskii's efforts at the negotiating table to extract concessions by tortuous interpretation of the ambiguous Moscow agreement. Nor did his veiled threats that the blockade might be extended to the air impress any longer.[42]

It was now the West's turn to raise ultimatums and refuse further discussions as useless, forcing Moscow on the defensive. "In every field," Marshall noted, "the Russians [are] retreating." He rightly perceived Berlin as their last foothold in a losing contest for Europe.[43] The note by the Soviet government, sent on September 25, conveyed vividly its predicament by complaining to the Western powers why they "do not confine themselves to their sovereign administration of the western zones of Germany but wish at the same time to administer also the Soviet zone of occupation in currency and finance matters by introducing their separate currency in Berlin, which is the center of the Soviet zone, thus disrupting the economy of the eastern zone of Germany and, in the final analysis, forcing the U.S.S.R. to withdraw therefrom."[44]

In its own distorted way, the document accurately identified the greater dynamism of the West's economic system—not to speak of the greater attractiveness of its political system—as the root cause of Moscow's inability to compete on open terms. Rather than go on playing a losing gambit, the Soviet government proceeded to smash the chessboard. The day after it had sent the note, it made public the confidential record of the futile four-power discussions, laying the blame for their breakdown to the other side.[45] By then, Washington had already decided to refer the Berlin question to the United Nations Security Council, where a month later a Soviet veto prevented it from being considered.[46] There was truth in Molotov's otherwise outrageous claim that "no 'blockade of Berlin' exists in reality,"[47] though the opposite of what he meant, for it was now the Soviet Union that had blockaded itself—not only in Germany but also in Yugoslavia.

The Dictator's Delusions

Like that of the Berlin airlift in its early stages, the failure of Tito's resistance had originally seemed all but certain. He was most unlikely to survive in power the solemn curse that Moscow had pronounced against him at the Cominform meeting convened for that purpose in Bucharest on June 19.[48] He had enough Soviet sympathizers within his own party, many of whom, especially diplomats, preferred exile to supporting his uncertain cause. The Yugoslav legation in Budapest, for example, fell apart after its chief political officer, Lazar Brankov, and eight of his colleagues had sided with the Cominform against their government and the remaining fourteen had been expelled.[49]

Tito desperately tried to avoid a complete break with Stalin. Swallowing his insults, Yugoslavia in July 1948 loyally supported the Soviet Union at the contentious international conference, held in Belgrade, on regulating navigation on the river Danube.[50] As late as September 26, Tito confided to the same British gadfly Konni Zilliacus in whom Stalin also had been disposed to entrust calculated confidences[51] that he considered Yugoslavia's place in Europe to be on the Soviet side, expressing confidence that sooner or later Stalin would realize his error and seek reconciliation.[52]

Instead Moscow grew if anything more intransigent. After Soviet military attaché in Belgrade Sidorovich had in August unsuccessfully conspired with two of Tito's generals to overthrow him, Stalin kept cultivating the one who had managed to escape, Pero Popivoda, as the prospective head of a Yugoslav government-in-exile.[53] But if Tito deluded himself about Stalin's having second thoughts, so did Stalin in thinking that he could merely "move his little finger, and there will be no Tito."[54] The longer the Yugoslav leader stayed in power, the more obvious was the resilience of a regime similar to Stalin's own which, like his during World War II, also proved its capacity to tap the patriotic sentiments of its people.

It was the ability of the Yugoslav communists to turn Stalinism against Stalin that prompted him finally to block their Eastern European comrades' "national roads to socialism" lest they follow suit. In the summer of 1948, Czechoslovak party general secretary Slánský could still tell the ubiquitous Zilliacus that his country was marching to socialism on its special road.[55] But soon Moscow made it clear that there was to be no other road than the Soviet one. Zhdanov's assistant Pavel Iudin reinterpreted the "people's democracy" no longer as a transitional stage to socialism of indefinite duration but as a system for all intents and purposes indistinguishable from the "dictatorship of the proletariat" practiced in the Soviet Union.[56]

This was the interpretation pioneered by none other than Tito as early as September 1947[57] but now turned against him by his former partner Dimitrov acting upon Soviet instructions.[58] The Bulgarian leader proclaimed its universal validity in an authoritative statement to his party's congress in December 1948. By then the forced merger of social democratic parties with the communist, begun in the Soviet zone of Germany as early as 1946, had been completed throughout the part of Europe Moscow controlled. At the same time, the many turncoats and opportunists who had jumped on the communist bandwagon before were now dropped off in a wave of purges during the fall of 1948.

As was Stalin's habit, someone had to pay for the policies discarded as unsuitable in an unexpected situation. The key figure in his miscalculations with Tito was Zhdanov. He had presided both over the Cominform's first meeting, which had chosen Yugoslavia as the vanguard, and over the second one, which had stigmatized its heresy. The reversal of roles did not reverse his fall from Stalin's grace, accelerated by the failure of the communist offensive in Western Europe, and did not prevent the rise of his rival Malenkov as the despot's favorite.[59]

The Soviet campaigns against both "cosmopolitanism" and its obverse, "bourgeois nationalism"—whose virtuous opposites were "proletarian internationalism" and "socialist patriotism"—reflected the accounting. The mismanagement of communism in the West, which was Zhdanov's responsibility, could be blamed on the heresy of cosmopolitanism, while the pitfalls of the national roads to socialism, highlighted by the recent developments in Yugoslavia, could be ascribed to the nationalist deviation. Until the Tito problem erupted, the bashing of cosmopolitanism took precedence over that of "bourgeois nationalism"—which was the failed Zhdanov's bête noire—only to subside after his sudden death on August 31.[60]

Zhdanov passed away soon after Malenkov's return to the party secretariat, from where he had earlier been removed by an intrigue.[61] Although the official obituaries extolling Zhdanov's accomplishments conspicuously omitted his Cominform assignment, he was given a party funeral with full honors. Rumors of foul play nonetheless surrounded his timely exit, dispelled neither by the insinuation in 1953 that he had been helped to die nor by its subsequent incomplete retraction.[62]

The Zhdanov-Malenkov rivalry has been a prodigy of Kremlinological reading from the tea leaves contours of a larger struggle, perhaps threatening Stalin himself.[63] Different authors have interpreted Zhdanov's demise as both "the defeat of moderation" and its opposite, and the concurrent policy turnabout as both crucial and nonexistent.[64] Whatever else may be said, despite all of Stalin's blunders, his authority remained overwhelming. Regardless of how much he allowed those around him to differ and bicker about strategies and priorities while he was still undecided, once he made up his mind the resulting policy was none other than his own.[65] The problem was his inability to make up his mind when so much was going wrong.

In trying to offset its setbacks in Europe, Moscow encouraged communist advance in parts of the world where the West had been retreating. It provided political direction, though precious little material support, to its followers participating in the anticolonial movements in Indonesia, Indochina, Malaya, Burma, and the Philippines.[66] At the international youth conference it sponsored in Calcutta in February 1948, it imposed a similar turnabout as that which the Cominform had required the Western European communists to perform the previous September[67]—with similarly disappointing results.

At issue was the rejection of an alliance with the "nationalist bourgeoisie"—shorthand for capitalists willing to work with communists—because of its alleged subservience to the colonial masters. Rather than follow a common leadership, the Asian communists were to wage armed struggle on their own.[68] Indonesia was an exception, presumably because the war for its independence, in which Indonesian communists had been taking part alongside the nationalists, was on the verge of being won. In Asia, Moscow envisaged the leading role for the Indian rather than the Chinese communists—whom Stalin saw as fighting an uphill battle against Chiang Kai shek's nationalist forces backed by the United States.[69] In contrast, after the demise of the British raj the Indians were left with no such formidable enemy to fight.[70]

Once the armistice with the Dutch brought to power in Indonesia a government not controlled by the communists, Moscow lent support, albeit insufficient, to an attempt at reversing this outcome. In August 1948 their leader Musso arrived in the country from the Soviet Union with the stated intention to implement there what he called the "Gottwald plan"—a seizure of power Czechoslovak-style. As had been the case in Eastern Europe, he dwelt on the need to capture the control of the "national revolution" in preparing the "proletarian" one. The proclamation on September 18 of a rival government by communist forces in the city of Maidun gave substance to his talk.[71]

The Maidun rebellion was quickly crushed. With few exceptions, turning to armed struggle proved costly for the Asian communists. It led them to political isolation and eventually defeat. The notable exceptions were China and Indochina, where they won enough support from not only the "national bourgeoisie" but also the rest of the population to keep advancing. By effectively defying the Soviet strategic concept, they did not endear themselves to Stalin, who withheld substantial support from their revolutions.

In contrast, the Soviet dictator helped both diplomatically and materially in the birth of the State of Israel, which he expected to become an anti-Western outpost in the Middle East. Ignoring the democratic and pro-Western thrust of Zionism, he encouraged, although he had not initiated, the considerable assistance provided to Jewish fighters in Palestine by Czechoslovakia—for profit rather than for political reasons —even before the communists took power in Prague.[72] Afterward Slánský and other party protagonists of Jewish origin used their extensive international connections to expand the assistance, which entailed not only massive arms deliveries but also the training of Israeli military personnel on Czechoslovak territory.[73] Following the Soviet approval on June 20 of the proposal by Shmuel Mikunis, the general secretary of the Israeli communist party, for the recruitment of Jewish military volunteers in Eastern Europe,[74] a unit was organized in Czechoslovakia.

The communists' efforts to infiltrate the unit, however, ended in failure. Nor did Moscow succeed in its attempts to win Western Jewry to its side by cultivating its gratitude for the Soviet contribution to the destruction of Nazism. Whether Stalin hoped to tap the supposedly fabulous Jewish riches[75] or, more likely, to enlist the U.S. Jewish lobby on his side, the weather vane of this unpromising enterprise was the Anti-Fascist Jewish Committee, created in World War II under the auspices of his security chief, Lavrentii P. Beriia. The front organization was supervised by Deputy Foreign Minister Solomon A. Lozovskii and presided over by the well-known Yiddish actor Shlomo Mikhoels.[76]

The murder of Mikhoels in January 1948 in a staged traffic accident, probably masterminded by Stalin behind Beriia's back,[77] was indicative of the dictator's growing skepticism about his ability to manipulate Jewish sentiments to his advantage. After doubts had been cast on the committee's loyalty by the junior party ideologist Mikhail Suslov,[78] Stalin contrived the absurd idea that its members had been colluding with the United States in trying to establish a Jewish republic in the Crimea.[79] Although he allowed Mikhoels to be buried, like Zhdanov, with full honors, he soon had more substantive reasons to ponder whether he had not miscalculated with his support of Zionism as well.

The arrival in Moscow at the beginning of September of Israel's first ambassador, Golda Meyersohn, touched off in the regimented city tumultous demonstrations by Jewish well-wishers. Although these demonstrations were not anti-Soviet they were worrisome to Stalin because they were spontaneous.[80] No sooner did they subside than his star Jewish propagandist, Ilia Erenburg, fired in *Pravda* the first salvo against Zionism as a variety of bourgeois nationalism. There followed a high-level decision to suppress Yiddish culture as another of its manifestations.[81] The Jewish Anti-Fascist Committee was disbanded, and its activists lost official favor though not yet their lives.[82]

Since Jews could conveniently be categorized as susceptible to both nationalism and cosmopolitanism, the campaign against the latter heresy surged again during December.[83] Lashing at Israel's American connection, Soviet media conspicuously publicized arrests of malefactors with Jewish names, reported with racial slurs. But the targets of the campaign, whose main goal was to help insulate the Soviet realm from nefarious Western influences, extended to alleged cosmopolitans in the arts and sciences regardless of race or religion.[84] By the end of January 1949, the anti-Jewish slant had disappeared as abruptly as it had appeared before.[85]

In those closing months of 1948, when Stalin was away on vacation, not everything that occurred could be attributed to his management, although much of what was happening reflected his delusions. On November 15, the Austrian communists secretly prepared a detailed plan for a coup d'état, astounding in its assumption that some resistance could be expected from the social democrats but none from the Western military forces stationed in most of the country.[86] The authors of the plan, which was wisely shelved, may well have been acting upon Moscow's prodding to probe the West's resistance in the remaining "gray" areas where the lines had not yet hardened as much as in Germany. The idea probably originated in the international department of the Soviet party central committee that was trying to read Stalin's mind in the absence of his orders.

This is what the department had been doing with more serious consequences about Greece, where the chief of the communist guerrillas, Markos Vafiades, was stripped of his party positions on the same day the Austrians finalized their plan.[87] The demotion on the preposterous charges of "opportunism" of the man who had been instrumental in the Soviet peace overtures to the Athens government followed Moscow's recent reactivation of the insurgency after a deceptive improvement of its prospects for success. A special committee of the Cominform was established to help meet "the needs of the democratic army of Greece,"[88] made urgent by the incipient termination of the Yugoslav deliveries, curtailed as a result of the Stalin-Tito break.[89] But the Soviet insistence on the reorganization of the guerrilla army into a regular one sealed its eventual defeat.[90]

Back in Moscow, Stalin presided in early January over the inauguration of the Council for Mutual Economic Assistance, later known as Comecon.[91] To the Eastern European officials invited there for the occasion, he confided his fanciful view that the Western European countries, particularly Italy and France, could be detached from the United States by making them critically dependent on the Soviet supply of raw materials. He fantasized that the Comecon, by creating "a raw material base for the whole of Europe ... will become more important than the Cominform."[92]

The new organization, designed to compete with the Marshall Plan, was its caricature rather than replica. Stalin himself showed a lack of faith in his creation by leaving it as little more than an empty shell for the rest of his life while exploiting his dependencies through imposed bilateral agreements without any pretense of partnership. In the end, this allowed the Soviet Union to extract *from* them for its reconstruction about the same fourteen billion dollars worth of deliveries that the United States supplied *to* its allies for their reconstruction through the Marshall Plan.[93] Here was the key difference between his empire by imposition and the American empire "by invitation,"[94] which was now about to take also a military form as Stalin's *cauchemar des alliances* was coming true.

Militarization of the Cold War

Having started as a political conflict, the Cold War changed its character as a result of the Berlin crisis. The risk of an armed clash created by Soviet action made the West more inclined to conceive of the Soviet threat as potentially a military one, thus giving the East-West rivalry a dimension it had been lacking before. Although neither side felt compelled to substantially boost its armies to prepare for war, they both came to see their relationship more in military terms.

The change was more pronounced, as well as more consequential, in Washington, which was prone to vastly exaggerate Soviet military might.[95] The Joint Chiefs of Staff believed the Russians capable of overrunning Western Europe within six months, bombing Alaska and the Puget Sound, and even stirring up "serious internal disorders by subversion and disruption" in the United States through their fifth columns.[96] To retaliate in kind against the "vicious covert activities of the USSR," Washington had created even before the German crisis the Office of Special Projects, later renamed Office of Policy Coordination, as an agency to foster trouble within the Soviet realm.[97] This was still a step ahead in political warfare only. Nor did the $4.4 billion increase of the $10 billion U.S. defense budget, approved under the impact of the Berlin confrontation, entail a significant expansion of the nation's armed forces, sharply reduced since the end of World War II.[98]

What changed because of the crisis was the importance attributed to military power as an instrument of policy. Truman later reminisced that the experience had taught him "that there was only one way to avoid a third world war, and that was to lead from positions of strength. We had to rearm ourselves and our allies and, at the same time, deal with the Russians in a manner they could never interpret as weakness."[99] The main reason the United States still did not feel compelled to substantially bolster its armed forces was the belief in the sufficiency of its nuclear monopoly. During the year preceding the Berlin blockade, its stockpile of atomic bombs had grown from thirteen to fifty. The blockade itself prompted an expansion of both the fleet of bombers needed for their delivery and the network of overseas U.S. bases encircling the Soviet Union.[100] Most important, it precipitated the decision to use the bombs if appropriate.[101]

On September 10, 1948, the National Security Council stated in no uncertain terms that the Russians must "never be given the slightest reason to believe that the U.S. would even consider not to use atomic weapons against them if necessary. It might take no more than a suggestion of such consideration, perhaps magnified into a doubt, were it planted in the minds of responsible Soviet officials, to provoke exactly the Soviet aggression which it is fundamentally U.S. policy to avert."[102] Since the United States was capable of delivering a nuclear strike against the Soviet Union but not vice versa, Moscow directed all its "intelligence resources ... to get to know ... concrete [American] plans."[103] To what extent it succeeded was an important, though not necessarily critical, determinant of its policy.

Although Stalin's spies were capable of penetrating the inner sanctums of Western governments, he was prevented from making the best use of their information. Impressed by the organizational prowess of his adversaries, the month after launching the Cominform as a response to the Marshall Plan he had also established a superintelligence agency in response to the recent reorganization of the U.S. intelligence apparatus. This Committee of Information at the Soviet foreign ministry, however, was so rigidly centralized that its staff could not effectively process the vast masses of incoming material. Stalin himself was receiving from Molotov only brief digests of what the foreign minister's overworked subordinates considered important, and given the dictator's erratic working habits, he hardly found enough time to read, much less absorb, even that.[104]

Whatever the limited value of what Stalin was thus capable of learning about the true intentions of his enemies, at least he could be quite certain that they were not about to attack him. At the same time, enough of what was being said publicly sounded sufficiently disturbing to lend support to the belief that, if push came to shove, they—like himself—would not be inhibited by scruples. The articles published in July and August 1948 by the retired U.S. Air Force Chief of Staff Gen. Carl Spaatz, for example, casually referred to Washington's intent to wipe out "a few hundred square miles of industrial area in a score of Russian cities" as part of its nuclear strategy.[105]

The tendency to expect the worst from the capitalists, which permeated the Marxist doctrine, reinforced the disposition of Soviet representatives abroad to overstate, for their own protection, the possibility of the worst scenario. It was an attitude more typical of generals preparing to fight a war rather than of diplomats, whose task is to prevent it. In the summer of 1948 Paniushkin, the former director of the foreign policy department of the Soviet party central committee now assigned as ambassador to Washington, was scaring his superiors by what he described hyperbolically as a "war psychosis" rampant in the U.S. capital.[106]

Further distorted by Stalin's ingrained insecurity, the estimates informing his policies were therefore shaped to an extraordinary degree by preconceptions and fantasies rather than hard facts. He was even more prone to exaggerate the Americans' strength than they were prone to exaggerate his, suspecting them of an ability to perform superhuman feats. If their top military men had visions of Soviet planes dropping bombs on the West Coast and of the Reds handily subverting the country, his nightmare was that of the United States encroaching on the Soviet homeland from the Crimea with Jewish help, mobilized by the actor Mikhoels![107]

While the specter of war was never far from the Soviet mind, the changing estimates of its probability nevertheless made a difference. In contrast to the reassuring opinion Molotov had given to the Czecho-slovak envoy in the fall of 1947,[108] a year later Stalin expressed to that country's communist boss Gottwald the somber view that war was inevitable, although he did not venture a prediction about its timing.[109] In response to Truman's dispatch to Europe of unarmed nuclear bombers, the Soviet Union reorganized its antiaircraft defenses—just in case the loads of the aircraft were more lethal next time.[110] Otherwise the Berlin crisis did not trigger any frantic Soviet military buildup either. Moscow continued to build its still mainly defensive submarine fleet, which was already the largest in the world, and streamlined its ground forces—which, too, had always been bigger than their Western counterparts—for greater efficiency.[111]

As long as the East-West conflict, despite its creeping militarization, remained primarily political, Stalin was encouraged to believe that the capitalists could be compelled even against their will not to act on their worst instincts, thus delaying the outbreak of the likely war. Yet his attempts to put them on the defensive by driving wedges between them and in other ways proved singularly counterproductive. Increasingly, he was left to rely upon the impersonal forces of history, which, though beyond his control, he nevertheless believed to be on the Soviet side and hence capable of changing the situation for the better before it could become seriously worse.

In the fall of 1948, the seemingly esoteric debate among Soviet academic experts about the allegedly imminent "general crisis" of capitalism was really about policy. At issue in this officially inspired disputation on the merits and demerits of the economist Varga's theory about the relative stabilization of capitalism—which he had in fact already abandoned in his 1947 memorandum on the Marshall Plan[112]—was nothing less than the correct assessment of where the U.S.-dominated part of the world was heading and what the proper Soviet response to its likely development ought to be.[113]

The forthcoming U.S. presidential elections made the subject topical. Stalin tried to influence the campaign by sending a flattering "open letter" to third-party candidate Henry Wallace, which for this advocate of gentler dealing with the Russians amounted to the kiss of death.[114] Moscow viewed with foreboding the generally expected victory of the Republicans, compared with whom even Truman appeared as a moderate. Shortly before the vote, the American press leaked out his uncharacteristically naive project to send Chief Justice Fred Vinson on a "peace" mission to Stalin, so that the sinner could "unburden himself" to the good judge, with beneficial results for U.S.-Soviet relations.[115]

Moscow's propagandists hailed Truman's victory as a blow to "reactionaries" bent on war, accounting for it by "his promise of a policy of peace and understanding with the Soviet Union."[116] Yet the president's reelection, having vindicated his handling of the Berlin crisis, made the U.S. posture in the Cold War if anything less rather than more accommodating. Unlike Stalin, officials in Washington did not see war impending. If it could be avoided now, the State Department estimated, it could possibly be avoided altogether, for if the deadlock with the Russians "threatens us with the loss of Berlin, it threatens them with the loss of Germany itself." In contrast with the failure of the Western appeasement of Hitler, the West's position now appeared to be the "opposite to that at the time of Munich."[117]

Not only was the Germany Stalin wanted already lost, but there were also other losses looming as preparations for the Western military alliance entered their final stages. In desperately trying to avert the inevitable, Stalin brandished both inducements and threats. In reply to questions by U.S. news executive Kingsbury Smith on January 30, 1949, he reaffirmed, after a half-year interval, his willingness to lift the Berlin blockade if the plans for the West German state were shelved, adding a proposal for the conclusion of a Soviet-American nonaggression pact.[118] Ignored, he amended the proposal into a four-power pact, while signaling a desire to meet with Truman for a tête-à-tête. Meanwhile he had communists in Western Europe stir up a gigantic "peace" campaign against the prospective alliance.[119]

Starting in a relatively low key, with French party boss Maurice Thorez invoking the threat of war but not its inevitability, the campaign grew progressively shriller.[120] The point at issue was, according to Gian Carlo Pajetta of the Italian party directorate, "to make it clear that we are in a position to create . . . a difficult situation for those who want war."[121] By the end of February, leading Western European communists were publicly serving notice that if war were to come, in their homelands the Red Army would be welcomed as a liberator.[122]

In an open letter to Truman, the Peace Council of France harangued against its membership in the projected North Atlantic Treaty Organization as unconstitutional because of the alliance's allegedly aggressive aims.[123] This tallied with Moscow's insistence that NATO would be incompatible with the UN charter as well. Threatening annulment of its still formally valid wartime treaties with the Western powers if the pact came into effect,[124] the Soviet Union tried to intimidate the still undecided candidates for membership. Challenging the Norwegians to explain their intentions, it precipitated their decision to join the alliance, clearing the way to its formal inauguration on April 4.[125]

Even before this happened, but when the outcome was already all but certain, on March 5 Moscow announced without explanation the replacement of its two top foreign policy officials: Molotov and Anastas A. Mikoian, the minister of foreign trade. They had been presiding first over the rise and then over the fall of, respectively, Soviet political and economic relations with the West during the past decade. The next day Nikolai A. Voznesenskii, until recently a Stalin favorite, lost his post of planning chief. And soon afterward the international department of the central committee was replaced by a foreign policy commission, headed by the editor-in-chief of the Cominform newspaper, Vagan Grigorian.[126]

In trying to explain the changes, Kennan noted a resemblance to the Soviet moves preceding the Hitler-Stalin pact ten years before. Expecting another such bombshell to follow, he wondered whether it might "be some sort of exploitation of the satellites against us, from which Russia herself would remain aloof? Somewhere Moscow must think that it has a means of bedeviling the West and promoting Soviet objectives which will not involve the Soviet Union directly."[127] This time, Kennan's famous intuition failed him. But at least he was right in suspecting that something was brewing in Eastern Europe.

The elevation to the post of foreign minister of Andrei Ia. Vyshinskii, the sinister prosecutor in the Great Purge trials of the 1930s, presaged symbolically what it would be. More to the point, the promotion of this yes-man, not to speak of the succession to the crafty Mikoian of his bland deputy Mikhail A. Menshikov and the other personnel changes, "reflected a growing lack of self-confidence about the correctness of the adopted course and apprehension about the uncertain future of the country."[128] At the same time, they marked a turn toward less structured and even more arbitrary ways of making policy.

Having bred an extraordinary series of foreign policy disasters, Stalin's harvest of blunders gave him good reasons to pause and take stock of what had been happening. The split within the once monolithic international communism, which he had gratuitously provoked by his row with Tito, threatened to get worse. Nor was the split in Germany, forced by his Berlin blockade, getting any better. Moreover, the ensuing militarization of the Cold War apparently brought a real war closer. Whether the appearance was deceptive or true, the time had come for the weaker Soviet side to start cutting its losses.

4

Retreat and Consolidation April 1949–November 1949

Once NATO had been established, there was little Stalin could do to reverse American ascendancy in Western Europe; what he could do to rectify the lopsided correlation of forces was to tighten the screws in Eastern Europe while retreating in the two places where he had played for high stakes and failed—Berlin and Yugoslavia. The manner in which he would proceed would be an indication of whether he preferred accommodation to further confrontation. The actual decision about which course to take, however, depended upon other factors as well, not all of them within Soviet control.

Cutting the Losses

Of Stalin's two entanglements, the German one was easier to disentangle. Unlike Tito's dissidence, which, if tolerated, threatened to set a dangerous example in Eastern Europe, the Berlin blockade could be disposed of without unacceptable damage to Soviet security. In mid-December 1948 Stalin began to prepare a retreat by admonishing the East German leaders to moderate their conduct, which had proved so plainly counterproductive.[1] Exempting their part of Germany, for the time being, from the sovietization drive he had been pushing elsewhere, he shared with them his confidence that the country could be reunified in accordance with his wishes once the Berlin problem has been settled.

Another six weeks elapsed before Stalin signaled publicly a willingness to terminate the standoff in his "interview" with Kingsbury Smith —actually a statement in which he chose to address some of the hundreds of written questions by Western journalists that he had been keeping on file for appropriate occasions. Commenting on Berlin, he pointedly omitted the currency issue that had originally triggered the crisis but had since lost its relevance because of the effective separation of the city's two parts.[2] Instead he hinted that the blockade might be lifted if the planned establishment of a West German state were held off pending the convocation of a conference of foreign ministers, which would address the German question in its entirety, and if the Allies simultaneously terminated their own counterblockade of the Eastern zone. Since this retaliatory measure had been increasingly painful for the Soviet Union,[3] while the siege of their Berlin sectors had become manageable for them,[4] the Western powers could afford to ignore the hints, waiting for Stalin's terms to be spelled out more clearly.

The replacement of Molotov by Vyshinskii did not presage, as Kennan feared, some new Soviet mischief on the order of the Hitler-Stalin pact, but was a prelude to doing business. On March 15 Soviet ambassador to the UN Iakov Malik invited his U.S. counterpart, Philip C. Jessup, to an informal meeting at his New York office, to convey the message that Stalin's omission of the currency issue in his latest interview had not been accidental.[5] When asked to be more specific, the diplomat turned evasive, expecting the other side to make the next move, but none came.

Unlike ten years earlier, in 1949 Moscow was playing a weak hand. When in 1939 Stalin dismissed Litvinov to signal a desire for rapprochement to Hitler, he had something substantive to offer—the promise of Soviet neutrality that Germany badly needed; now all he could come up with was retracting his own blunder. Although the makings of a deal were missing, Malik desperately tried to bargain. Six days after his first secret talk with Jessup, during another meeting he delivered Vyshinskii's more specific, though merely oral, message that the blockade could be ended even before the ministers would meet if there were an agreement on their meeting.[6]

The absence of anything in writing that would show what had actually been discussed so far enabled the Soviet diplomat to bid for a U.S. concession by pretending to believe that Jessup had said something that he had not. Malik averred that Vyshinskii had understood Jessup's remark that the date for creating the West German state was not definite as a promise to postpone it indefinitely. Disabused of such a notion, Malik finally produced a real proposal: to make the end of the blockade coincide with the conference and hold off the inauguration while the foreign ministers negotiated.[7] Having dropped all its substantive demands, Moscow still cared about saving face.

Unable to agree among themselves, the Allies were not helpful. The Soviet Union prodded them by releasing to the press an account of the secret Malik-Jessup talks.[8] But it no longer insisted that West Germany's statehood must wait until the ministers had concluded their deliberations; a mere understanding on the date on which they would start talking would be good enough. The offer was irresistible; yet it still took another ten days of haggling to achieve an agreement.

Promised not to bring the West German state into being before the ministers met, Moscow now sought to postpone for even a few days the date of their meeting; it finally accepted May 23. For their part, the Allies had been pressing for the termination of the blockade as early as May 9 but eventually settled for May 12.[9] On that date, both sides claimed victory, although only the Soviet Union had suffered a defeat. It placed its last dwindling hopes in its ability to thwart the Allies' West German project at the eleventh hour by exploiting their possible disagreements at the conference.

Shortly before the ministers got together, Soviet hopes had been encouraged by the press leak of Kennan's secret memorandum advocating a united, demilitarized, and neutral Germany.[10] Yet the leak, never to be clarified, was a deceptive harbinger of a change in U.S. policy. It made Washington if anything even more determined to proceed strictly in accordance with the letter of the agreement—if not with the spirit that Moscow wanted to read into it. On the same day the foreign ministers convened in London, May 23, the West German constitution was adopted in Bonn—to become effective at midnight.

Acting as if the German division could still be averted, Vyshinskii at the conference pleaded for a "return to Potsdam"—the principle of joint responsibility for the whole of Germany rendered obsolete by recent developments. The new Soviet formula envisaged the conclusion of a peace treaty with an all-German State Council—to be elected in the four occupation zones in a manner yet to be decided, to be followed by the withdrawal from there of foreign troops by a date yet to be determined. Rebuffed, as could be predicted, Vyshinskii, in trying at least to undo the damage that the blockade had done to Soviet interests, proposed to reinstate unitary administration in Berlin—with the same result.[11] Since the discord among the capitalists that Moscow hoped to turn to its advantage had not materialized, the conference would have been a total failure had they not been more successful in exploiting a discord between communists—Stalin and Tito.

In an attempt to break the deadlock preventing the conclusion of a state treaty with Austria, the Allies offered to back the Soviet economic claims against it if Moscow abandoned its support of the claims by Yugoslavia.[12] Vyshinskii obliged only too readily, thus opening the way to an agreement whereby the four ministers entrusted to their deputies the preparation of the treaty's final draft by September 1.[13] As the Soviet bank in Vienna stopped issuing credit and Soviet-administered factories in the country's eastern part ceased to accept long-term orders, it was reasonable to assume that the Austrian settlement was on hand.[14] It was not so reasonable for Soviet officials to draw from the outcome of the conference the conclusion that another gathering of foreign ministers would result in the German settlement as well.[15]

Such wishful thinking was all the more unwarranted considering the rapid sequence in which the Federal Republic of Germany was proclaimed, its parliament elected, and its government installed in office in Bonn. Yet the creation in East Berlin on October 8 of Moscow's own German Democratic Republic showed that the illusion was genuinely held. Contrary to retrospective appearances, the Soviet Union did not create this artificial state as the second best solution of the German question for the foreseeable future but as a temporary expedient, which was to provide an institutional base for its continued unification efforts.[16]

Professing to speak for all Germans and ignoring the state that had arisen in the West, the constitution for East Germany, promulgated under Soviet auspices in their former capital, described the country as indivisible, with a single citizenship, and comprising one customs and trade area.[17] The authoritative explanation of this fiction, intended for insiders, proclaimed programmatically that "the origin of real Germany is in the Soviet occupation zone. Consequently, this is not the creation of an East German state or an East German government but of a government for all of Germany."[18]

A similar fiction underlay the Bonn constitution, likewise presumed to be expressive of the whole nation's will and refusing to recognize the existence of another German state—but for different reasons. Heeding the wishes of their constituents, the West German politicians were those who kept the door open for reunification. They made the constitution temporary and wrote into it stronger provisions to safeguard national unity than originally favored by the Western occupation powers[19]—which for all intents and purposes envisaged an indefinite division.

Otherwise, however, significant differences distinguished the attitudes of the three Allies. The Americans were the most confident in their belief that the vigor of West German democracy would eventually compel Moscow to concede defeat and allow for the absorption of its German state by the Western one as a result of free elections—but regarded the consolidation of the Bonn republic the necessary prerequisite. The British still had their fears that a united Germany might go communist while many, perhaps most, French—as the author François Mauriac quipped—loved Germany so much that they would rather have two of them.

In the reversed situation in the East, the communists could afford to ignore the preferences of the people while trying to convince Stalin that the state they wished to rule was in his best interest. Their desire to build it up before extending it to the West mirrored the American hopes about the Federal Republic but, more pertinently, drew inspiration from his notion of "socialism in one country," advanced in the 1920s as a precept for gathering strength for later expansion of Soviet power. Yet Stalin, still hoping that he could have his way in the whole of Germany, was not to be easily convinced about the viability of communism in the part he controlled. He was wrong and its local proponents were right in their banking on the creation of a state that Moscow would eventually find useful—although in the end its absorption by Bonn would vindicate, if only after forty years, the foresight of the founding fathers of the German democracy and their American supporters.

Stalin's exaggerated optimism about Germany facilitated progress toward the state treaty with Austria. After its draft had been finished, Vyshinskii reported having instructions from Moscow to conclude an agreement providing for the departure of all foreign troops from the country in exchange for hefty Austrian reparations to the Soviet Union.[20] Following the acceptance of its economic demands by the Western Allies on November 18, the deal was all but ready to be signed.[21] Regardless of the stabilization in central Europe, however, Stalin's readiness for further accommodation hinged also upon the outcome of the other crisis he had brought upon himself—the Yugoslav one. Its aggravation coincided with his presiding over a new wave of repression showing how much his internal insecurity influenced the Soviet conduct abroad.

The Trials of Past Errors

As early as the aftermath of Zhdanov's providential death, U.S. ambassador to Paris Jefferson Caffery reported having heard from a "trustworthy former Comintern agent" that a reenactment of Stalin's prewar Great Purge might be in the offing.[22] What followed indeed resembled the model that had unfolded after the death in 1934 of another of Stalin's close associates, Sergei M. Kirov (allegedly murdered by enemy agents but actually liquidated with the despot's connivance).[23] The purge first targeted "saboteurs and wreckers" from among noncommunist loyalists, then extended into the party's own ranks, and finally culminated in the spectacle of staged trials during which some of its most senior leaders were confessing to incredible crimes before being dispatched to their deaths. Yet not everything was planned in advance or was Stalin's doing.

This time the initial impetus did not come from Moscow. The purges began in Yugoslavia, where in April 1948 a show trial in Ljubljana was the first to cast a shadow on the integrity of veterans of the international brigades in the Spanish civil war, insinuating their involvement after their defeat with Nazi agents in French internment camps, besides their more recent liaison with Western secret services.[24] When Tito, after his break with Moscow, began to clamp down on its Yugoslav sympathizers, these were similarly accused of having been "opportunists, Trotskyites, and careerists" already during World War II—traitors then and now again.[25]

Inspired by Stalinism rather than by Stalin personally, other Balkan purges helped his local minions to preventively protect themselves from his fury and settle their own scores as well. No sooner did Moscow start blackballing Tito in April 1948 than Romania's party boss Gheorghe Gheorghiu-Dej, deeply implicated in past contacts with Belgrade, had his rival, Lucreţiu Pătrăşcanu, jailed on the not entirely false charge of nationalism, thus identifying a prominent potential Titoist other than himself.[26] Half a year later, his Albanian counterpart, Hoxha, similarly sacked his rival, Xoxe, from the ministry of the interior, branding him as guilty of Titoism—as, until recently, had been almost everyone in the Tirana leadership. Denouncing him was for Hoxha "almost like denouncing himself but preferable."[27]

The concurrent purge in Poland, which zeroed in on Gomułka, evolved from the intraparty debate about the viability of the nation's special "road to socialism" after Moscow had voiced its displeasure with such roads. The debate entailed jockeying for Soviet favor but not any real or suspected Titoism, which would have required a turnabout that none of the intensely unpopular Polish communists could afford to risk. Not an admirer of Tito, Gomułka differed from the party chairman, Bolesław Bierut, in his sensitivity to the political costs of the dictated change of course but not in his fundamental loyalty to Moscow. When Bierut convened a special session of the central committee to ostracize him as a nationalist, he acted on his own account rather than on Stalin's orders, though within his presumed intentions.[28]

As the crowning piece of evidence against Gomułka, his accusers unearthed a 1944 article in the underground press by his aide Władysław Bieńkowski, written to comply with Soviet criticism of the doubts Gomułka had expressed about the feasibility of collaboration with nationalists.[29] Since the article had articulated the Moscow line that was now finally being discarded as erroneous, the ploy had the familiar Stalinist twist. Yet after no more than perfunctory repentance Gomułka suffered the loss of his secretaryship but not of Stalin's grace. Invited to visit Stalin shortly afterward, he was received in the Kremlin benevolently, with expressions of regret about the "misunderstandings."[30]

The Czechoslovak communists, too, found Moscow "keenly interested in how we conduct the purge in our country,"[31] but were not always able to read its signals correctly. In June 1948 they had to be advised "not to overdo it."[32] Three months later, their ideologue Arnošt Kolman nevertheless took it upon himself to publicly attack the Prague leadership for nationalism. Since he had close connections with Moscow and was even a Soviet citizen, panic ensued. Yet he was himself purged because of his "Trotskyism" after the party chief Gottwald went to see Stalin to seek clarification.[33]

While providing the inspiration and the pressure, the despot initially let his Eastern European underlings determine the pace and manner of the purge. Only in October 1949 did the first Soviet security advisers come to Prague to help—at Gottwald's request.[34] By then Tito's political survival had created the urgent need to preventively immunize the region against the spread of his heresy, prompting Moscow to take the lead in enforcing a "purgative encirclement of Yugoslavia" from its "Macedonian periphery" outward.[35] As early as January 1949, the Soviet hand could be detected in the expulsion from the party of Vafiades, the already disgraced commander of the Greek guerrillas, whose now moribund insurgency had been mainly supported by Tito.[36]

In the same month Dimitrov, Tito's other former partner in the ill-starred Balkan federation, was secretly spirited from Sofia to Moscow in a Soviet plane—according to an official report (issued only three months later) for medical treatment. He never returned. In July he passed away amid similar rumors of foul play as those surrounding the death of Zhdanov the year before; even the Kremlin physicians who certified it were the same.[37] Dimitrov, like Zhdanov, had been prominently associated with a Soviet policy that failed, yet was not a good choice for being publicly dishonored. Having earned international reputation by putting to shame the Nazi kangaroo court that had tried to frame him as a culprit in the 1933 Reichstag fire, he was especially unsuited to be hauled before another such court on Stalin's orders.

Instead the choice fell on Traicho Kostov, like Xoxe a former minister of the interior though, unlike him, no creature of Tito's but rather his antagonist in having resisted most vigorously his idea of a federation that would have reduced Bulgaria to a Yugoslav province. Expelled from the party in March, Kostov was now given a push by the same Tito smearing him as a police agent in a public speech at a time when the charges against him in Sofia were not nearly so serious.[38] They had so far consisted mainly of insinuations that he had been "insincere" in dealing with Soviet comrades, particularly too "secretive" in matters concerning trade between the two countries.

Reminiscent of the topics that had started Stalin's row with Tito, these were nevertheless portentous accusations, all the more so since at the Cominform conference in Bucharest the preceding June Kostov had cited for the benefit of the Yugoslavs precisely the respect shown to Soviet advisers by Bulgarians, whose "communist conscience is clear," with "no secrets to hide from ... comrade Stalin."[39] Although anything but Tito's man, Kostov resembled Tito in his revolutionary disposition—a liability rather than an asset in Stalin's eyes. His record as the chief organizer of Bulgaria's bloody postwar repression of the communists' political enemies was now turned against him.[40] The description of him as "the standard bearer of all reactionary and restoration elements ... *whether he wished it or not*"[41] encapsulated the purge's modus operandi.

The purge gathered momentum after Belgrade had started hitting back. In April, a court in Novi Sad sentenced eight Yugoslav citizens as "Cominformist" agents, and Hungary expelled Tito's diplomats in retaliation.[42] The next month, his accused spies were displayed at trials in both Bulgaria and Albania, where the imprisoned Titoist Xoxe also was dragged to court.[43] This was the first time a top communist official faced criminal charges in Soviet Eastern Europe. Yet his trial differed from the later pattern in its being secret rather than public and the publicized charges against him in being essentially true.[44] He was sentenced and put to death for having conspired with Belgrade to seize power and make his country part of Yugoslavia. This was something that Stalin had once approved of but later regretted, so he all the more willingly gave his fiat once Hoxha's had set up Xoxe's downfall for reasons of his own.[45]

The Hungarian László Rajk was the third former interior minister for whom the bell tolled—though not in Budapest but in Prague. On May 12, the day the Berlin blockade ended, one of the Cold War's more bizarre episodes began to unfold in the Czech capital with the arrest of the American citizen Noel Field at the request of the Hungarian leader Mátyás Rákosi. Attesting to the importance Moscow attached to the case, it sent Gen. Fedor Belkin with a team of police experts, assisted by Hungary's security chief Gábor Péter, to mastermind the elaborate operation. Czechoslovakia did its part by delivering another implausible character: the director of its official travel agency, Gejza Pavlík. Shipped to Budapest for questioning, he was subjected to a treatment that could not fail to yield the names of as many as sixty accomplices of Field's putative conspiracy, including Rajk.[46]

A quintessential Western fellow-traveler, Field was the parody of a superspy. During World War II, as a manager in Europe of a philanthropy using U.S. government funds to support mainly left-wing refugees from Nazism, he had made enough friends among people who had since risen to prominence in communist Eastern Europe. With such friends he could hardly help getting involved in small-time intelligence gathering—for Moscow rather than for Washington, which rightly suspected him to be a Soviet agent.[47] Indeed, had he not been arrested in Prague, he might well have been upon returning to the United States, although he was never formally charged with any wrongdoings either place. Though useful to the Soviet managers of the imaginary conspiracy as its centerpiece, the pathetic Field could not be displayed in court to any good effect; kept in prison, he would later emerge from it as a convinced communist and end his days in Hungary.[48]

On one of his previous Eastern European peregrinations, Field had been overheard in Prague as saying that the Stalin-Tito rift might cause unspecified "difficulties" to his friend Rajk.[49] Jotted down by the police, the remark provided the thread from which a plot could be woven, although it was not the real cause for Rajk's demotion soon after Field's disappearance. Nor was Rajk, like Kostov, the most obvious choice for a show trial; according to Tito, he was "the very one we had the least to do with. Most of our contacts were with Rákosi, Farkas and the others, while Rajk was always as silent as a grave."[50]

As another Hungarian purge victim later reminisced, however, "assumable and potential deeds were more important than those which had been committed."[51] While for Stalin "*which* of the leaders was arrested was unimportant ... , so long as *some* were,"[52] for Rákosi, who came to be suspected by Soviet ambassador Georgii M. Pushkin of hindering the investigation,[53] attributing the contacts with the Yugoslavs to Rajk was an alibi. Taking charge of the purge together with Tito's other former interlocutors—Mihály Farkas, Ernő Gerő, and József Révai—Rákosi invited Soviet police advisers, supervised the selection and interrogation of the defendants as well as the drafting of their indictment, sent it to Moscow for comments, and, when Stalin evaded his question of whether Rajk should be executed, made the decision himself.[54]

Since, unlike Rajk, the foursome also happened to be Jewish, as were Péter and a disproportionate number of high security officials in Hungary and elsewhere in the realm of Stalin, the purge assumed the appearance of a Jewish undertaking.[55] In reality, Jews were conspicuous among both Stalin's accomplices and his victims. As a dismal aftereffect of the Nazi Holocaust, those of its few Eastern European survivors who had not chosen to emigrate as Zionists were for that reason particularly eager to assimilate and hence embraced the ascendant communist regimes with zeal. As renegades who would sometimes treat other Jews even worse than non-Jews, they were often Stalin's preferred choice for especially dirty tasks, but were then also more vulnerable to being purged once they had served their time.[56]

In concocting the case against Rajk and his supposed confederates, the substitution of the original charge of Trotskyism with that of Titoism[57] was suggestive of the Yugoslavs' success in striking back with Stalin's own weapons. "Now we treat Stalin's followers the same way he treated his enemies," Tito's alter ego Đilas proudly noted.[58] There were trials of real or presumed Bulgarian spies in Skopje and Hungarian ones in Maribor.[59] The Stalinist methods used included frame-ups and forced confessions, besides confinement of alleged Cominformists in the notorious concentration camp of Goli Otok on a desolate Dalmatian island, where they perished by the hundreds.[60] If this was a difference from the millions in Stalin's gulags, it was in scope rather than in kind.

In August, the Soviet-Yugoslav confrontation peaked when Moscow, in a menacing note protesting the detention of more of its agents by Belgrade, threatened it with unspecified countermeasures.[61] Stalin reportedly had decided to invade Yugoslavia;[62] in any case, an intervention force was assembled in Bulgaria, Romania, and Hungary. It included anti-Tito émigrés, some six thousand Eastern European "volunteers," and a contingent of airborne troops in the city of Blagoevgrad in Bulgarian Macedonia.[63] Instead of invading, however, the Soviet Union and its allies merely abrogated their still formally valid alliance treaties with Yugoslavia and severed economic ties.[64]

Moscow could have hardly ignored that by then Washington had adopted the policy of supporting Titoism to undermine Soviet rule in Eastern Europe. Persuaded by Kennan's argument that "everything possible should be done to increase the suspicion between the Kremlin and its agents abroad,"[65] U.S. officials had come to the conclusion that the only practical way to reduce and eventually eliminate "preponderant Soviet power without resort to war [is by] fostering Communist heresy among the satellite states, encouraging the emergence of non-Stalinist regimes as temporary administrations, even though they be communist in nature."[66] To put the policy into effect, the chief U.S. diplomats in Eastern Europe met in London in October.[67]

Although the threat to Soviet interests was real, the way of combatting it in the Budapest and Sofia courtrooms bordered on the surreal. The show trial of Rajk in September, followed three months later by that of Kostov, amounted to grotesque accounting for much that had gone wrong with international communism during the past decade. The script, which in the Bulgarian case Moscow supplied verbatim,[68] sought in a Marxian fashion to overcome an unsatisfactory reality by reaching a superior one through a "dialectical" interaction of the truth and its opposite.

In this distorted mirror, the allegedly irresistible march of communism turned into a tragicomedy of errors. The Spanish international brigades appeared as a nursery of enemy agents, the communist resistance in World War II as a German cover operation, Moscow's wartime alliance with the Western powers as an elaborate deception by their intelligence services, the communists' postwar coalition strategy as a capitalist scheme for the infiltration of their ranks by bourgeois subversives masquerading as leftists. The point of the exercise was to explain away why Stalin's policies had so often turned out differently than intended; since no doubt could possibly be cast on their "scientific" management, conspiracy and treason could be the only explanation.

The trials addressed, with the benefit of hindsight, some of Stalin's main misjudgments on Yugoslavia during the war. The prepared scripts attributed to Tito the encouragement of an Allied landing in the upper Adriatic in 1944—something he had dreaded but Stalin had favored as being conducive to a quicker defeat of Germany;[69] had he had his way, the later achievement of Soviet control of the area would have become that much more difficult. Further accusations against Tito included his having contrived handing over control over parts of Yugoslavia to the Western powers, although it was again Stalin who had made his "percentages" deal with Churchill to divide with them influence in the

Balkans—retrospectively another faux pas. Even the decision that Stalin had the strongest reason to regret—the Red Army's postwar withdrawal from Yugoslavia—was depicted as having somehow been Tito's doing.[70]

The more the revision of history in the courtrooms proceeded toward the present, the more Tito stood out as the real defendant in absentia. He was blamed for plotting the Balkan federation in league with its enemy Kostov, while the credit for foiling the scheme went to none other than Dimitrov—the one whom Stalin had chided for promoting it.[71] Tito was further accused of having at the end of 1947 slyly alerted his accomplice Kostov to an imminent change of Belgrade's policy toward Moscow—as if the former rather than the latter had subsequently changed its policy.[72] And the moment when this happened in January 1948 was supposedly the one that the villain had chosen for his most despicable act of treachery—an attempt to overthrow the governments of both Hungary and Bulgaria with the help of local confederates, thus delivering the two countries into the hands of Western imperialists.[73]

It was this scary scenario, which only Stalin's futile quarrel with Tito had made plausible, that the purges were most urgently staged to avert. To make it more plausible, defectors from the Yugoslav foreign service, notably the Lazar Brankov who had led their exodus at the legation in Budapest the year before, were now displayed at the trials, confessing to having actually spied for Belgrade—which, true to form, had promptly disowned them as traitors all along.[74] According to the indictment of Rajk, they had been instrumental in a plan masterminded by Tito's security chief, Aleksandar Ranković.

The "plan," which conveyed accurately enough the U.S. intent to use Tito for subverting Soviet Eastern Europe, rationalized Stalin's failure to subdue him by attributing it to an ingenious enemy ploy. This was to have consisted first in antagonizing the Yugoslav people against the Soviet Union, thus exacerbating international tension to cause them economic hardship, then using the hardship as a pretext for Tito's turning to the West for help, and finally aligning the country with it. Kostov's rehearsed testimony gave this sequence of events that had actually happened the appearance of premeditation by recounting his alleged conversation with Tito, during which the Yugoslav was said to have bragged about how the Americans had at last "recognized that only through him would they be able to attract the countries of the people's democracy" on their side.[75]

The Soviet economic blockade of Yugoslavia had indeed prompted Tito to move closer to the West, albeit reluctantly, and Washington to extend to him—more willingly—the material assistance needed to ensure his independence from Moscow.[76] Having unsuccessfully tried to crush Tito by all means short of the outright use of force, Stalin effectively had to write off Yugoslavia, although he would never acknowledge his defeat or reduce his hostility. Yet by insulating it from the remainder of his empire and intimidating any potential imitators there, he succeeded in containing the spread of the heresy.

No End to the Cold War

Despite his concern about NATO, Stalin did not initially act as if he worried too much about its presumed aggressive intentions. The Soviet military response to its creation was moderate rather than alarmist: a 20 percent increase in defense spending, mainly calculated for public effect,[77] a bolstering of the troops stationed in East Germany, the establishment of an office to supervise the modernization of the armed forces of the Eastern European allies. Having previously preferred to keep them weak as potentially disloyal, Moscow now pressed forward a purge of their officer corps, to replace holdovers with more reliable party men.[78]

Whatever the Western intentions, Stalin had no need to be disturbed about NATO's military capabilities. Although the alliance was formed to discourage Soviet aggression and reassure the Western Europeans as a result, its ability to defend them—much less take offensive action against the enemy—was for the time being severely limited. Rooted in the experience of World War II, the U.S. contingency plans envisaged first strategic retreat from the continent and its eventual liberation after landing troops from overseas—a prospect so dismal that Washington thought it prudent to keep its plans hidden from its allies.[79] They were unlikely to remain hidden from the Russian enemy, whose spies supplied accurate enough information about America's fighting potential, including the number of atomic bombs in its arsenal.[80]

Those bombs, rather than the modest conventional forces, were NATO's main deterrent, as well as possible offensive weapons. Successive U.S. strategic plans reconfirmed the intention to use them if required by circumstances—according to the "Trojan" plan of January 1949 for attacking seventy industrial cities, with the expected loss of 2,700,000 lives.[81] Five months later, however, the Joint Chiefs of Staff concluded that the Soviet Union could not be defeated by using nuclear weapons alone.[82] This was the conclusion of its own experts as well, in accordance with which its war plans envisaged the destruction of U.S. bases and troop concentrations in Europe before reinforcements could arrive from across the Atlantic. This was to be followed by rapid advance of Soviet forces into Western Europe and possibly the Middle East.[83]

The defensive nature of both sides' military planning ensured the essential stability of their relationship, but not their security. When Stalin in July 1949 received the new British ambassador David Kelly, he predictably complained about NATO, ruminating about its aggressive character and contrasting it with the ostensibly benign Soviet alliances, formally directed only at Germany.[84] Although the comparison was disingenuous, his apprehension was not feigned. If to one of his aides—his emissary to China, Ivan V. Kovalev—he confided the view that war was not imminent, he qualified his estimate by saying that this was only because it it would not be "advantageous for the imperialists. Their crisis has started; they are not ready to fight. They scare us with the atomic bomb but we are not afraid. There are no material preconditions for an attack, for launching a war. The U.S.S.R. is strong enough to defend itself."[85]

Thus Soviet confidence in avoiding, or at least postponing, a military showdown with the capitalist powers hinged upon the premise that they could not *afford* to strike rather than that they would not *want* to. Accordingly, the United States' supposedly impending economic breakdown remained an intense Soviet preoccupation in 1949.[86] The proper assessment of the direction in which capitalism was moving and the consequences of the trend for policy had already been at issue in the Varga case[87] and now figured again in the evolving Leningrad purge. This convoluted affair was different in both origin and thrust from the Eastern European purges, although its exact rationale, if any, has so far defied a satisfactory explanation.

Having arisen from a corruption scandal surrounding the December 1948 trade fair in the city,[88] which implicated many of its officials, the ensuing purge assumed deeper significance because of the simultaneous downfall of the Soviet planning chief and politburo member Nikolai A. Voznesenskii, until recently a Stalin favorite.[89] He began his descent when he lost his party posts in the month of the Leningrad fair, which was also when the campaign against cosmopolitanism surged. But he was never charged with that particular deviation, and when in March 1949 he lost his key government job as well, the campaign had already subsided. His presumptive guilt was only hinted at by the dismissal four months later of the editorial board of the leading party journal *Bolshevik,* although the reshuffle was not made public until September and the reasons for it until three years thereafter.[90]

At issue in the upheaval was the editors' susceptibility to the supposedly erroneous ideas Voznesenskii had expounded a year earlier in his book on the wartime Soviet economy—a work previously judged so excellent as to be awarded the prestigious Stalin Prize. The ideas complemented Varga's theory about the relative stabilization of capitalism by addressing the other side of the equation germane to the crucial question of the Cold War rivals' competitive advantages, namely, the strength of the Soviet Union's own system. Voznesenskii's thesis that Soviet planning practices—which were his particular responsibility—assumed the infallibility of "laws" expressed in Marxist jargon the conviction that the country's economic progress after World War II was significant enough a safeguard of its security to justify more attention to be paid to the needs of the consumers. Conversely, Stalin's critique of Voznesenskii's "subjectivism" conveyed the skeptical message that in a menacing world "objective" necessity dictated more sacrifices—including the human ones.[91]

Besides that of the presumptous planner, the sacrifices included an unknown high number of officials from Leningrad, once the fiefdom of Zhdanov. Accused of high treason without further specifications, some of them were arrested in Malenkov's office on August 13, and many more—including Voznesenskii—were subsequently executed in deep secrecy.[92] The difficulty of finding out exactly what happened—reminiscent of the shroud surrounding the 1934 slaughter by Hitler of the followers of Erich Röhm in the Nazi party—was later officially attributed to the destruction of the evidence by Malenkov. He, rather than Stalin's chief security aide, Beriia, seems to have organized the bloodbath with the assistance of Viktor S. Abakumov, the minister of state security variously considered as Beriia's stooge or archfoe, possibly both.[93]

Whatever the alignments in the intrigues within Stalin's entourage, the Leningrad purge with its distinctive secrecy highlighted his insecurity precisely at a time the Soviet Union was on the verge of a feat that could make or break its security in ways not easily to be predicted. The purge climaxed just as its first atomic device was ready to be tested. When Stalin on August 17 met with the new U.S. ambassador Adm. Alan G. Kirk, he no longer mentioned his worries about NATO; in fact, he was reluctant to speak about anything.[94] Unlike Truman, who four years before had alerted him in advance about the American bomb ready to be dropped on Japan, he tried to keep his own blast concealed.

Contrary to later speculations, Stalin had grasped the importance of the new weapon immediately after its potency had been demonstrated by its leveling Hiroshima in August 1945, though not before. Soviet scientists had started working on the bomb four years earlier, yet only then had a crash program for its production been initiated and entrusted to Beriia. The availability of a detailed design of the first American bomb, spirited out by Soviet spies, helped, as did the expertise of captured German physicists, although decisive for the outcome was the totalitarian state's ability to purposefully mobilize its vast resources—not only the traditional excellence of Russian physicists but also Beriia's contingents of forced labor.

Stalin's appreciation of the bomb's importance did not preclude simplistic notions about it, further distorted by Moscow's inadequate knowledge of the consequences of the two nuclear explosions during the war against Japan. On the one hand, he disparaged the weapon as something "meant to frighten those with weak nerves"; on the other hand, he admired it as a "powerful, pow—er—ful" thing.[95] Yet either way, he did not consider it enough of a reason to discard his banal theory of "permanently operating factors," which attributed the decisive importance in warfare to such unsurprising matters as the number of troops, their firepower, leadership, equipment, and morale. In any case, the Soviet military doctrine ignored the revolutionary strategic consequences of nuclear weapons as long as he lived.[96]

In fairness to Stalin, his view of the atomic bomb as simply another weapon, only more powerful, did not substantially differ from the prevailing American view at that time;[97] the realization that a nuclear war would be an unmitigated disaster for all concerned only came as a result of new technological developments after his death. Lacking the bombs anyway, Stalin had all the more reason to pretend he was unimpressed,[98] as long as the United States chose to abstain from using them to blackmail him.[99] Nor did Washington ever approve their use in a possible preventive war, recurrent recommendations in its favor by high-ranking U.S. military notwithstanding.[100] But since Stalin could not be certain about the future, it made sense trying to nullify the adversary's advantage by acquiring the weapons, too, thus discouraging any blackmail also in the years to come.

As the first Soviet test, scheduled for August 29, was approaching, Stalin could not be confident about its outcome and the possible American reaction to its success or failure. The success might conceivably shock Washington into considering a preventive strike after all, thus nipping in the bud Moscow's fledgling nuclear capability, which U.S. experts did not expect for several more years.[101] Even if the test failed, finding out about it could still prompt the Americans to accelerate and expand their own programs, thus increasing their head start. To prepare for either contingency, before letting out the news Stalin wanted his scientists to build a second bomb as a deterrent. If it is true that the night before the test he went to bed and slept through the historic event, he must have been confident enough that he could keep it secret as long as he wished.[102]

Once the United States detected the blast and Truman was the first to announce it to the world on September 23, Stalin reportedly scurried about in search of traitors until he was satisfied that even without them the Americans had the technical means to make the discovery.[103] In any case, Moscow's first public reaction was little short of panicky. While ludicrously alluding to gigantic public works that necessitated much blasting, the official Soviet statement did not deny the American announcement, but played it down as if nothing new had happened. It cited Molotov as having stated in 1948 that for the Soviet Union the "atomic secret" was no secret any more.[104]

Washington's immediate response to the loss of its nuclear monopoly was reassuring. No new American programs started as a direct reaction to the Soviet test. The U.S. resolve to expand the production of atomic arms had antedated it, and Truman's January 1950 decision to produce a hydrogen bomb as well, while taking into account Moscow's newly acquired nuclear capability, had been mainly conditioned by domestic politics, particularly the need to compensate for the congressional cuts in the financing of conventional forces.[105] Nor did the revised U.S.

emergency plan for war, adopted in December 1949, differ from its predecessors in anticipating the use of atomic weapons if necessary.

In the longer term, however, the diminishing American reliance on conventional forces fostered the nuclear arms race[106]—a development more alarming for Moscow than for Washington. The Soviet Union could not hope to match the U.S. production, not to speak of the already stockpiled two hundred bombs, any time soon. Once its test had become public knowledge, it therefore took the high moral ground by proposing at the UN to ban all nuclear weapons.[107] In the improbable case the Americans agreed, their advantage would vanish; in the meantime, the proposal helped communists to rally by their "Stockholm appeal" worldwide opposition against any intention by Washington to utilize its arsenal.

The high priority Stalin assigned to the development of the hydrogen "superbomb," initiated as early as 1946, was not a response to the American program either.[108] While both programs were plagued with conceptual and technical difficulties, Moscow's capacity to deliver nuclear arms further suffered from the cancellation of its plans for a new long-range heavy bomber in favor of a lighter model, usable for conventional warfare only. Nor did Soviet construction of strategic missiles, started in 1950, result in operational models during Stalin's lifetime.[109] Yet since he cherished nuclear weapons primarily for their political effect, the Soviet nuclear inferiority was a secondary matter as long as the overall correlation of forces was improving.

Invoking the Soviet nuclear capability, Malenkov's keynote address on the November 7 anniversary of the Bolshevik revolution exuded a new tone of self-confidence. He expatiated on the enviable security the Soviet Union derived from its being surrounded by friendly states for the first time in its history. His assertion that "the forces of democracy and socialism are growing, while the forces of capitalism and warmongers are suffering a loss"[110] concerned especially the recent victory of the communists in China, which more than offset their defeat in Greece.

The overall situation no longer favored Soviet concessions. The conference of deputy foreign ministers in New York, convened to settle the Austrian question, did not produce the expected result. At the last moment, the Soviet negotiators failed to receive from their superiors the necessary instructions to proceed toward the conclusion of the Austrian State Treaty, thus leaving it in limbo for another five years.[111] The setback added to the multiplying signs that Moscow's posture was hardening.

At the secret third conference of the Cominform in the Hungarian Mátra hills on November 16–19, Mikhail A. Suslov, Zhdanov's successor as Moscow's chief ideologist, warned that "we must not conclude from the weakening of the anti-democratic, imperialist camp that the danger of war has diminished. . . . Historical experience teaches that the more desperate the position of imperialist reaction, the more furious it becomes and the greater is the danger that it would launch into warlike adventures."[112]

Here once again, projected abroad, was Stalin's thesis that the approaching defeat of the enemy makes class struggle not less but more imperative. Whether the difference between Suslov's and Malenkov's views reflected their personal opinions competing for Stalin's attention or—more likely—his decision to escalate the confrontation despite his country's improved security, there could be no end to the Cold War.

During 1949, after having cut his Berlin losses, Stalin had proved willing enough to negotiate about Germany and Austria. At no time, however, did he show a readiness to entertain a German settlement on terms other than his own, although in the less important Austria his representatives had temporarily indicated a greater readiness for concessions. The unfolding Eastern European purge, driven by a desire to both contain Titoism and provide a forum for the badly needed reassessment of past Soviet policies, succeeded in consolidating Stalin's grip on his dependencies, yet the manner in which this was achieved attested to his abiding insecurity. Nor did the bomb he had finally acquired make him secure enough—only more inclined to act on his insecurity. In the ensuing adjustment of policy, defensive and offensive considerations merged, enouraging him to resume advance.

5

Resuming Advance
November 1949–June 1950

Moscow regarded 1949 as a watershed year during which the West turned more aggressive than before.[1] This was true—not so much because of the creation of NATO, which for the time being lacked important operational consequences, as because of the American intention to destabilize Eastern Europe. As in 1947 by launching the Marshall Plan, the United States again had begun to move to a higher level confrontation before the Soviet Union did. Previously Stalin had used the communists in an attempt to destabilize Western Europe but had failed; would the West be more successful in his own backyard? Or, would he again—as in 1947—overreact to the challenge by anticipating its results before they could materialize, thus driving the spiral of conflict farther upward? In either case, the dynamics of East-West relations presaged more conflict.

Real and Imaginary Subversion

The Office of Policy Coordination, created in Washington in 1948 to supervise a counteroffensive against "Soviet-directed world communism,"[2] proved notably successful in mobilizing resistance against it in Western Europe, particularly Italy. Made part of the Central Intelligence Agency and entrusted with covert operations, it subsequently became the main tool to achieve, as the National Security Council put it November 1948, "gradual retraction of undue Russian power and influence from the present perimeter areas" and the restoration of the Soviet satellites as independent states.[3]

A year later, Washington concluded that the time appeared to be "ripe for us to place greater emphasis on the offensive to consider whether we cannot do more to cause the elimination or at least a reduction of predominant Soviet influence in the satellites."[4] Although

"Stalinist penetration" seemed to make Soviet control of Eastern Europe "well nigh invulnerable," it also fostered there "self-stultification and demoralization—the recurring necessity to purge personnel —and ... nationalist resistance which constant Soviet interference partially generates and inflames."

At the initiative of the British Russia Committee, whose mission was similar to that of the American agency, Albania was chosen as the first target of the joint counteroffensive.[5] Separated geographically from the rest of the Soviet bloc, but close to Italy and bordering on hostile Yugoslavia and Greece, Albania was an obvious choice. The British had experience with supporting guerrilla operations there during World War II, as well as a special grudge against its government since the sinking in 1946 of a British destroyer by a mine in Albanian waters, with considerable loss of life.

London's main Albanian expert, Julian Amery, advocated the overthrow of the pro-Soviet government in Tirana as a fair revenge for the support Moscow had given to the Greek communists in their trying to topple the pro-British government in Athens.[6] The United States welcomed the idea to help discourage a resumption of the Greek communist insurgency and also to get rid of the Soviet military base at the Albanian port of Vlorë, where missiles of World War II vintage capable of reaching Western Europe were reportedly installed.[7]

Sensitive to the country's vulnerability—not to speak of its dubious value as an ally—Moscow had refrained from concluding with it the same kind of mutual defense treaty that protected other Soviet dependencies. In 1948, Stalin's representatives in the Cominform had as well ruled out Albania's request for membership in the organization, lest "possible difficulties in Greece and also in Yugoslavia"[8] arise as a result. By 1949, however, Stalin had abandoned the Greek communists to their fate while the Yugoslav communists had abandoned him, and Albania's value as a Soviet outpost had accordingly increased.

In September 1949, British foreign secretary Ernest Bevin and his U.S. counterpart, Dean Acheson, agreed at their Washington meeting to "try to bring down" the Hoxha regime in Tirana "when the occasion arises."[9] By then preparations for the occasion had already started, and soon the two Western powers' secret services were sending into the country agents, mostly Albanian émigrés, to link with anticommunist guerrillas in the mountains. The Americans provided transportation by air via parachute mainly to the north, the British by sea to the south. After the first of several groups had landed in October, a reported mutiny in the Albanian army, followed early in the next year by a purge of two hundred of its officers, seemed to suggest that the conditions for inciting a rebellion were right.[10]

The conditions were not necessarily wrong. The reason most of the infiltrators were almost immediately caught was not any demonstrable popularity of the Hoxha regime but rather the presence of the notorious

Soviet double agent Kim Philby on the very committee in Washington that supervised the operation. Hence the information about their itineraries could travel to Tirana through Moscow even faster than they could their destinations.[11] Despite the unfair advantage Hoxha enjoyed as a result, however, Stalin was obviously worried.

When the Soviet dictator and his Albanian retainer met in January 1950 at the Black Sea resort of Sukhumi, Stalin not only promised Hoxha more arms,[12] but also took his side in his dispute with the chief of the Greek communists in exile, Nikos Zakhariades, about the future of southern Albania, claimed by Greeks as northern Epirus.[13] With Soviet backing, the Albanians disarmed the remnants of the Greek communist guerrillas on their territory, whose presence there risked attracting additional Western attention.[14]

If Stalin happened to have the right man in the right place to frustrate the Anglo-American scheme in Albania, he could not count on the same good luck in more important places where the Western agencies might try to hit him. And they were trying not only in Poland, where an underground civil war was still going on, but also in the parts of the Soviet Union where nationalist guerrillas were operating, particularly the Baltic republics and Ukraine.[15] Although there, too, most of the agents sent from abroad were soon caught—in Poland, 152 "bandits" were killed and 1,989 arrested during ten months in 1949[16]—more were coming. Even to Albania the British and Americans kept sending them, despite their losses, till as late as July 1951.[17]

The most exposed parts of Soviet Eastern Europe, besides Albania and East Germany, were Bulgaria and especially Czechoslovakia. They bordered on countries from which Western secret services could operate most easily—Greece, Turkey, West Germany—and had no Soviet troops stationed in their territories. They were particularly vulnerable to low-flying U.S. aircraft, which in those days could overfly much of the region with impunity. Their governments usually abstained from protesting, lest they publicize their impotence. Instead they publicized with self-congratulatory indignation the attempts at infiltration they said they had uncovered and foiled.

The publicity of such cases mounted suddenly and dramatically as soon as the Cominform had sounded the alarm at its meeting in Hungary in November 1949. The Czechoslovak party general secretary Slánský set the tone by outdoing other speakers in painting in garish colors the perils his country was supposedly facing. He mentioned as many as three attempts at a coup d'état and the discovery just recently of a subversive network of opposition politicians directed from abroad.[18] Soon after the Cominform's call for vigilance, early in the next year the Soviet Union, seconded by its satellites, formally reintroduced the death penalty "for traitors, spies, and saboteurs."[19]

Did the communists really believe that they were so threatened, or were they merely posturing? Stalin confided to Khrushchev, whom he had brought to Moscow from the provinces to rejuvenate the politburo, his anxiety that even the Soviet capital was "teeming with anti-party elements."[20] His Eastern European lieutenants shared his anxiety. According to the confidential files prepared in February 1950 for the meeting of the Czechoslovak party presidium, some of the persons recently arrested for belonging to "terrorist and conspiratorial groups" were actually party members.[21] Having revealed the shocking discovery to the select audience, Minister of State Security Ladislav Kopřiva proceeded to harp on it in his published speech calculated to dramatize the seriousness of the situation.[22] During the subsequent months, the official media reported arrests and trials of purported subversives almost daily.

At the same time, the Eastern European regimes conducted a concerted campaign of harassment against Western diplomatic missions, prosecuting their local employees and sometimes even the diplomats themselves on assorted espionage charges. In Poland, French diplomats were tried and sentenced amid extensive publicity.[23] Bulgaria expelled the American minister, severing its ties with Washington.[24] The American embassy in Prague suffered the worst after three Czechoslovak pilots had defected to the U.S. zone in Germany along with their planes.[25]

Yet the evidence offered to substantiate the reported upsurge of subversive activities fell notably short of justifying the emergency invoked, thus leaving no doubt that the looming threat, though genuinely perceived, was also deliberately exaggerated. No figures showing the number of enemy agents captured during this period have been found in Czechoslovak archives. The figures seem never to have been compiled, presumably because they were irrelevant to the authorities trying to avoid making a distinction between the real and imaginary subversion.

Significantly, the size and capability of the counterintelligence department of the Czechoslovak ministry of state security were far from commensurate with the scope of alleged foreign infiltration. Nor was any crack effort undertaken to match the recent expansion of the agencies in Washington. The Office of Policy Coordination, which had started in 1948 with a staff of 302 and an annual budget of $4.7 million, grew in three years to employ 2,812 persons in the United States and 3,142 more on overseas contracts, while its budget rose more than seventeen times—to $82 million.[26] In contrast, as late as June 1950 the department in the Prague ministry by its own admission employed no more than "207 poorly directed undercover agents," who in the whole of that year managed only "in two cases to collect compromising material about capitalist diplomats."[27]

Rather than any freshly uncovered conspiracies, the publicized cases concerned activities supposed to have taken place much earlier—in 1948 and even before—and consequently posing no threat any more. Of the three attempted "coups d'état" Slánský mentioned in his report to the Cominform, one probably referred to the pathetic assault on the building of the general staff in Prague by a group of students on May 17, 1949.[28] There had also been a "plot" by army officers, which contemporary U.S. observers had considered authentic but which had in fact been concocted by the Czechoslovak counterintelligence chief, Gen. Bedřich Reicin.[29] Its "discovery" had led to a purge of the officer corps—made topical by the progressing militarization of the Cold War—though not to the execution of the supposed main plotter but instead of the agent provocateur employed in the ruse.[30]

Similarly, the "conspiracy" of the opposition politicians was an incongruous group of people, ranging from Catholics to "Trotskyites," who hardly knew one another but were linked together by the police. As their alleged ringleader, former parliamentary deputy of the Czech socialist party Milada Horáková was indicted for having expressed, in private conversations with friends, the hope for a military conflict between East and West that would bring the communists down. After a show trial before a handpicked mob, she was dragged to the gallows and hanged—the only time in Soviet Eastern Europe such an outrage was committed against a woman.[31]

The atrocities perpetrated since the end of 1949 under the pretext of combatting a surge in subversion could not possibly be justified by an internal threat at a time when active domestic opposition in Eastern Europe, if not already crushed, had been diminishing, and any potential imitators of Tito had been effectively deterred. Yet the U.S. covert operations, such as there were, proceeded from the assumption that significant underground resistance existed and needed to be encouraged. The Free Europe Committee, created in June 1949 under CIA auspices to encourage the resistance with the help of anti-communist émigrés, publicized the belief that opposition networks in their homelands were ready for action.[32]

If Stalin knew that Washington believed this to be true—and he was certainly in a position to know from his informers—that alone would not have allowed him to rest until finding out where the networks were. He pressed the Czechoslovak communists to unmask in their country a conspiracy that would make it possible to stage there a monster trial of their Kostov or Rajk. He was drumming up the threat of war although the challenge he was facing was not a military one. He stepped up preparations for a war before the Americans did, thus anticipating a new stage in the confrontation.

Concurrently with the wave of repression since the beginning of 1950 the Soviet Union undertook the major military buildup that it had not considered necessary to undertake in response to the creation of NATO the year before. Although no challenge of the same order had intervened, Moscow pushed forward an expansion of the armed forces of its Eastern European allies and a systematic militarization of their economies. It formed a Unified General Staff in Bucharest and installed Soviet Marshal Konstantin Rokossovskii as the defense minister of Poland.[33] The establishment of a special Soviet ministry of the navy symbolized a widening of Stalin's strategic horizons at a time when the focus of the Cold War was shifting from Europe to Asia.[34] The far-reaching consequences of the shift can now been extensively documented from newly available evidence in all their fascinating details.

The Second Front in Asia

Since mid-1947 the Soviet Union had been on the defensive not only in Western Europe but also in the Far East, and in both places it had largely itself to blame. In March 1947 it had missed an opportunity to influence decisions on the future of Japan. It could have had a say in shaping them if it had been responsive to the proposal for the country's demilitarization under international control submitted by the United States at a conference of the victor nations represented in the Far Eastern Commission. But rather than discuss the subject within the framework of this body, whose two-thirds voting rule would have compelled the Soviet representatives to compromise, Moscow insisted that a peace treaty with Japan must be prepared by the Council of Foreign Ministers, whose decisions it could block by exercising the right of veto.[35]

Following the deterioration of East-West relations later that year, Washington abandoned its proposal. "Because of Russia's conduct,"[36] it began to seek Japan's transformation into an anti-Soviet buffer state that would "function as an American satellite."[37] Although the conduct in question had been in Europe rather than in Asia, in the American mind the two theaters of the Cold War became interconnected. In November 1947 Secretary of State Marshall could note with satisfaction that the containment of Soviet expansionism had brought closer the necessary restoration of the "balance of power in Europe *and* Asia."[38]

In vain did Moscow strive to tilt the balance in its favor by manipulating nationalist movements against Western colonial powers in Southeast Asia. Its prodding the local communists to wage armed struggle did not save them from painful setbacks in Indonesia, Malaya, Burma, and the Philippines. Their Soviet-inspired rejection of the nationalist middle classes as enemies rather than potential allies proved a precept for political isolation. Alone in Indochina did the communist guerrillas manage to hold their own. Yet their successes there were more a nuisance than a boon to Stalin by making it more difficult for communists to beat the patriotic drum in France—for him a more important battlefield than an Asian jungle.

The developments in Southeast Asia gave Stalin little reason to revise his skeptical opinion about the prospects of the Chinese revolution, the communists' military advances in early 1948 notwithstanding. Although he began to provide them useful nonmilitary assistance in

Manchuria,[39] he kept their leader, Mao Zedong, at arm's length, repeatedly sidetracking Mao's demands to receive him in Moscow to coordinate policies. Among the flimsy excuses Stalin gave to delay their talks was "the incipient grain harvest," on account of which "the leading comrades will be travelling to different parts of the country and staying there."[40]

Moscow continued to bestow legitimacy on Generalissimo Chiang Kai-shek's nationalist government. Even when the retreat of its forces turned into a rout, the Soviet ambassador still followed the government from one place to another after other foreign envoys had already given up.[41] It was, after all, the treaty Stalin had concluded in 1945 with this government that enabled him to secure at China's expense his most valuable gains on the Asian mainland—the control of Outer Mongolia, Xinjiang, and Manchuria, besides the naval bases of Port Arthur and Dairen.

By early 1949, however, the situation had changed. The Chinese communists had done so well that, barring direct U.S. intervention to prop up Chiang, the collapse of his regime was all but certain. Still, in forwarding its request for Soviet mediation to Mao Zedong on January 10, Stalin advised the communists to negotiate; only on the next day did he hastily clarify that he really wanted them to set such conditions that would make the negotiations fail.[42] To avoid any more misunderstanding, promote good will, and assess the probable future course of development, he secretly dispatched his seasoned emissary, politburo member Anastas A. Mikoian, to the communist headquarters at Xibaipo in Hebei province.[43]

Agreement on strategy was all the more desirable since Mao soon had to make the crucial decision of whether he should order his forces across the Yangzi River and probably finish off the enemy as a result—unless, of course, the U.S. intervened militarily to save Chiang. There is no evidence that Mikoian tried to prevent the communists from crossing the river, although the implication of Stalin's earlier suggestion favoring their negotiations with the nationalists was that he would have preferred a compromise settlement of China's civil war, as he had ever since the end of World War II and even before. Having by this time become acceptable to Chiang, the compromise would have amounted to a division of power in the country along the Yangzi line, thus denying the United States a cause to intervene.[44]

Mao thought that the American intervention, which could reverse his march to victory, was quite likely.[45] He could have tried to avert it, but did not, by secretly working out a deal with the Americans through people in his entourage who had had contact with them in the past. He loyally kept Stalin informed about the few current contacts, and the Soviet leader showed no signs of alarm.[46] Yet being what he was, he had to be worried; unlike most officials in Washington, he considered an arrangement between it and Mao possible. Reminiscent of the situation in the Balkans two years before, the strife between the local communists and their domestic opponents threatened to attract American involvement in the region. The dissimilarity of the two situations, however, outweighed their similarity.

The intense contemporary speculation about Mao's becoming for Stalin another Tito was totally unwarranted.[47] Not only had Tito's dissidence been of Stalin's own making, and therefore in his power to prevent, but he could also better afford to accept Mao as a junior partner than he could an Eastern European. China's strategic significance for Moscow, though considerable, was not as overwhelming as Eastern Europe's, and consequently did not require as pervasive a Soviet control as Stalin thought indispensable there. That sort of control would have been in any case inconceivable in a country of China's size. Most important, unlike Tito, Mao was sensitive to Stalin's concerns and acted to allay them by going out of his way in supporting the Soviet condemnation of the Yugoslav heresy.[48]

Once the communist army in April 1949 crossed the Yangzi and captured the nationalist capital of Nanjing while the United States stood by, Mao demonstrated a readiness to burn bridges with Washington. He had American diplomats in Manchuria arrested and tried as spies, ignoring U.S. protests.[49] He solemnly proclaimed his desire to "lean to one side," namely, to the Soviet one, from which alone he expected effective support for his revolution.[50] With its victory within sight, he sent his second-in-command, Liu Shaoqi, to Moscow to ascertain what the Soviet views and intentions were and adapt his policies accordingly.

Having proved themselves successful revolutionaries and reassured Stalin about their loyalty as well, the Chinese communists became for him an asset not to be neglected. Nor, to be sure, could he allow being outflanked by them on the left if he were to keep intact his authority as the supreme leader of international communism. By showing that their interests could coincide with his, they helped to revitalize his interest in its expansion, which he had so often been prepared to sacrifice to the demands of Realpolitik.

In anticipation of Liu's trip, *Pravda* paid the Chinese the unusual tribute of prominently displaying a translation of his seminal article on their winning strategy of collaboration with willing members of the nationalist bourgeoisie—the strategy Moscow had previously decried as erroneous.[51] Now it was sanctioned as particularly suitable for Asia, though not for Europe. Stalin volunteered to Liu the confession that he had underestimated China's revolutionary potential, gloating that "the center of the world revolution is transferring to China and East Asia."[52] After this promising beginning, their conversations revealed a substantial identity of views on policy.

Since the Chinese took pains to show that they respected Moscow as the center of world revolution, Stalin's remark about its eastward shift concerned not so much the responsibility for its direction as the opportunity for its exploitation. The proclamation of the communist state in Beijing on October 1, 1949, provided not only an inspiring example for other countries to follow but also an accomplished fact that Washington could recognize only by acknowledging defeat. For Stalin, the prize now was Japan, which not only had eluded him after World War II but had also been recently evolving into the main U.S. bastion against the further spread of communism in the Far East. The American "loss of China" opened up for Stalin the prospect of reversing his "loss of Japan."

Until then Moscow had been guiding the Japanese communists to accept U.S. supremacy while trying to mitigate its impact by policies calculated to win them broader support in the country[53]—the same main goal that their Western European comrades used to pursue under Soviet direction in the aftermath of World War II. In December 1949, however, the staging in the Far Eastern Russian city of Khabarovsk of a show trial of captured Japanese army officers accused of having conducted bacteriological warfare during the war was the harbinger of change in Soviet policy.[54] While the accusations were plausible enough, they were timed to back up Moscow's simultaneous campaign for the trial as war criminal of Emperor Hirohito himself—on whose keeping the imperial mantle hinged so much of the Japanese deference for the U.S. occupation power.

In the same month, Mao traveled to Moscow for the first time. What happened after he had received there a hero's welcome on December 16 could until recently be gleaned mainly from self-serving accounts selected for publicity by the Beijing authorities[55] and from unverifiable memoirs by surviving Russian witnesses. Most of these sources tend to retrospectively project the later Sino-Soviet rift into a period when a fundamental conflict of interest between the communist giants was neither present nor anticipated. Although not everything was sweet and smooth between the two ruthless and devious dictators, this was the time Mao's revolutionary ambitions and Stalin's power interests converged the most closely.[56]

The contemporary Soviet minutes of the historic conversations hardly bear out the Chinese descriptions of Stalin's imperious boorishness and his guest's insulted dignity, images that Mao himself—no doubt retrospectively embarrassed by the extent of subordination he had once been willing to accept in relation to Moscow—later tried to disseminate.[57] Although his initial reception was marred by displays of Soviet bad manners—such as Molotov's curt refusal, on alleged protocol grounds, to attend the impromptu banquet the Chinese delegation had prepared on its train[58]—no insults were intended. Once Mao reached the Kremlin, Stalin was effusive, rhapsodizing that "the victory of the Chinese revolution will change the balance of the whole world."[59]

If after the initial excitement the pace of Mao's visit slowed down, this was partly because he got sick and badly needed a rest,[60] partly because of its coincidence with the mammoth celebrations of Stalin's seventieth birthday. Otherwise, according to his interpreter Nikolai T. Fedorenko, the leader of world communism was calm and attentive to his younger guest, who in turn deferentially let the senior statesman choose the conversation topics.[61] The Russian eyewitness has described their talks as long and relaxed, wide-ranging and comprehensive—a fair description considering the mutual interest of the two unequal partners in forging a working relationship.

Mao indicated that he wished "something that should not only look nice but taste delicious"—a subtle allusion to a close alliance that may have escaped the cruder Stalin as being part of the small talk.[62] Once he got down to business, he told his guest that he wanted to preserve the treaty he had concluded with China's previous government rather than replace it with a new one. He described the treaty as an outgrowth of the Yalta agreement, indispensable to safeguard Moscow's title to its territorial acquisitions in the Far East, particularly southern Sakhalin and the Kurile Islands. He was willing to consider amendments favorable to China, such as the withdrawal of the Soviet garrison from Port Arthur, but Mao welcomed its continued presence, presumably to help deter U.S. military moves against his country.[63] Whatever he may have thought, he chose not to push the issue of a new treaty.

The conversations could have taken a turn for the worse when Ivan V. Kovalev, Stalin's chief liaison with the Chinese communists, prepared for him a memorandum that imputed an anti-Soviet and pro-American disposition to their ranking officials.[64] This was the same sort of smearing of a friendly government by a supercilious Soviet representative that in 1948 had helped trigger Stalin's row with Tito. But this time the otherwise suspicious Stalin resisted the temptation to act on the accusations, while Mao was able to control whatever frustration he may have felt about the slow progress of their talks. They both proved an impressive ability as well as a willingness to accommodate each other's very different personalities and dispose of potentially contentious issues.

On January 1 Mao broke the question of future Sino-Soviet relationship to the public in a press interview expressing his preference for a speedy conclusion of an alliance that would supersede the Stalin-Chiang treaty.[65] To the central committee in Beijing, Mao at the time described the interview as his idea; seven years later, however, he would tell Soviet ambassador Pavel Iudin that Stalin had supplied him with the text.[66] Be that as it may, the statement expressed accord between the two leaders. Mao plausibly surmised that the recent diplomatic recognition of his regime by India and Great Britain was what swayed the cagey Soviet dictator to scuttle the treaty with its predecessor. With the stage set for negotiations, Mao sent for Premier Zhou Enlai to come from Beijing to take charge.

The talks assumed added significance because of the pending proposal to Stalin by North Korean party chief Kim Il Sung for the forcible unification of the divided country by communist military action. Until documents from Russian archives proved otherwise, Stalin's and Mao's very knowledge of Kim's project could, however implausibly, be questioned and the origins of what became the Cold War's hottest confrontation attributed to a dispute between the Koreans themselves.[67] In reality, North Korea was dependent on Moscow much the way its Eastern European satellites were and China was not,[68] and solicited both Soviet and Chinese backing from the outset. The backing, however, evolved tortuously.

While in Moscow in March 1949, Kim presented his plan to Stalin, but was rebuffed. Stalin ruled that "it was not necessary to attack the south," although "in case of an attack on the north of the country by the army of [South Korean President] Rhee Syngmann" it would be "possible to go on the counteroffensive to the south of Korea."[69] In view of the bellicose talk and border provocations by the South's government in Seoul, the possibility of its attacking was not far-fetched at that time. In any case, both the Northern regime and its Soviet sponsors took the threat seriously.[70]

Calculated as it was to persuade the Americans to maintain their support for Seoul, however, the threat was exposed as a bluff once their military forces, to visible South Korean distress, completed their withdrawal from the peninsula in June.[71] Although Kim Il Sung henceforth had little reason to worry about the South's saber rattling, instead of calming down he exploited the new situation to renew his proposal to Stalin. In September he asked for Soviet permission to seize a portion of territory beyond the dividing line between the two parts of the country and, if the going were good, advance farther.[72] He argued that South Korea's economic plight and the unpopularity of Rhee's regime would make the invasion easy and even welcomed by its people.

Stalin was not easily to be convinced by an argument bearing an uncomfortable resemblance to that which the Bolsheviks had used in trying to justify their disastrous invasion of Poland in 1920. Nor could Washington's acquiescence in an attack on its client in close vicinity of Japan be taken for granted. On September 24 Stalin sensibly ruled out the proposed invasion as liable to result in a war for which he rightly judged the North to be unprepared both militarily and politically.[73]

Mao was more willing than Stalin to take the risk of challenging the Americans. After his fears of provoking them by crossing the Yangzi earlier that year had proved groundless, he became inclined to discount the probability of U.S. military action in Asia. In May, he met Kim Il

Sung's demand to bolster North Korea's armed forces by releasing two Chinese army divisions consisting predominantly of its nationals.[74] Although by doing so he did not endorse Kim's aggressive plans, neither did he discourage them. Mao told Kim's emissary, Kim Il, that "in case of necessity we can throw in Chinese soldiers for you; all black, no one will notice."[75] But after ejecting the nationalist forces from the mainland, Mao first wanted to complete his victory by the conquest of their last stronghold on Taiwan.[76]

The success of Kim's scheme depended ultimately on Moscow rather than on Beijing. In addition to Stalin's consent, it required Soviet material support and military expertise, which in turn presupposed reassuring the wary Stalin about the soundness of such an investment. Before the crucial negotiations for a Sino-Soviet alliance started, the United States unwittingly provided the necessary reassurance, thus facilitating the collusion between the two communist powers in support of the aggressor.

The Design for Aggression

On January 5, 1950, President Truman announced the termination of U.S. military assistance to Chiang's regime in Taiwan. On January 12, Secretary of State Acheson in a public speech described America's "defensive perimeter" in the Far East as including, besides Japan, the Chinese offshore islands under nationalist control but not any part of the mainland. He also told the Senate that in Asia, as in Europe, what was to be countered was not so much the threat of military aggression as that of political and economic chaos.[77] Yet Congress subsequently voted down the economic assistance bill for South Korea, thus giving the impression of not caring about its political future either.

How did these signals, possibly confirmed by Moscow's knowledge of the secret decisions in Washington that had preceded them, influence Soviet policy? Molotov read a translation of the Acheson speech to Mao, thus calling his attention to its importance.[78] By that time, however, Moscow had already begun acting on the assumption of American disengagement from the Asian mainland. Not only had Stalin indicated to Mao his readiness to sign the new treaty the Chinese wanted, but he had also made the Japanese communists abruptly change their course.

Accusing them of a lack of militancy, *Pravda* on January 4 pointedly published an article by Liu Shaoqi from previous November, in which he had exalted the "Chinese way" as the way for other countries, including Japan.[79] Two days later Moscow followed with a reprimand similar to that which in 1947 it had given the French and Italian communists at the founding session of the Cominform in preparation for their campaign to block the Marshall Plan. Now what was to be blocked was the conclusion of a separate peace treaty with Japan by the United States. The violent demonstrations staged by Japanese communists, reminiscent of the tactic used in Western Europe four years earlier, substantiated Acheson's prediction of looming chaos.[80] The moment was well chosen, for in Washington "no consensus could be reached on the timing and nature of a Japanese peace settlement."[81]

Meanwhile Moscow had been trying to force the UN to accept Mao's rather than Chiang's regime as the legal government of China. It was two days before the Acheson speech that the Soviet delegates walked out of the Security Council and other agencies of the world organization where nationalist representatives were seated. Since such tactic was hardly suited to persuade the United States to tolerate under pressure their replacement by communists, more devious motives—including Stalin's alleged desire to isolate Mao internationally—have sometimes been suspected.[82] Yet the Soviet conduct, agreed upon in previous consultations with the Chinese,[83] was not so much devious as foolish.

As shown by the sequence of events, the signals from Washington did not make a critical difference in prompting Stalin's aggressive turn, which he had initiated, gradually rather than suddenly, even before they were sent. Once they had been sent, however, he was not discouraged from going farther than he had previously shown a willingness to go in testing the limits of American power. What did make a difference at this critical point was Kim Il Sung's new initiative. On January 19, while "in the state of some intoxication" at a luncheon party in his capital city of Pyongyang, he let Soviet ambassador Terentii F. Shtykov know that he just had to see Stalin again, to "receive an order and permission for offensive action," for lately he could "not sleep at night, thinking about how to resolve the question of the unification of the whole country."[84]

Now Stalin was beginning to change—and lose—his mind. He sent word that he would be glad to receive Kim although he cautioned him that "such a large matter in regard to South Korea such as he wants to undertake needs large preparation. The matter must be organized so that there would not be too great a risk." He was ready to help but asked the North Koreans to pay for his trouble in kind on the order of "a yearly minimum of 25,000 tons of lead."[85] He still made his final approval of the attack contingent on Kim's guarantee that it would be successful. Meanwhile Pyongyang's request to Beijing for the release of an additional fourteen thousand Korean soldiers serving in the Chinese army, along with their equipment, had already been favorably received.[86]

In view of these happenings, the State Department's Policy Planning Staff director Paul H. Nitze was right when he observed in early February that "recent Soviet moves reflect not only a mounting militancy but suggest a boldness that is essentially new and borders on recklessness."[87] They also proved right—albeit by reverse proof—his belief that Moscow's concept of correlation of forces, which took into account the perception of the balance of power as well as its reality, imposed upon it restraint which the United States could promote by projecting an image of strength rather than weakness.[88]

The conclusions Nitze drew in the celebrated document NSC-68, which provided new guidelines for the American grand strategy, presumed Moscow's relentless drive for aggression that the United States must oppose with all the resources at its disposal. Attributing to the Soviet regime a "peculiarly virulent blend of hatred and fear," the text described it as "inescapably militant" because it "possesses and is possessed by a world-wide revolutionary movement, is inheritor of Russian imperialism, and is totalitarian."[89] None of these sweeping assumptions was entirely correct, but neither were they entirely incorrect. Unwittingly Stalin was preparing to substantiate them even before the NSC-68 was ready for official approval in April 1950.

Even in the unlikely case Stalin did not immediately tell Mao about his decision to support Kim—and the decision was, after all, still provisional—the prospect of the adventure made the conclusion of a Sino-Soviet alliance all the more imperative to clarify the two governments' respective commitments in case of an emergency. In his own way, Stalin showed his benevolence to Mao by revealing to him the slanderous Kovalev report from the month before and sacking its author.[90] When the two got together again on January 22, Stalin confirmed his change of mind in favor of a new treaty with China and, contrary to later insinuations, there were no serious disagreements about its terms. Again, Stalin offered to withdraw the Soviet troops from Port Arthur but deferred to Mao's wish to keep them there to help protect China from the Americans.[91]

If, as conspicuously emphasized in the retrospective Chinese accounts, disputes occurred during the subsequent talks between Molotov and Zhou Enlai, more remarkable was the willingness of both sides to resolve them. It is quite possible that the Soviet Union balked at giving China an unequivocal promise to help if it were invaded.[92] In the end, the two parties settled for the formulation that obliged each other to "render military and other assistance by all means at its disposal" against an attack.[93] Such a provision in a published treaty was designed less to be tested than to deter their American adversary from testing it. The treaty implied their common expectation that the war in Korea would not necessitate their direct military involvement.

The expectation was also implicit in the timetable chosen for transferring to China the strategic facilities on its territory that were currently under Soviet control: the Chang Chun railroad in Manchuria and the naval bases of Port Arthur and Dairen.[94] The agreements stipulated that the first two were to be turned over after the peace settlement with Japan, but not later than 1952, by which time their military significance would have presumably diminished. The discussions for the transfer of Dairen were likewise to be initiated after that settlement, which the signatories thus indicated they hoped to be able to accomplish on satisfactory terms within that time limit. The only plausible reason why they could possibly expect to achieve a Japanese peace treaty to their liking was the crushing effect that a successful unification of Korea by the communists would have on the United States.

Otherwise, as befitted a relationship between two unequal allies, the stronger one made greater concessions for the sake of common good. They agreed to a joint management of the mining operations in Xinjiang while recognizing the territory as an integral part of China. Beijing also received the Japanese property that the Soviet troops had seized in Manchuria—or what was left of it after they had looted whatever they could.[95] The Soviet Union further extended to China a three-hundred-million-dollar loan at 1 percent interest—the modest amount that Mao, who entertained exaggerated notions about his country's self-sufficiency, had himself requested.[96] In addition, the Chinese obtained Soviet technical assistance at the time of their greatest need.[97] For all this, their recognition of Outer Mongolia as Moscow's satellite state, which merely confirmed the status quo, was not a high price to pay. All things considered, never before or after did Stalin go so far in treating a client as a partner—or the Chinese communists in willingly submitting to a foreign patron.

The treaty Moscow had concluded the year before with North Korea excluded, unlike its Chinese counterpart, even a hint of military assistance. On its first anniversary, in March 1950, the Soviet press deceptively exalted the "friendship and economic cooperation" between the two countries as a fountain "of universal peace."[98] At the same time, Stalin authorized the delivery of arms and supplies requested by Kim Il Sung—though not before securing from him not only all the lead he wanted but also gold and other precious metals and raw materials to boot.[99] Unlike in his dealings with the Chinese, whom he had learned to respect, the Soviet dictator was prepared to squeeze out of the Koreans whatever they were worth, and more.

As the military buildup continued, the Soviet Union tried to cover up what was in the making. It berated the Indian communists for still adhering to the strategy of armed struggle it had imposed on them two years before.[100] In a *tour d'horizon* of the international situation, Malenkov harped on the supposed threat to peace in Europe but kept all but silent about Asia.[101] And the Cominform assigned a high priority to promoting its Stockholm appeal for the peace that was soon to be at risk in Korea.[102]

During his visit to Moscow in April, Kim Il Sung requested additional support. He received one thousand trainloads of military equipment—enough reason to feel in "good spirits" when he returned home.[103] From the Chinese he secured the implementation of the previously approved release of their soldiers of Korean origin, thus bringing the total of those put at his disposal between fifty and seventy thousand.[104] Before the end of the month, a high-ranking Soviet military team arrived in Pyongyang to finalize what it euphemistically called the "Preemptive Strike Operations Plan," for even insiders had to pretend that it was responding to an expected invasion from the south.[105] On May 1 the Soviet government transferred the responsibility for Korean affairs to the Ministry of Defense.[106]

When Kim came to Beijing on May 13, reporting that he had Stalin's green light for the invasion,[107] the startled Mao requested clarification from Moscow, which Stalin supplied in the following manner:

> In a conversation with the Korean comrades Filippov [Stalin] and his friends expressed the opinion that, in the light of the changed international situation, they agree with the proposal of the Koreans to move toward reunification. In this regard a qualification was made that the question should be decided finally by the Chinese and Korean comrades together, and in case of disagreement by the Chinese comrades the decision on the question should be postponed until a new discussion.[108]

In effect, Stalin was saying that he was prepared to open a second front in the Cold War, thus exploiting the opportunity apparently presenting itself as a result of America's retreat from positions in Asia, but only if Mao, whose track record in estimating U.S. behavior in that part of the world was better than Stalin's, shared the responsibility for the decision, which under no circumstances Kim must be allowed to make on his own.

Although Mao later fretted that he had not been sufficiently consulted,[109] he had no objections to the aggression as such. He had already the month before conveyed to the North Koreans his opinion in no uncertain terms: "A peaceful unification of Korea is impossible; Korea may only be united by military means. As far as the Americans are concerned, there is no need to be afraid of them. They will not go into a third world war for the sake of such a small territory."[110] If for Stalin the attack in Korea was still conditional upon favorable circumstances, Mao saw it coming in any case.

Although the preparations for war were unfolding in deep secrecy and deception was used to divert the enemy's attention, surprise could not be guaranteed. On April 12, the U.S. Communications Intelligence Board Watch Committee in Washington recorded a report received through the headquarters of the supreme commander in the Far East, Gen. Douglas MacArthur, stating that the North Korean army was poised to invade the South in June.[111] The report was accurate, for this is what Kim Il Sung told ambassador Shtykov on the very same day.[112] Western intelligence services, which evidently had their informers well placed, received numerous additional warnings about the offensive preparations but dismissed them as not immediately topical.[113] Still, the nearer the invasion, the greater was the possibility that the warnings would be heeded and acted upon.

By this time Washington's axioms about Moscow's relentless expansionism had become public knowledge as a result of the congressional hearings preceding the approval of NSC-68 as the basic guide for policy.[114] As a result, Stalin was in a position to know that the aggression he was abetting would confirm the worst American assumptions about his intentions. In particular, it could consequently provide justification for the massive U.S. military buildup recommended in the document, thus turning the victory he expected into its opposite.

As the North Koreans and their Soviet advisers were putting the final touches to the invasion plans, the despot showed signs of nervousness. In early May he expressed to the visiting UN Secretary-General Trygve Lie his "profound belief" in "the possibility of long-continued coexistence between the two systems, communist and capitalist."[115] He also put his name at this time to the most bizarre of his writings—devoted to, of all subjects, the theory of linguistics. His treatise struck a curiously reassuring note precisely as the Cold War threatened to grow into a real one. He eulogized the language as a common human asset overcoming even the class struggle, which, "however sharp [it] may be . . . cannot lead to the disintegration of society." Musing about the mutual dependence of workers and capitalists, as well as the importance of continuity in human affairs, he rejected the inevitability of a revolution, insisting that the necessary changes could take place without violence.[116] He nonetheless allowed the invasion plan to be put into effect.

On May 27 the entire staff of the Soviet diplomatic mission to Japan, headed by Gen. Kuzma N. Derevianko, left for home.[117] Three days later Shtykov reported from Pyongyang to Moscow that the North Koreans were ready to attack before the end of the next month.[118] He conveyed their demand for still more hardware, along with Kim Il Sung's estimate that the enemy was ignorant of what was going on but that any delay could cause this favorable situation to change. Before replying, Stalin asked Mao to reaffirm his support for Kim, and again it was the bold Chinese leader who reassured the hesitant Soviet one.[119]

Grigorii Tumanov, in 1950 the official in charge of Korean affairs at the Soviet ministry of defense, recalls a crucial meeting convened in Moscow on June 10 to consider Pyongyang's request for the "approval of its proposed action toward the unification of the Korean people."[120] His testimony gives an air of authenticity to a document obtained three years later by U.S. intelligence, allegedly from Stalin's archives, according to which Stalin and several of his close associates listened at a meeting to reports by Derevianko, Shtykov, Chinese politburo member Li Lisan, and Soviet expert on Pacific affairs Georgii Voitinskii.[121] Shtykov and Li Lisan were described as maintaining even at this gathering of insiders the pretense that the forthcoming invasion was needed to preempt a contemplated South Korean attack, although Derevianko pertinently observed that the South was in no position to resist.

Voitinskii is said to have added his confident estimate that U.S. public opinion would in any case never tolerate giving Seoul military support.

Only after the conference, according to Tumanov, did Stalin make the final decision and dispatch diplomat Semen K. Tsarapkin to Pyongyang as his special envoy to personally deliver the great news to Kim.[122] As a precaution, Soviet advisers were withdrawn from the proximity of the future war zone, where their possible capture might expose Moscow's complicity and invite U.S. retaliation.[123] As a result, the North Koreans were deprived of badly needed expertise when they needed it most.

Incredible as it may seem, the exact timing of the invasion was possibly left up to the reckless Kim. Otherwise Shtykov would have hardly reported to Moscow on June 20 with surprise and disbelief what was Pyongyang's ruse to justify its aggression, namely, its supposed interception of South Korean army orders to invade the North in an hour.[124] Even if the choice of the hour was theirs, when the North Koreans attacked on June 25, they had not yet fully mobilized and much of their Soviet matériel was still in transit.[125] Nor was their superiority overwhelming, despite their outnumbering the enemy in both troops and equipment. Their success depended on their ability to exploit the element of surprise and win within twenty-two to twenty-seven days at most, as their plan envisaged.[126] All things considered, "it was reckless war-making of the worst kind."[127]

In authorizing the design for aggression in Korea and supporting it materially, Stalin responded to a challenge as well as an opportunity. The challenge was not so much Washington's stepped-up subversion of Eastern Europe, the efficacy of which he overestimated, as the victorious advance of the Chinese revolution, whose vigor he had underestimated. To compensate for this error and ensure Mao's continued respect for his authority, he became more favorably disposed to advancing communism by force. Toward that goal, the North Koreans provided the opportunity by their eagerness to advance it even at their own risk. Their expected success promised to more than offset the recent Soviet setbacks in Western Europe by making America's retreat in the Far East all but irreversible. So enticing was this prospect that no provisions were made for a failure.

6

The Test of Strength
June 1950–January 1951

The outbreak of hostilities in Korea injected a measure of unpredictability into the Cold War that had not been there before. The surprise attack was to make the war short; its prolongation required improvisation from both sides, for which neither had been adequately prepared. The waxing and waning of military fortunes amid fast-moving developments and abrupt changes on the battlefield added to the difficulty of predicting what was going to happen next. Assessing the likely course of the war, and adapting the policies accordingly, was a major Soviet preoccupation from its beginning.

An Adventure Gone Wrong

The initial Soviet response to the North Korean attack on June 25 was dilatory and noncommittal. For thirty-six hours, Moscow offered no public support to the announcement on the Pyongyang radio that had tried to justify the aggression by accusing the victim of starting the war.[1] Even after it endorsed the fiction, factual reporting by the Soviet media remained notable for its scarcity.[2] No responsible official could be found in Moscow to act on Washington's frantic attempts to contact the Soviet government in hopes of persuading it to intercede to stop the fighting. Letting events take their course, Vyshinskii absented himself from his office while his deputy Gromyko evaded a meeting urgently requested by the U.S. ambassador on June 27.[3]

Already on that day, however, the original calculation started going wrong as the UN Security Council, boycotted by Moscow since January, acted swiftly on the U.S. request to condemn the aggression, appealing to UN members for the necessary assistance to repel it. In trying to account for the Soviet failure to take part in the crucial vote that made this turn of events possible, Gromyko offered the lame explanation that his government had not refused to participate but had been unable to do so because of the exclusion from the world organization of the Chinese communists, insisting that this had rendered its decisions null and void.[4] Yet this was the same explanation that Stalin gave confidentially to Czechoslovak communists.[5] Gromyko would later suggest that he tried to convince Stalin that the boycott was a mistake but did not succeed.[6]

It was not so much the UN call for action as its speedy implementation by Washington that was bound to both alarm and confuse Moscow. In less than twenty-four hours, the United States sent not only its air force and navy but also its ground troops to defend South Korea. In addition, it announced the dispatch of the Seventh Fleet to the Straits of Formosa to protect Taiwan against an attack from the mainland. Although these actions were initially intended more for political than for military effect—a demonstration to impress both friend and foe that the U.S. government was willing and able to act decisively—against the background of its previous reticence they could not but appear as more than what they were.[7]

If Stalin had been accurately informed by his spies that the Americans had been unaware of the impending attack, then so purposeful a deployment of their military might could hardly fail to generate grave concern about the reliability of the intelligence he had been receiving and about the enemy's ulterior goals, concealed by deception. In the first major Soviet editorial comment about the clash in Korea, *Pravda* on June 28 charged an American ploy calculated to provoke the war in order to exploit it.[8] Mao Zedong as well suspected Washington's unfolding aggressive design in Asia.[9]

Stalin soon had reasons to regret having left too much of the operational planning in the incompetent North Korean hands. He tried to find out from his envoy in Pyongyang what the high command there was really up to. "Do they intend to push forward? Or have they decided to stop the advance? In our opinion, the attack should be continued under any circumstances; the sooner South Korea is liberated the smaller are the chances of the [American] intervention."[10]

In reply, Kim Il Sung pleaded organizational difficulties, whining about his misfortune of having to fight the Americans in addition and begging for still more weaponry and advisers.[11] Stalin must have temporarily developed second thoughts, for at a meeting with British ambassador David Kelly on July 6 Gromyko hinted at a Soviet desire for a peaceful settlement. When he again met the envoy a few days later, however, he turned evasive, ignoring Kelly's demand that the invaders must withdraw back beyond the 38th parallel, the dividing line between North and South Korea.[12] Before committing himself one way or another, Stalin evidently preferred to wait for the military situation to clear up.

While the extent of the Soviet setback was yet to be determined, the Chinese had already suffered a painful one. Even though the U.S. flotilla could not be physically positioned between Taiwan and the mainland for some time, the prospect that it would be there soon ruled out any easy seizure of the nationalist stronghold. If this is what they had hoped to accomplish after the expected success of their Korean comrades —Beijing's plans envisaged the invasion of the island in the summer of 1951[13]—then the sudden risk of a direct clash with the United States ruined the timetable.

The North Koreans had planned for a campaign of no more than one month; indeed, as early as June 25 they prematurely celebrated victory with a display of fireworks in Pyongyang.[14] Expecting that the capture of the capital, Seoul, would suffice to make the South Korean government collapse,[15] they were not prepared for its survival and continued armed resistance, which compelled them to fight their way down the peninsula. Their advance was still easy enough to justify Kim Il Sung's belief that "within a short time the whole territory of Korea would be cleared,"[16] but having to fight U.S. troops as well raised serious questions about the price of victory—not only for Pyongyang but also for its Soviet and Chinese supporters.

By the first week of July, Stalin got nervous enough to urge the Chinese to speedily dispatch nine divisions to the border of Korea, to be ready for intervening there as "volunteers" should the enemy be able to retaliate by crossing the 38th parallel. He vaguely indicated a willingness to provide them with air cover.[17] Mao Zedong, again rating the chances of U.S. military action against his own country high,[18] gave no reply. Stalin repeated his demand, sweetening it with the more specific promise of a division of 124 jet fighters, later to be donated to China.[19]

Although no clear understanding ensued—particularly not about the air cover—Mao was led to believe that this would be available if needed. On this uncertain premise, by August he had already redeployed the bulk of the Chinese army near to the war theater.[20] At the same time, he decided to delay the planned invasion of Taiwan until 1952, but in effect indefinitely.[21] If the prolongation of hostilities in Korea thus led to the postponement, though not yet the abandonment, of Beijing's most cherished goal, it created a potentially even greater problem for Moscow.

The longer the fighting continued without conclusive victory for the North, the more worrisome for Stalin was the possibility that Americans might try to compensate for their rout in Asia by an offensive in Europe. Khrushchev later recalled Moscow's "considerable alarm that the US might send its troops into Czechoslovakia."[22] Indeed, no sooner did the Korean war break out than U.S. observers in West Germany noted a "very considerable" increase of military activity on the Czechoslovak side of the border, including the construction of *defensive* installations.[23]

Stalin may not have known at the time that the day after the war began, Truman had secretly ordered preparations for a nuclear strike against the Soviet Union should it attempt to intervene militarily.[24] But the president's subsequent warning that the communist aggression "might well strain to the breaking point the fabric of world peace" was made publicly.[25] On July 11 the press reported the incipient transfer across the Pacific of U.S. bombers capable of carrying atomic bombs.[26] This was the same show of strength that Truman had made in response to the Berlin blockade, only more credible. Although the first set of planes had not been configured to carry the bombs, those that followed three weeks later were.[27] By then the beleaguered American forces in Korea were on the verge of being pushed out into the sea at Pusan.

If Stalin worried about a possible U.S. military diversion in Europe, so initially did Washington about the possibility that the North Korean operation might be a mere sideshow to camouflage preparations for a Soviet attack there. Some of its European allies feared that the attack might be invited by the Americans responding too forcefully to the provocation in the Far East. French high commissioner in Germany André François-Poncet thought that such a response, compounded by their plans for an accelerated rearmament of West Germany, could persuade Moscow that its time was running out, thus prompting it into "premature action."[28]

By the end of June, however, Washington concluded that the test of strength was not intended to extend beyond Korea.[29] Acheson believed that a firm response was called for precisely to keep it so limited. The Joint Chiefs of Staff concurred in their assessment that the aggression in Asia was not the beginning of a global war, in which case Korea would have had to be abandoned, so that Western Europe could be defended.[30] The National Security Council determined that sacrificing U.S. positions in Europe would leave global war as the only alternative.[31]

Despite the communists' demonstrated readiness to wage a war of conquest, the Soviet behavior left little doubt that there was no immediate danger of one in Europe. During the critical first week of the fighting in Korea, no unusual military movements by Moscow or its allies were noticed along the sensitive Yugoslav, as well as Turkish and Iranian, borders.[32] While Americans were facing probable defeat in the Far East, the Soviet Union had good reasons to avoid giving them an excuse to retaliate elsewhere. While verbally supporting the North Koreans, it did not interfere with U.S. shipping and other military operations near the peninsula, although it could have done so easily using submarines and by other means.[33] Stalin also rejected Kim's request for sending Soviet naval personnel lest their presence give the Americans "grounds for interference."[34]

U.S. ambassador Kirk considered Moscow's attitude flexible enough to allow for a North Korean retreat if a face-saving formula could be found. He urged Washington to keep the door open by blaming the war on communism rather than on the Soviet government or Stalin.[35] Indian diplomatic intermediaries reported the Chinese as well to be favoring an armistice in Korea and restoration of the status quo under Security Council supervision—in return for their admission to the UN.[36] But if Beijing thus seemed ready to see the hostilities ended in return for merely the achievement of its own single goal, Moscow continued to hold out for more. As long as the North Koreans kept advancing and the prospect of their ejecting the Americans from the peninsula remained open, it gave no signs of a readiness to support a return to the prewar situation.

By the end of July, however, a quick communist victory became problematic as, in Khrushchev's memorable phrase, the "air went out" of the North Korean offensive, allowing the U.S. forces to consolidate their defenses around Pusan.[37] Amid these new uncertainties, on August 1 Soviet representative Iakov Malik unexpectedly turned up at the Security Council to assume its rotating chairmanship. He did so despite the UN's failure to meet the condition his government had set for discontinuing its boycott, namely, China's admission. But Malik brought no conciliatory proposals—indeed, the purpose of his returning may have been to be on hand, in the still conceivable event that the Americans were driven into the sea, to strike a deal that would bar them from ever returning to Korea.[38]

In Europe, meanwhile, Moscow let the tension rise. At the beginning of August, Western Europeans had more reason to feel threatened, albeit indirectly rather than directly and not so much by the Soviet Union as by some of its satellites. Rumors of an impending military action against Yugoslavia spread again—later to be substantiated by Marshal Georgii A. Zhukov, recalling Stalin's having toyed with the idea of an armored thrust and airborne landings in Bosnia.[39] Moreover, East Berlin party boss Ulbricht made himself heard that the Bonn government would suffer the fate of the Seoul one. His party's new "theses" referred ominously to a forthcoming "armed uprising" in West Germany.[40]

In trying to cajole Moscow to support his ambitions, Ulbricht had a propensity for pushing harder than he had been authorized to do, thus putting Moscow on the spot. This may have been the reason for his agents playing into Western hands misinformation about supposed delivery of Soviet armored vehicles, designed for offensive operations, to East Germany's fifty-thousand-person militarized police units, which in reality remained lightly armed.[41] Their Soviet supervisors did not sufficiently trust them to be assured that these "Teutons" would not "stab them in the back."[42] Ulbricht's antics served Moscow's interests poorly by strengthening Washington's resolve to rearm the West Germans without making it more pliable in Korea.

On August 4, Malik proposed bringing the North and South Koreans to the UN for negotiations as equals.[43] Since this would have enabled the aggressors to negotiate from strength, the proposal was unlikely to be entertained by the losing side unless there were no more hope of turning them back. Mao had come to the conclusion that "there would probably be complications and a reversal." Unprepared to fight the United States on their own, the Chinese warned it against any expansion of hostilities that might make their intervention topical, belatedly protesting Truman's decision two months earlier to send warships to the Taiwan Strait.[44] While Beijing was thus trying to discourage Washington from fighting on, Moscow encouraged Kim Il Sung to push for victory.

Malik's intransigent speech at the UN on August 22 dashed hopes for a negotiated settlement.[45] Three days later more oil was poured onto the fire by Secretary of the Navy Francis Matthews, who advocated publicly a preventive war against the Soviet Union—a rare such statement by a ranking U.S. official.[46] But Stalin, in a confidential message, reassured the worried Czechoslovak leaders that the United States, forced to divert its attention from Europe to Asia, was losing the Korean war and, should the Chinese be drawn into it, would collapse altogether. He predicted that this would make possible the consolidation of "socialism" in Europe.[47]

The Chinese were not so sanguine about the Korean situation. Gao Gang, the regional party chief in Manchuria, estimated that the time for a North Korean victory had passed. Gen. Deng Hua, the army commander in the area, correctly anticipated that the United States would try an amphibious landing deep behind the enemy lines. At the end of August his warning about it was forwarded to Kim Il Sung and Stalin, but discounted by both.[48] Yet the Chinese had already done whatever they could to prepare their possible intervention.[49]

The day after Stalin had predicted American defeat in his message to Prague, Truman approved MacArthur's plan to land U.S. forces at Inchon. Its spectacular success on September 15 turned the tables on the North Koreans with a vengeance. Suddenly they were facing total defeat, the Chinese a likely threat to their own territory, and Stalin his worst setback since the Berlin blockade. Predictably, his foremost concern was blaming others for what had happened.

In upbraiding his people in Pyongyang, Stalin found the main fault in the "strategic illiteracy" of Soviet military advisers and their "blindness in intelligence matters."[50] But in assessing the situation for the benefit of the Chinese leaders, he preferred rather to shift the onus onto the "Korean comrades."[51] On September 29 Kim Il Sung, faced with the imminent collapse of his army, desperately asked for an outright Soviet military intervention or, "if for some reason this is not possible," a Chinese one.[52] Since Stalin indeed had his reasons to avoid such an involvement and, except for sending supplies and equipment for six divisions, delayed a reply, on October 1 the tottering dictator turned to Beijing to prop him up.[53]

War Expanded

At this dramatic moment, Moscow's and Beijing's responses to Kim Il Sung's plight were not as diametrically different, or straightforward, as they were later to be misrepresented. Trying to shift the burden of the risky rescue operation onto Mao, Stalin on October 1 urged him to immediately dispatch his troops to North Korea and deploy them as far as the 38th parallel—as he had previously indicated he would be willing to do.[54] At the same time, the Soviet Union proposed at the UN a resolution calling for the withdrawal of all foreign—that is, American—troops from Korea, after which its two governments would negotiate about holding elections under UN supervision.[55] This was not a very good proposal after the North Koreans had been all but beaten, and is was predictably rejected.

Mao may have been prepared to comply with Stalin's request promptly—if the document dated October 2 and printed in an official Chinese publication but not made available in original is authentic. It consists of his purported message to Stalin announcing the decision to intervene and asking the Soviet Union to supply the necessary equipment and air cover.[56] Yet the message was never received in Moscow; what was received there was another Mao message, dated on the same day, which stated the opposite. Accompanied by Soviet ambassador Nikolai V. Roshchin's annoyed commentary, it listed all the right reasons why the intervention would not be in China's interest and suggested that the North Koreans accept defeat for the time being.[57]

Since the second message mentioned that a final decision had not yet been taken—for Stalin's demand had provoked severe disagreements within the Chinese leadership—Mao's positive reply was probably written but never sent. His estimate that "the decisive duel between China and the U.S. imperialists being inevitable, Korea as a battlefield chosen by the imperialists is favorable to us"[58] was unlikely to be shared by most of his colleagues. While the central committee in Beijing held a marathon session to consider the matter further, Mao summoned from the provinces influential officials who could be expected to support him.[59]

Acting on the Chinese reluctance to intervene, Stalin pushed harder. Describing the intervention as a way of forcing the Americans to abandon Taiwan, thus allowing Beijing to achieve its foremost goal, he wrote Mao in congenial language that "if war is unavoidable, then let it be now, and not after several years, when Japanese militarism will have been restored." Stalin tried to assure Mao that if China were drawn into war so would be the Soviet Union as its ally.[60] He let China's leaders believe that if they moved troops into Korea at the very least he would give them air cover.[61]

While the result of Stalin's entreaties remained to be seen, he began to explore—more convincingly than in the past—the possibility of a cease-fire. Following the rejection of the Soviet proposal at the UN, his diplomats proceeded to outline a more reasonable Korean settlement. Its terms, which they hinted at in private conversations with their Western counterparts between October 5 and 9, coincided closely with those formulated by the State Department's Policy Planning Staff short-ly after the war's beginning. The coincidence may have been another example of Moscow's ability to peek into Washington's top secrets. In any case, the situation was grim enough for the Soviet Union to start evacuating its personnel from North Korea,[62] while signaling a readi-ness to broker an armistice along the prewar border, to be followed by UN-supervised elections in both parts of the country.[63]

Such a scenario would have meant the end of not only the war but also the wretched Pyongyang regime that had started it and lost. This outcome seemed to be forthcoming anyway once MacArthur's forces had crossed the 38th parallel on October 7—unless the Chinese stepped in. Since they were thoroughly undecided, Washington's giving them firm assurances of its readiness to respect the integrity of their territory would have strengthened in the divided Beijing leadership the hand of the opponents of the intervention, thus possibly sealing the fate of Kim Il Sung and his state.[64] The failure to give such assurances, rather than MacArthur's invasion of the North with the goal of reunifying the country, was what was wrong with U.S. policy at that time.

The day after MacArthur had crossed into North Korea, U.S. aircraft strafed a Soviet airfield sixty-two miles from its border in broad daylight, eliciting no more than a mild protest.[65] Stalin was obviously at pains to placate the Americans at such a delicate moment. In con-trast, Mao responded to the invasion by reiterating his readiness to go farther than the rest of the Beijing leadership had so far been willing to go. On October 8, he notified Kim Il Sung that the Chinese army would be coming to his help, although no firm decision to that effect had yet been reached.[66] He stepped up preparations of the expeditionary force and even set a date for its entry, October 15.

Because of the persisting disagreements in the central committee, however, the conditions of Sino-Soviet collaboration in the campaign, provided it would take place, needed to be clarified first. This was the purpose of Zhou Enlai's mission to Stalin which, in the absence of a contemporary record of what really transpired during their difficult talks at Stalin's Black Sea villa on October 10–11, has allowed for sharply different interpretations, hinging on the issue of air cover.

According to the recollection of Shi Zhe, Zhou's interpreter known to have taken liberties in narrating his experiences on other occasions, the Chinese emissary was shocked to find out that Stalin had changed his mind, giving the excuse that the Soviet air force was not quite ready.[67] In trying to bargain, Zhou then supposedly pretended that the Chinese would have to postpone the operation they were actually set to undertake, but the ploy backfired by prompting Stalin to renege even on his promise of military supplies. According to this account, he later sent Molotov to tell Zhou that he no longer wanted them to intervene at all.[68] The evidence from the Soviet side, however, has inspired the alternative interpretation that Stalin in trying to induce the Chinese to intervene had never refused to provide the air cover and that Zhou and his entourage had invented the refusal as an excuse to justify their opposition to Mao's plan to move in.[69]

The subsequent developments tend to support the familiar view of Stalin as the wary, even craven one, while Mao's reckless revolutionary bravado eventually carried the day. The message to Beijing that Stalin is said to have composed with Zhou to summarize the outcome of their talks, reportedly stated that "it will take at least two or two-and-a-half months for the Soviet air force to be ready."[70] This did not necessarily mean that the planes would actually support the Chinese army in Korea but rather that Stalin was going to wait and, depending on its progress and the American reaction, would take the specified time to ponder the situation before deciding whether the deployment would be in his interest. His reticence may have also been calculated to dissuade Beijing from a large-scale offensive liable to provoke the Americans with unpredictable results.[71]

Khrushchev later reminisced that Stalin "showed cowardice. He ... had his nose to the ground. He developed fear, literally fear, of the U.S."[72] Reconciled to the prospect of having the Americans as neighbors in the Far East, he decided to let the North Korean regime sink. On October 12 he directed the heartbroken Kim Il Sung to evacuate the country and take the remnants of his army to China,[73] where he might establish a government-in-exile.[74] But on the next day he suspended the directive, characteristically attributing his own vacillation to "the Chinese comrades" who, "after some wavering and various provisional decisions, in the end made the final decision to help Korea by sending troops."[75] In reality, the main reason the Korean war did not end at that time by American victory was Mao's refusal to tolerate it regardless of cost.

At first, having received the distressing news from Moscow, Mao on October 12 ordered his military machine to stop moving and cancelled the date he had set for starting the intervention.[76] After an agonizing debate on the next day, however, he persuaded the Beijing politburo to go ahead, even without the air cover, for otherwise the "arrogance of reactionaries at home and abroad would grow."[77] Invoking the credibility of China's commitment to communism, he insisted that "the Chinese had no right not to send the troops,"[78] and reinstated the date of their entry into Korea.

Mao may have believed that the decision would shame Stalin to reconsider the question of Soviet assistance, for he notified Moscow that he expected to "receive the air cover as soon as possible and in any case not later than in two months."[79] Zhou kept pressing Stalin for at least bombers, if not fighters, and tried to pin him down on other forms of assistance, but to no avail.[80] The willingness of the Chinese communists to absorb the resulting frightful casualties changed the course not only of the war but also of their relations with Moscow. Although their loyalty to Stalin remained unimpaired, his failure to deliver on his promise at a critical moment predetermined the later Sino-Soviet rift more than anything that had happened so far.

When Mao ordered his troops across the Yalu River into Korea on October 19, he did so with the intention of uncompromisingly leading them to complete victory. Stalin, while impressed by the Chinese audacity, was not so reckless. Once the intervention started—though not before—he compensated for his broken promises by expanding deliveries of hardware but never committed himself to seeking an all-out American defeat. With a keen eye on the fluctuating military situation, he instead tried to influence the conduct of the belligerents to his best advantage while keeping his own country out of the fray.

Stalin let the Chinese use the Port Arthur naval base, sent two air force divisions to protect the Yalu bridges, and moved a third one into Manchuria, thus providing a deterrent against U.S. attack on Beijing's home territory. He went so far as to have Soviet fighter pilots in Chinese uniforms fly aircraft with North Korean markings, but not over parts of Korea where they could be shot down and captured.[81] In the twelve months following November 1, fifty-nine Soviet air and artillery personnel were reported killed in action.[82]

According to different documents from Russian archives, the American losses up to December 1951 ranged from 569 to over a thousand aircraft shot down, compared with which the Soviet air force supposedly lost seven to nine times fewer planes—a stunning ratio in its favor.[83] While the reported kills, inconsistent with the more reliable American sources,[84] were almost certainly shamelessly inflated, the war provided Moscow a welcome opportunity to test its own air power and take a close look at some of the most advanced American models that were in fact shot down.[85]

In trying to prevent the Americans from retaliating in Europe, the Soviet Union tended to be more conciliatory whenever they seemed to be winning in Korea, only to take the opposite tack as the tide of the war turned in the other direction. The usual weather vane was Germany, where since the outbreak of the war the Soviet fortunes had been deteriorating rapidly. Having already decided to create a West German army, the Allies were also prepared to lift restrictions on the country's military production. The Korean events prompted them to issue a guarantee of its territorial integrity, including West Berlin, and reinforce their pledge to defend it by demonstratively dispatching additional troops into their part of the city.[86]

As the Chinese intervention unfolded, but while its outcome was still uncertain, Moscow revived its campaign for a German settlement on a seemingly conciliatory note. On October 21 it convened in Prague a conference of its satellite foreign ministers to make them pass a declaration proposing the establishment of a joint council composed of representatives of both German states. They were to prepare the country's reunification, to be followed by the conclusion of a peace treaty with its new government and eventually the departure of all foreign troops from its soil.[87]

Molotov told the participants in the Prague meeting that since the declaration was primarily aimed at West Germany, it should omit any invectives against its government, as especially the Polish delegate wanted to include. "If we succeed to detach West Germany from the [U.S.] alliance," East German foreign minister Georg Dertinger paraphrased the Soviet wishes, "it is probable that this would shatter the Anglo-American war conspiracy in Europe at its roots."[88] Signaling a belief that this might be possible, Moscow followed the Prague appeal with a proposal to reconvene the dormant four-power Council of Foreign Ministers.[89]

The Soviet Union could hope to utilize the Council, where its voice would count, by linking the possible settlement in Germany with the one in Korea. This is what the "peace" congress that it subsequently sponsored in Warsaw attempted to do on November 16 by calling for a grand gathering of all great powers, including China. The proposed agenda—ending the war in Korea by withdrawing all foreign troops, stopping the "remilitarization" of West Germany and Japan, forcing the U.S. to terminate its support of Taiwan, investigating American "war crimes," and much more—was preposterous.[90] But its Soviet authors evidently did not think so after the Chinese had completed the first stage of their offensive and positioned themselves to deal a crushing blow to the Americans.

Moscow also tried to take advantage of the situation to influence the prospective peace settlement with Japan which Washington had been actively pursuing with its other wartime allies since the Korean war had highlighted its urgency.[91] State Department consultant John Foster Dulles, a veteran of the 1919 Versailles peace conference in whose hands the project was entrusted, thought it "most unlikely" that Moscow "would ever agree with the kind of a peace treaty we want."[92] Hence he was surprised when on October 17—just as the Chinese were about to cross the Yalu—Malik expressed to him a desire to discuss it. Dulles assured him that "if the USSR were a party to the treaty, Japan would, by the treaty, cede South Sakhalin and the Kuriles to the Soviet Union,"[93] thus legalizing its possession of the territories it had conquered in 1945 and controlled ever since.

When Malik returned a month later—by which time the Chinese were getting ready to deliver their blow—he made a "very definite attempt ... to create a friendly atmosphere,"[94] thus indicating that he expected the message he had brought from his government to be taken seriously. Yet the message not only ignored Dulles's offer to confirm the 1945 Soviet conquests but also raised insidious questions about the American conquests. It questioned the propriety of continued U.S. military presence in Japan—indispensable for waging the Korean war—and protested the exclusion from the negotiations of the Chinese communists—the other party in that war.[95] Even without Moscow's releasing in addition the record of the confidental Malik-Dulles talks to the press, Washington's rejection could be taken for granted.[96]

The attempts to simultaneously press in both Germany and Japan outrageous demands that the Soviet Union was not in a position to enforce may be explained only by its expectation that the Americans would be compelled to yield because of their looming defeat. After the failure of MacArthur's brief counteroffensive that was to "bring the boys home by Christmas," the great Chinese onslaught started on November 27, resulting in an unprecedented U.S. rout and heavy casualties. Suddenly the specter of a global war, which the CIA concluded Moscow believed necessary for the attainment of its objectives,[97] assumed a more definite shape than before.

On November 29 Acheson warned in his "Strategy for Freedom" speech that "no one can guarantee that there will be no war. The present crisis is very serious. Whether reason prevails depends only partly on us. We have to hope and try for the best, but at the same time prepare for the worst."[98] What the worst might mean was suggested the next day by Truman in his statement to the press that the use of atomic weapons in Korea was under "active consideration."[99] Most Americans supported the idea, but their European allies panicked.[100]

Less inclined to believe in Moscow's design for aggression, Western Europeans had, if anything, been reassured by its behavior.[101] British Prime Minister Clement Attlee hurriedly flew to Washington to calm Truman.[102] The president subsequently explained that he had not meant to say that the U.S. policy had changed, which was true. Having considered the use of nuclear weapons in Korea before, Washington had by then ruled it out as politically unacceptable, except in the last resort to avert an imminent military disaster.[103]

Barring such an extreme situation, the Korean emergency thus did not make a nuclear war in Asia more likely; what it did was to make the United States more attuned to the possibility of a conventional war in Europe, which might or might not turn nuclear. In August the Joint Intelligence Committee had already expressed concern that "the Soviets may estimate at any time that they possess the capability to deliver a crippling blow to U.S. military capabilities. The situation may become critical at any time within the next two years."[104] As the Chinese pressed relentlessly forward, the critical time seemed at hand, prompting Washington to prepare for a nuclear defense of Europe.

The secret "Reaper" plan of November 29 envisaged defending the Continent by dropping more than a hundred atomic bombs on the

Soviet Union[105]—information that at least one of its spies in the U.S. capital, Donald Maclean, was in a position to pass on to his Moscow handlers.[106] The NATO "Agreement on Berlin Security," concluded on December 9, also provided for military action against East Germany if the Western positions in the city were endangered.[107] A week later, the decision was reached to integrate the prospective West German army into the European Defense Community, intended to supplement the transatlantic alliance. The alliance's chief in World War II, Gen. Dwight D. Eisenhower, was appointed to the new post of supreme allied commander in Europe.[108]

The memory of the last war inspired America's massive rearmament program, whose expected cost was estimated by chairman of the Joint Chiefs of Staff Gen. Omar Bradley to be "heavy, but not as heavy as the war which, we are now convinced, would follow our failure to rearm."[109] Proclaiming a "national emergency," the president announced the so far greatest peacetime military buildup in U.S. history. It eventually added 1 million soldiers to the nation's armed forces, making their total 3.5 million; the production of aircraft and combat vehicles almost quintupled.[110]

America's rush to organize to meet the communist challenge echoed the hectic weeks preceding the adoption in 1947 of the Marshall Plan—except this time the challenge to be met was military. Was the threat that Moscow was seen as posing as China's provider a mere figment of the imagination or something more tangible? Soviet behavior at the time when U.S. defeat seemed most likely suggested the answer.

At the Brink

On November 30, the day Truman made his bellicose press statement, the Soviet delegation in the UN vetoed there a resolution condemning the Chinese for their intervention.[111] Voicing unqualified support for its successful progress, Moscow at the same time signaled renewed interest in a settlement of the German question. It commissioned from East German premier Otto Grotewohl an amiable open letter on that subject to his Bonn counterpart, Chancellor Konrad Adenauer,[112] thus complementing the Asian sticks with European carrots.

As before, however, the prospective German settlement was to be on Soviet terms only. The Grotewohl letter merely restated the contents of the previous Prague declaration, already rejected by the West as unacceptable.[113] Nor could any inducements be found in Moscow's repeated call for the convening of the Council of Foreign Ministers or the appeal by the East German rubberstamp parliament for the convocation of an all-German constituent council.[114] Seconding the wishful thinking of their Soviet sponsors, the leading communists in East Berlin nevertheless continued to believe that "the Bundestag won't say no" and that, despite obstruction by the Bonn government, the council would come into being.[115]

Such beating around the bush showed how badly Stalin overestimated Washington's disposition to bow under pressure. The United States, having earlier given the wrong impression of a willingness to cut its losses in Asia, had been trying that much harder to avoid giving the same kind of impression ever since the North Koreans had attacked and even more after the Chinese had joined them. "Policies dictated by caution," the three secretaries of the U.S. armed services concluded, "are as likely, or more likely, to involve us in war with Russia than those which indicate firmness of purpose."[116] Symmetric with Mao Zedong's uncompromising drive for victory, the American attitude contrasted with Stalin's own greater willingness to retreat and cut his losses, demonstrated in the 1948 Berlin crisis. But that crisis had been of his own making, whereas the Korean war was not of U.S. making.

At the end of 1950, Moscow expected American resistance to falter. On December 4, the day before the Chinese recaptured Pyongyang, their ambassador Wang Jiaxiang solicited Gromyko's opinion about whether Washington might be ready to negotiate, suggesting doubts about the political wisdom of advancing beyond the 38th parallel. He got the advice to "beat the iron while it is hot."[117] At the UN, Vyshinskii received from Moscow instructions that any Soviet encouragement of an armistice would be "incorrect in the present situation, when American troops are suffering defeat and when the proposal about the cessation of hostilities in Korea is more frequently being put forward by the American side, in order to win time and prevent the complete defeat of the American troops."[118] Back in Pyongyang, ambassador Shtykov urged the Chinese to go for an all-out offensive to take the whole Korean peninsula.[119]

Although Mao refused to do so on sound military grounds, he did order advance beyond the 38th parallel.[120] And once his troops crossed it, he announced his conditions for an armistice: the complete withdrawal of U.S. forces from Korea, Washington's disengagement from Taiwan, China's admission to the UN—in short, abject American surrender.[121] This was the moment he and Stalin, who had approved these terms in advance,[122] believed to be the closest to victory, and they were preparing to exploit it to the utmost.

After the inconclusive end of the fall session of the UN General Assembly, Vyshinskii, before sailing home from New York, brought together a group of Eastern European diplomats for a secret briefing where he told them that the situation in Asia was "already beyond American control."[123] At the same time, he assured them that there would be no world war because the Western Europeans resented the U.S. rearmament program and the Americans would be unable to afford its cost anyway. Concluding on an optimistic note, he described the correlation of forces as tilting in favor of communism—the reason why the Soviet Union, though capable of overrunning Western Europe, need not do so.

The notion that communism could prevail without war permeated Soviet efforts to activate and strengthen the Cominform. Shortly after the Chinese had cast the dice by entering the Korean War, thus enhancing the prospects of communism's victory but also the risks of a wider war, Stalin indicated to Grigorian, the chairman of the central committee's foreign policy commission, his wish to expand the Cominform's functions and appoint its permanent general secretary.[124] Its meeting scheduled for December 23 was to implement what Stalin wanted.[125]

The projected measures highlighted the attention Stalin had recently been paying to Western Europe's key communist parties, the French and the Italian. In the fall of 1950, he brought their respective chiefs to the Soviet Union, apparently intending to keep them near. The first to come was France's Thorez who, having suffered a stroke, had been treated by French physicians until a Soviet one suddenly turned up and spirited him to Moscow in a plane sent specially for that purpose.[126] This was a similar departure for Soviet medical treatment as the one from which Dimitrov had not returned alive the year before. Thorez survived—but was prevented from returning home until after Stalin's death. Meanwhile, in Paris the party leadership was purged on Moscow's orders.[127]

The Italian party secretary Togliatti, too, came to the Soviet Union to recuperate, in his case from an automobile accident. Stalin himself visited the patient at the hospital before having him to his home on New Year's Eve.[128] Ruminating on the threat of war, he tried to enlist Togliatti for the job of the manager of a Cominform that, much like the Comintern before, would again serve as something of Europe's general staff of international communism, now set to resume a perilous but promising advance thanks to the latest Chinese military feats. His prospective appointment was important enough to warrant the postponement of the organization's planned December meeting—as it turned out indefinitely.

While the Rome party directorate welcomed the news of the likely promotion of its chief, the nimble Togliatti himself hedged and in the end evaded the honor.[129] In a letter which was a masterpiece of dissimulation, he wrote to Stalin about the indispensability of his remaining in Italy at such a critical time.[130] Having managed to slip back home and stay there despite Soviet entreaties to return, he used his presence in the country to preside in April 1951 over the Italian party congress distinguished for restoring to good standing prominent communists who had been purged before—a daring action at a time when everywhere else the trend was the opposite.[131]

The rehabilitations took place regardless of Moscow's reprimand to the Italian as well as the Czechoslovak communists for their alleged lack of "attention to raising vigilance and developing criticism and self-criticism."[132] All too familiar with the deadly workings of Stalinist purges, in which he had assisted as one of Stalin's stooges in the 1930s, Togliatti managed to spare his party the devastation they were increasingly inflicting on the French, Czech, and other parties. While the Italian communists prospered as a result, the unreformed Cominform declined and eventually faded away after Stalin had given up trying to appoint its director.[133]

The differences between Stalin and Togliatti about the price worth paying for advancing communism in Europe followed on the heels of the Stalin-Mao disagreement about the price of saving it in Korea. Although the disputes remained latent rather than open for the time being, they showed how the war bore the seeds of the later estrangement of both the Chinese and the Italian communists from Moscow. While in Europe the correlation of forces had by the beginning of 1951 failed to turn in Soviet favor, in Asia its improvement depended entirely on the Chinese. At no previous juncture of the Cold War had the stakes been so high yet the outcome so dependent upon forces beyond Stalin's control.

The longer the Chinese were winning battles in Korea without being able to win the war, the more reason Stalin had to be concerned that the Americans might strike in Europe and—to paraphrase in reverse British Prime Minister George Canning's famous boast—in redressing the balance of the Old World would try to save the New. At the end of 1950, Moscow alerted its Eastern European allies to be prepared for war by the end of 1952.[134] The target date was tantalizingly identical with the one recurring in the contemporary secret American estimates of the period of the presumed greatest danger of a Soviet attack[135]—estimates accessible to Stalin's spies in Washington.

The resulting transparency would not have necessarily been reassuring for Moscow. As a knowledgeable veteran of its intelligence agency pertinently observed, "mistaken though genuine Western fears of a Soviet attack were, when reported to the Kremlin they were almost certainly interpreted by Stalin as a cover for the West's own aggressive designs."[136] Thus, if he knew that his enemies expected him to attack, he had all the more to worry that they might want to preempt him, thus giving him a reason to strike first.

The defensive measures undertaken at the peak of the Korean War in Central Europe and the Soviet Union proper—whose capital city Stalin had surrounded by heavy antiaircraft artillery[137]—did not rule out preparations for an offensive. These were reportedly on the agenda of his three-day secret meetings with top Eastern European military and party officials in Moscow in mid-January, the details of which were later recounted by one of the participants, Czechoslovak defense minister Gen. Alexej Čepička. They have not been confirmed, but neither have they been contradicted by other sources.[138]

In speaking to the gathering about his favorite subject of the inevitability of war, Stalin is said to have explained to his Eastern European dependents that advantageous conditions for fighting it would last only three to four years. Referring to the Korean conflict as a probe, he then supposedly directed them to prepare for an invasion of Western Europe. In the end, they were to agree to place their armed forces under Soviet command in case of war.

American threat assessments reached their highest level of urgency at the same time. On January 15 the Joint Chiefs of Staff cautioned that "the Korean War could be the first phase of a global war between the United States and the USSR" after all.[139] The Joint Intelligence Committee concluded that this might be started by either side:

> The conflict between the increasingly hostile action of the Soviets and the objectives and policies of the Western Powers will in the end determine the beginning of global military warfare. Whether it is the Soviets or the Western powers who may formally initiate such action cannot be determined at this time. Nonetheless, increasing Western resistance to communist expansion tends to bring closer rather than to retard the outbreak of global military warfare.[140]

Indeed, "in light of the grave world situation," the National Security Council anticipated using atomic bombs against China and warning the Soviet Union that its aggression would be answered by using them against it as well.[141]

The evidence of Stalin's putative design for an aggression in Europe has similar strengths and weaknesses as the notorious "Hoßbach minutes" of the remarks Hitler made to his top military men in November 1937, cited by the Nuremberg war crimes tribunal as the most important, though still inconclusive, proof of such a design by the German dictator. Although both he and Stalin were reported to be expatiating on the inevitability of war and setting deadlines by which they had to strike in order to win, neither offered an accurate preview of what he would actually do next.

Hitler did proceed by resorting to offensive military moves, though not in accordance with the scenario he had outlined to his audience, thus giving evidence of an aggressive intent rather than design. Stalin, however, is not known to have followed up with any specific military preparations apart from the general military buildup he had subsequently accelerated. If he had a design for invading Western Europe in 1951, it was notable for a lack of directives necessary for its implementation.

The absence of a follow-up does not necessarily cast doubt on what Stalin was presumably saying, but it does give his words a different meaning. If he indeed feared the American attack so much that, according to Khrushchev, he "trembled with fear," putting "the whole country ... on military alert,"[142] then he had every reason to ensure that his Eastern European subordinates would not similarly tremble, too. And nothing would have scared them more than his telling them that the war he described as forthcoming would start by their being attacked first. If he wanted them to prepare for it, only by pretending that the timing was in his hands could he hope to achieve the desired effect.

There was a danger of a self-fulfilling prophecy, however, in the belief that Stalin and his American adversaries held in common. For "to believe in the likelihood of war, rightly or wrongly, means in some degree to behave in a manner that will actually enhance that likelihood."[143] William Strang, the permanent secretary in the British Foreign Office, therefore implored his government that it should do what it could to "deflect the Americans from unwise and dangerous courses," avoiding particularly "any action which might convince the Soviet government that an attack from the West was inevitable."[144]

The test of strength in Korea nevertheless did not grow into a wider war because neither Stalin—who in the end controlled Mao—nor Washington was willing, much less eager, to go over the brink. Although their symmetric estimates of the propensity of the war to expand fostered instability, they simultaneously provided a critical margin of stability by placing the expected showdown far enough into the future to prevent giving either side sufficient incentive for preemption. By the end of January, the Chinese offensive ground to a halt, after which the Korean belligerents reached a stalemate for the rest of the war. But they could not know in advance that they did, so that even though the conflict was contained, international tension remained high, and within the Soviet bloc the perception of both external and internal threat kept increasing.

7

On the Defensive
January 1951–March 1952

The incipient Korean stalemate, reassuring though it may have been for Stalin in the short run, was hardly suited to allay his concern about the war's possible long-term repercussions. From his perspective, even if the Americans felt too weak to expand the hostilities beyond the Far East, for that very reason they might want to compensate for their weakness in that area by increasing pressure on his vulnerable Eastern European periphery. In anticipating their challenge, should he try to appease them by facilitating a negotiated peace in Asia? Or should he rather do his best to lock them in a prolonged confrontation with the Chinese, while preparing for a perhaps inevitable showdown in Europe? These were Stalin's unappealing strategic choices in early 1951.

Eastern Europe under Siege

As long as the Chinese appeared to be capable of winning despite having been pushed out of South Korea in early February, Stalin shunned his choices. On the 16th he told *Pravda* that the United States was going to lose and should accept the terms demanded by Beijing. He added that "at least at the present time war cannot be considered inevitable" if "the people take the cause of peace into their own hands."[1] When the people assembled for that purpose under Soviet auspices met a week later at the international peace rally in East Berlin, no alarm was sounded.[2] But neither did the shrill rhetoric of the gathering hint at any Soviet interest in winding down the Asian war. With the likelihood of its expansion receding as a result of the Chinese setback, Soviet attention could be shifted to Eastern Europe.

Three days after Stalin's statement to the press, a bomb explosion wrecked the Soviet embassy in Tirana.[3] The Albanian authorities instantly rounded up thirty persons to be summarily shot, hauling hun-

116

dreds more to prison. Whether the bombing was a sabotage or a provocation, staged to provide an excuse for the crackdown as more Western
agents were on their way into the country,[4] it highlighted Albania's
special vulnerability, accentuated by the recent rapprochement between
neighboring Yugoslavia and the United States. The incident bore on the
topical question of whether the Eastern European security services, so
far successful in combating the modest U.S. covert actions, would be
able to keep up with their rapid expansion.

Washington's new amity with Belgrade prompted a Soviet outcry
about alleged U.S. plans for a Yugoslav-Greek attack on Albania and
perhaps Bulgaria.[5] Even if the outcry was feigned, it could be no con-
solation to Stalin that it was Tito, rather than the Americans, who
determined the pace and extent of his country's expanding military collaboration with the West. After the first shipment of armaments from
Britain had reached Yugoslavia in May, foreign minister Gen. Koča
Popović requested and received more equipment from the United
States, while the two countries' military representatives met for talks
in Washington.[6]

As official Western statements sounded more and more as if they
were meant to guarantee Yugoslavia's integrity against a Soviet attack,
its spokesmen, with their customary recklessness, advertised their intention to turn any such adventure into a world war. Tito's ideological
aide Kardelj served Stalin the notice that "we shall not allow anyone to
stage a new Korea in Yugoslavia, that is to say, to throw this or that
satellite or several of them against Yugoslavia while he himself is
supposedly protecting peace."[7] Molotov responded by another appeal to
its people to get rid of Tito and his gang.[8]

In its venom, the Moscow-Belgrade hostility surpassed even the
Soviet-American one. Hurling exquisite insults at each other, the communist adversaries paid no respect to the diplomatic immunity of their
respective agents. In retaliation for the mistreatment of its personnel in
Eastern Europe, Belgrade put on trial members of the Czechoslovak,
Bulgarian, and Hungarian diplomatic missions, along with their local
helpers.[9] Yet the standoff between Tito's and Stalin's versions of Stalinism continued.

Washington's support for Tito increased despite its having concluded
that his example was unlikely to be followed elsewhere.[10] Rather than
banking on the anti-Soviet potential of independent-minded communists, the United States instead turned to encouraging anticommunist
resistance. Hence the more centrally located East Germany and Czechoslovakia replaced Albania as the prime targets. A report prepared in
December 1950 for the U.S. high commissioner in Germany envisaged
luring away important officials from East Germany to undermine its
economy and political system while using RIAS, the widely popular
U.S.-run radio station in Berlin, to keep up the people's spirit of resistance.[11] Yet Washington kept distance from the West German Fighting

Group against Inhumanity that encouraged sabotage and favored guer-rilla warfare.[12]

While not enough is known about how far the intentions were translated into actions in Germany, in Czechoslovakia the United States tested its regime's vulnerability by staging border incidents. Khrushchev recalled that "not a single day went by when American planes didn't violate Czechoslovak airspace."[13] In January the Prague government charged that fifty-six such violations had occurred recently, including drops of radio transmitters for undercover agents. The next month, its protest against penetration by two U.S. aircraft as far as the capital elicited first Washington's denial but then its acknowledgment of the incident. Three months later, a group of American soldiers strayed into the country, soon followed by two jet fighters crash-landing there. By June the communist government had counted a total of 116 violations.[14]

While most of the U.S. data remain classified, those from the Czech archives reveal a different picture from that of the previous year, giving substance to the official claims of sharply rising foreign infiltration. Of the 2,224 persons listed as captured in 1951 after having crossed the borders, 62 were identified by the police as Western agents; during that year and the two that followed, a total of 1,200 were reported arrested or killed. From 1951 to 1956 the communist authorities recorded 79 political murders (not counting the judicial ones they themselves had perpetrated), listing 35 party officials as assassinated by enemy hands in 1952 alone.[15]

In May 1951 Washington created the Psychological Strategy Board as the nerve center of its counteroffensive against Soviet Eastern Europe.[16] Aiming to promote there passive—or not so passive—resistance, the Munich-based Radio Free Europe had already been inaugurated the month after the Korean war had broken out.[17] Its programs had since been expanded to cover five countries, with particular attention given to Czechoslovakia.[18] Putting into effect Kennan's ideas, the ostensibly private organization of concerned citizens that sponsored the broad-casting complemented the official Voice of America to bring its lis-teners particularly the local news that their rulers had reasons to hide.

Judging from the pains the communists took to silence it, the broad-casting was effective. In trying to make Washington stop transmitting the Voice of America Hungarian program through Germany, the Budapest government swapped a U.S. businessman it had previously jailed—only to have the program continued through Greece.[19] Indignant at the Munich station's practice of personally attacking individual offi-cials, Czechoslovak minister of state security Kopřiva demanded from the foreign minister, Viliam Široký, that something be done about "this scandalous method, which incites to murder."[20]

In response, the Prague regime took an American citizen—Asso-ciated Press correspondent William Oatis—hostage for better behavior of his government. Besides raising the usual charges of spying, it linked his arrest with the broadcasts "openly inciting to revolution" as well as the recent border incidents.[21] Although Washington refused to negotiate under blackmail, its inability to prevent the indignity of a public trial, where Oatis was made to confess what the prosecution wanted him to confess, made U.S. "prestige seriously jolted as a result."[22]

Široký, the author of the ploy, further envisaged allowing the captive's American wife to visit him in jail, whereupon the couple were to be persuaded to request asylum, in return for which Oatis would be pardoned and forgiven his long prison sentence. "Progressive lawyers" would then have a field day conducting interviews and collecting material to embarrass his purported Washington paymasters.[23] Yet the scheme was shelved after the Americans were the ones to have a field day by being able to give asylum to a whole trainload of Czechs, whom the engineer had crashed through the border into Bavaria.[24]

While the war of the secret services was real enough, some of the feats it entailed were, as before, both unintentionally or deliberately exaggerated. The "Summary of Our Knowledge about the Espionage Services of the U.S.," prepared for Široký's office, showed that the knowledge was pretty rudimentary and was further distorted by ideological blinkers.[25] Unable to determine with enough accuracy just how much of Washington's subversive intent was actually translated into action, the communists preferred to err by overestimating it. Since their doctrine presupposed a formidable capitalist threat, in order to document it more incidents had to be fabricated than actually occurred, thus simultaneously demonstrating the prowess of the security forces in coping with the threat. The notorious "Babice case" was named after the village where the Czechoslovak police "foiled" a terrorist attack they had themselves organized, complete with the murder of one of their own men, allegedly by assassins dispatched from the West.[26]

Unlike before, Washington was indeed trying hard to destabilize Eastern Europe after the outbreak of the Korean War, and the justifiable fear of the local regimes that it might succeed added to the ferocity of their repression. In providing them with guidance, Moscow warned that "the establishment of People's Democracies" had not ended the "capitalist encirclement," which was not to be understood as "a purely geographical concept" but also the source of a threat from within.[27] The resulting persecution of the already suppressed class enemies was coordinated throughout the region, singling out the once independent farmers and the clergy of the no-longer-to-be-tolerated Catholic church. In the spring of 1951, the deportations for forced labor in the Czechoslovak uranium mines coincided with the evictions of the dispossessed upper classes from Hungarian cities and the "unmasking of Vatican agents" in countries from the Baltic to the Danube.[28]

Some of the American policies that may seem ludicrous in retrospect were taken very seriously at the time. In the summer of 1951, Moscow and its satellites were most perturbed by the debate in the U.S. Congress preceding the vote on the Kersten amendment to the Mutual Security Act. The sponsor of the bill, Congressman Charles J. Kersten, was on record as having said that "unless the Soviet regime is undermined, subverted, and overthrown it will bring us war," and his colleague Sen. Alexander Wiley as favoring the dispatch of commandos to incite insurgencies in Eastern Europe, thus impairing Moscow's ability to wage war elsewhere.[29] The bill envisaged spending one hundred million dollars to create military units of exiles from the region, ominously declaring the intention to recruit "any selected persons who are residing" there as well.[30] Gen. Władysław Anders, the commander of the Polish army in the West during World War II, claimed that in addition to the estimated 239,000 exiles, he could mobilize as many as 6,250,000 fighters in the countries under Soviet control.[31]

The passage of the Kersten amendment on October 10, preceding the conclusion four days later of the treaty providing for U.S. military assistance to Yugoslavia, elicited formal complaints by Moscow and its client states to the UN. Citing numerous cases of Western subversion, the Soviet Union invoked the 1933 document the United States had signed in trying to prevent communist subversion—the Roosevelt-Litvinov agreement whereby the two governments had pledged to abstain from interference in each other's internal affairs.[32]

While none of the cases mentioned in the complaints could be substantiated, those that could were not mentioned. This concerned particularly the Western agents captured over the years in Albania, several of whom were put on trial on the day the Kersten amendment was passed; with similar pointed timing, another batch of them followed the signing of the U.S.-Yugoslav treaty.[33] A hundred more infiltrators were officially reported killed soon after their arrival. Calculated to impress the outside world with the effectiveness of combating subversion, but not to spread alarm about its scope in the other countries that were exposed to it, the trials publicized in Albania were ignored elsewhere in the Soviet bloc.[34]

Although the National Security Council went on to authorize "the intensification of covert operations" to help "develop underground resistance and facilitate covert and guerrilla operations,"[35] there was more bark than bite in the Kersten amendment. Its implementation was delayed, and in the end petered out after the formation of but a few symbolic detachments without military value. To the House Committee on Foreign Affairs, CIA director Allen Dulles conceded, "We want to keep alive in these countries the desire for liberty, and the hope of liberty. But until we are much better prepared, do we want to do anything as provocative as having legions sitting there ready to go in but not able to go in? ... And I must admit there are times when I feel we have almost stirred up Czechoslovakia too much, without the military ability to do something."[36] This was too narrow a view, however, to give a balanced picture of the results of Eastern Europe's siege by the U.S. secret services.

Some of Washington's European allies, particularly the British and the French, considered its subversive efforts irresponsible, unwise, dangerous, and liable to backfire by provoking Soviet military response.[37] But there was no need to be worried. Although the efforts proved not as successful as their architects had hoped, they were successful enough to make the communist regimes palpably less secure internally but not any more dangerous externally. If they fueled more domestic repression, this was by then increasingly directed against the communists' own ranks rather than the remnants of the already defeated opposition. And as far as war was concerned, that remained limited to Korea, gradually making Moscow more rather than less helpful in trying to bring it to an end.

The Inauspicious Negotiating

The Soviet readiness to support a negotiated settlement in the Far East evolved slowly, in interaction with the German question. Ever since the the communists had attacked in Asia, in Europe the U.S. plans for the rearmament of West Germany had been making strides, prompting Moscow to push harder on its proposal for a conference of foreign ministers to initiate a discussion that could stop them.[38] But any hope that the Americans could be forced by their distress in Korea to become more forthcoming in Germany was misplaced. Refusing to discuss the German problem unless the larger issues that had created it in the first place were addressed as well, Washington countered the Soviet campaign for the foreign ministers' conference by proposing a meeting of their deputies that would first prepare the agenda for their later talks.[39]

There was a similar correlation between the prospects on the battlefield and the Soviet attitude toward the American efforts to speed up the conclusion of a peace treaty with Japan. Since Moscow had in December 1950, when the prospects were good, spurned Dulles's invitation to share in its preliminaries, Washington had been busy consulting with its other prospective signatories to hammer out its text. Pending the clarification of the trends in the Korean War, the Soviet government stood by. Behind the scenes, however, Stalin held secret meetings with a delegation of Japanese communists, instructing them how to intensify their struggle against U.S. occupation to prevent the American-made treaty from coming into effect.[40]

Moscow resisted the U.S. proposal for a conference of deputy foreign ministers until the fading prospects of communist victory in Korea gave it a reason to reconsider. On March 1 Mao Zedong notified Stalin that he was switching to "rotational warfare," supposedly to wear down the Americans by attrition, but in effect acknowledging that his plan for defeating them had failed.[41] Five days later, deputy foreign minister

Gromyko met with his Western counterparts at the Paris Palais Rose. They had expected him to be mainly interested in Germany but were surprised to find out that the Soviet priority had meanwhile become the "arms race."[42]

French ambassador to Moscow Yves Chataigneau suspected that the Russians were afraid Washington would rearm to the point where it could either compel them to negotiate on its terms or else "force them to submission by resorting to arms."[43] No substantive discussions ensued. The Americans were not prepared to talk about disarmament when fighting a war, while the Soviet Union was not yet ready to help end it. As the conversations were limping on, Gromyko hinted to Trygve Lie, the UN Secretary-General, that his government might possibly be interested in helping to reach a cease-fire in Korea, but mentioned no terms. Nor were any specified by the Soviet ambassadors in London and New Delhi, who, too, dwelt casually on the need to end the war. When U.S. chargé d'affaires in Moscow Charles Bohlen raised these hints with Vladimir Semenov, the chief Soviet political adviser in Germany, he got nowhere.[44]

By the end of March, both Soviet and American interest in the Paris discussions had all but disappeared. Both sides tabled extraneous topics they knew were not negotiable—Moscow the Trieste dispute between Italy and Yugoslavia where it had no leverage, Washington the even more intractable question of human rights in Romania, Bulgaria, and Hungary. This was the time the Chinese were attempting their last big push in Korea, while in the United States the debate about extending the war into their homeland was nearing a denouement. If Moscow had reasons to stall while awaiting the results of the push, so did Washington as long as the outcome of the debate remained undecided.

Despite its nonbelligerence, the Soviet Union had by then become deeply involved in the Korean War. Besides providing the Chinese with air and naval bases on its territory, it supplied them with the bulk of their armaments, including up-to-date MiG-15 fighters superior to any in the U.S. arsenal at that time.[45] But the Soviet aid, though enough to keep the war going, was not enough to win it. To be paid for in hard cash with interest added, it earned its donors Beijing's later epithet of "merchants of death."[46]

Stalin was not above husbanding his aid to prolong the war—and China's dependence on him—as he had previously done in Spain with its republican government after its fight had already been lost.[47] In the spring of 1951, however, delaying the termination of the war was not yet in his interest. Washington was debating the use of atomic bombs against China, and the longer the fighting continued without the prospect of a resolution, the greater might be the American temptation to force it by that drastic step. This, if successful, could tilt the whole East-West balance irreversibly against the Soviet Union, making it lose the Cold War.

On April 7, the United States prepositioned nuclear bombs to Guam
—though not as far as Okinawa, from where they could reach China
more easily.[48] Truman tried to keep their movement secret—except
from his military chiefs and the congressmen whose support he needed
for sacking MacArthur, whose unauthorized public calls for bombing
China made him intolerable. If Moscow had first been alarmed by
finding out about the transfer of the bombs from the leaky Congress or
its spies, four days later it was bound to be reassured by the news that
the general was fired for favoring their use.[49]

Interpreting China's new offensive as a possible trigger of global war,
Washington did not abandon its nuclear option.[50] At the end of April,
additional U.S. aircraft capable of carrying atomic bombs flew across the
Pacific. With an eye on Soviet presence at Chinese air bases in Manchuria, besides the growing traffic at the Soviet ones in Vladivostok and
Sakhalin, MacArthur's successor, Gen. Matthew Ridgway, was given
discretion to respond by atomic weapons against any air attack on Korea
launched from outside its territory. U.S. reconnaissance of their possible
targets in northern China preceded the still more unnerving experiments simulating nuclear strikes against North Korea.[51]

By late May the failure of another—the most costly[52]—Chinese offensive ended the last serious attempt to win the Korean War by military
means. Gen. Peng Dehuai, the commander of Beijing's expeditionary
force, painted a grim picture to his superiors, contrasting the surging
morale of U.S. troops with its decline among of his own.[53] Fearing more
enemy landings behind his lines, Mao Zedong rightly judged June to be
the critical month.[54]

Expecting the Americans to press their advantage, Stalin implored
Mao to reverse the morale problem by ordering quick and powerful
strikes against their selected units that could dissuade them from advancing beyond the 38th parallel.[55] Yet what Stalin suggested was easier
said then done. Drawing the appropriate conclusions, Mao at the beginning of June sent his Manchurian lieutenant Gao Gang together with
Kim Il Sung to Moscow to discuss there what was to be done next.[56]

At this point in time, the United States would have been in the best
position to try ending the war by using its military superiority if its
own generals, in another proof that wars are too important a business
to be left to them, had not opposed pursuing the North Koreans into
their homeland. Unlike later, the U.S. public opinion was supportive of
the campaign, which would have created the optimal preconditions for
a peace settlement favorable to the West.[57] Whether the deal could
have entailed the elimination of the obnoxious Phyongyang regime,
which both Stalin and Mao had on previous occasions been disposed to
sacrifice, is a question of abiding interest, as only the manner of its
eventual demise will determine whether allowing it to survive in 1951
rather than trying to get rid of it was merely a pardonable folly or a
fatal mistake.

In any case, the previously secret Russian documents now show that Mao, having abandoned his dream of total victory, was at that time more ready than ever to end the war on reasonable terms. On June 13, he submitted for Stalin's approval armistice terms very different from those he had advanced in January. They amounted to no more than the restoration of the prewar status quo along the 38th parallel, to be safeguarded on both sides by a neutral buffer zone. Omitting the question of Beijing's admission to the UN on the ostensible grounds that the organization was not worthy of its membership, he was also willing to leave the Taiwan issue for later discussion.[58]

Unaware of how much the enemy was prepared to compromise, the United States, instead of using its military assets while they lasted, chose on May 31 the slow way of sounding out Malik at the UN about Moscow's willingness to mediate. Five days later, even before Gao Gang and Kim Il Sung began their consultations in Moscow, came an encouraging reply, thus indicating that Stalin shared Mao's desire to bring the war to an early end.[59] As a result of the consultations, they all agreed to pursue the American feeler to initiate armistice negotiations. "If they were to begin," Mao wrote to Stalin, it was "absolutely necessary for them to be directed personally by him, so that we would not find ourselves in an unfavorable position."[60] By deferring so willingly to Stalin's presumably expert guidance, the Chinese thus allowed him to manage the initiation of the talks to serve his own interests—for better or for worse.

Judging the time propitious to again tackle the German question, the United States on May 31 proposed to hold a conference of foreign ministers that would meet in Washington in five weeks' time.[61] Although their deputies at the Palais Rose had not even been capable of agreeing on its agenda, the State Department considered the progress of the talks sufficient to warrant their continuation on a higher level. The Soviet response, however, showed that this was not the case. Reaffirming its real priority, Moscow made its acceptance of the American invitation conditional upon raising at the conference the twin issues of NATO and U.S. military bases abroad.[62] The first pertained to its worries about West Germany's increasingly likely integration into the alliance, the second to those about America's growing capacity to project power and influence in Europe.

By setting such conditions, Moscow overplayed its hand. On June 22 the Americans broke up the Paris meeting, and nothing ever came of the proposed other conference. But despite Gromyko's spending two hours pestering them for spoiling the talks,[63] the Soviet government was not to prevented from taking on the next day the decisive step toward mediating armistice in Korea, as Stalin had promised the Chinese to do.[64] Malik's radio statement on June 23, favoring cease-fire negotiations, omitted any of the previous unacceptable preconditions for their initiation.[65] He still lamented the failure of the Paris talks, complaining a few days later to deputy U.S. representative to the UN Ernest A. Gross about the growth of NATO, America's air and naval power, its stockpile of atomic arms.[66]

Within two weeks, the belligerents agreed to start negotiating. By that time, however, the self-imposed U.S. restraint had already taken its toll. Having shed his moderation, Mao reverted to his demand for the total withdrawal of American troops, and before the negotiators met, Stalin had approved his idea of advancing it as a condition of the armistice, thus making its attainment that much more difficult.[67] The condition contradicted Malik's and Gromyko's assurances to American diplomats that the talks would be limited to strictly military matters, excluding political and territorial issues.[68]

Stalin's approval of Mao's expanded negotiating platform entailed a wrong, though understandable, estimate of U.S. readiness to retreat after the Americans had defied military common sense by their reluctance to exploit their strategic advantage. At the same time, Stalin was signaling his anger and frustration at the progress of preparations for the Japanese peace treaty which, railroaded by Washington while Moscow demonstratively stood aside, had by then reached a critical stage. When the U.S. government, on July 6, forwarded to the Soviet foreign ministry the revised version of the text it had sent there five months earlier, it created something of a panic.[69]

The earlier version, still drafted with the intention of inducing Moscow's adherence, had clearly spelled out the transfer of the former Japanese part of Sakhalin and the Kurile Islands to the Soviet Union provided its government would sign the treaty.[70] The new version merely envisaged the renunciation by Japan of its title to those territories without specifying in whose favor.[71] Moreover, the text was presented to Moscow as final, along with the invitation to sign it, without any further discussion, at the conference in San Francisco scheduled to meet two months later.

As Soviet foreign ministry officials were frantically drafting and re-drafting replies to Washington besides the text of a substitute peace treaty—only to have Stalin dismiss their papers as inadequate—the armistice negotiators tackled their business at Kaesong. The Chinese pressed their untenable demand for total American retreat—only to make the United States dig in deeper. They did not fare any better after adding the insistence on the 38th parallel, instead of the current front line, as the future border between the two Koreas—the division promoted by Kim Il Sung as a way of prejudging the prospective peace settlement to preserve the integrity of his rump state and his own power.[72]

All this time, Mao Zedong kept Stalin meticulously informed about the progress of the talks, but the Soviet leader, if he bothered answering at all, merely approved the inauspicious demands Beijing chose to push.[73] Thus, contrary to Mao's original desire, Stalin let him personally direct the negotiations with the goal of extracting from the enemy concessions that he was unlikely to get. On the fallacious assumption that the Americans had become so tired of the war that they were ready to quit,[74] Mao specifically instructed his representatives to be obstreperous and rude; on one day, the two delegations stared at each other in silence for 135 minutes.[75]

Although Mao Zedong gradually abandoned his other unpromising demands, his intransigence about where the armistice line was to be drawn ensured that the conference would remain stalled. After his negotiators reinforced his belief that their American counterparts would not risk its breakdown, bombing incidents blamed on South Koreans but possibly masterminded by Mao without consulting with Moscow,[76] supplied him with a welcome pretext to suspend the talks six weeks after they had started. Stalin approved, seconding the Chinese line that "the Americans have the greater need to continue" them.[77]

Stalin's view was consistent with the otherwise baffling Soviet conduct at the San Francisco conference. Only his belief that the Americans had lost their nerve could possibly justify his dispatching there his representatives with the assignment to disrupt the peace process at the eleventh hour by insisting on amendments that would kill the treaty which was about to be signed.[78] These ranged from the ban on U.S. military presence in Japan and the demilitarization of the sea straits around it to the inclusion of Beijing in the peace settlement—tenets that the presiding Dean Acheson had no trouble shooting down.[79]

Khrushchev later wondered whether it was Stalin's vanity or pride that made him leave his Japanese territories in limbo by refusing to cooperate with Washington while their confirmation was still on offer.[80] After all, this was the issue about which he had cared so much during his first meeting with Mao in December 1950. It is unlikely that a sophisticated explanation is needed; his behavior was simply stupid.

In contrast to Stalin's fumbling, American self-confidence was high, and growing. Washington officials believed that it was the progress of Western rearmament which deterred Stalin from risking a global war, prompting him to seek instead to downgrade his intended showdown with the West from military to political.[81] He remained impervious to persistent Chinese demands for more arms and ammunition; after four months of frustrating Moscow meetings, Mao's chief of staff Xu Xiangqian not only failed to get what he wanted but had to acquiesce in getting less.[82]

No sooner was the Japanese peace treaty done and not to be undone than the Soviets became more helpful again in trying to restart the armistice negotiations. As when cutting their losses in the Berlin blockade, however, they wanted to save face. On October 5 Vyshinskii first answered evasively U.S. ambassador Kirk's request for his government's mediation.[83] The next day saw Stalin boasting about his country's second nuclear test, expatiating on its readiness to flex its military muscle.[84]

Only afterward came the breakthrough, as the Chinese and North Koreans reversed themselves by accepting Washington's demand to resume the cease-fire talks at the new Panmunjom site they had been ruling out as unacceptable.[85] Still, the conference did not start again on October 25 until the Soviet government had made public its doctored version of the secret Kirk-Vyshinskii parley, to make it look as if the American had tried to be threatening but the Russian put him down.[86]

The lull that subsequently descended on the Korean battlefield was deceptive of progress toward peace, unexpectedly blocked by the un-willingness of large numbers of Chinese and North Korean prisoners of war to return to their homes.[87] In this affair, deeply embarrassing for the communists, at issue was not merely prestige but the very credi-bility of their rule; for the United States, apart from the propaganda boon, in resisting involuntary repatriation in principle the credibility of Western values was at stake.

Washington also encouraged the mass desertion because of its belief in the deterrent effect the sight was having on Moscow's penchant for aggression. Whatever the penchant, however, it had been sufficiently deterred already. Despite the continued growth of the U.S. defense budget fueled by the Korean War, the Soviet military expenditures after their upsurge in early 1951 declined substantially in the winter of 1951–52.[88] Meanwhile, Stalin stood to benefit from the impasse that made the Korean conflict simmer without threatening to explode, keeping the United States locked "in the wrong war, at the wrong place, at the wrong time"[89] and his Chinese clients dependent on his good graces for support.

Unlike in the spring of 1951, by the end of that year Stalin acted as if he no longer wanted the Korean War to end any time soon. He praised "the Chinese and Korean side [for] using flexible tactics in the negotiations while it has continued to pursue a hard line, not showing haste and not displaying interest in a rapid end to the negotiations."[90] At the same time, lest an impression of weakness be given to the enemy, Stalin blocked the appeal Pyongyang intended to make to the UN to accelerate the resolution of the war.[91]

In the long run, however, an indefinite prolongation of the war was wrought for Moscow with the danger of stimulating even more the growth of Western military power it had triggered. By 1951 the United States had acquired a fleet of intercontinental bombers and the capabil-ity of their in-flight refueling that could carry the four hundred nuclear bombs available in its arsenal to Soviet targets, as well as enough over-seas bases and trained crews to ensure that the ordnance would reach there.[92] In trying to narrow the lag in the arms race, Vyshinskii in vain tabled one disarmament proposal after another at the UN.[93] Meanwhile, in Moscow political scholars and party officials had been grappling with the prize question of whether or not war was inevitable. Having failed to reach a conclusive answer, they left it to Stalin to decide.[94]

What made the question more difficult to answer for Stalin, too, was the loss earlier that year of the premier Soviet source of information about American intentions—the spying pair of Guy Burgess and Donald Maclean. Their cover about to be blown off, since April 1951 they had been unable to provide any top secret information, and the next month they had to be spirited to Moscow in a hurry.[95] Nor did the reorganization of the Soviet intelligence apparatus, already overloaded because of its excessive centralization, make the proper processing and evaluation of data easier and more reliable.[96] All this was bound to make Stalin even more insecure than he had always been.

"I Trust No One"

In the summer of 1951, Khrushchev overheard the autocrat as fretting that he was "finished. I trust no one, not even myself."[97] Stalin's daughter, Svetlana, later recalled having been shocked by her father's saying that he no longer trusted his security man, Beriia, either.[98] According to Albania's Enver Hoxha, Stalin was afraid that "the enemy will even try to worm his way into the Party, indeed into its central committee."[99] Whatever the accuracy of these testimonials of the despot's seemingly progressing paranoia, they were amply borne out by the dimensions of the purge he had set into motion and expanded when, following the failed communist offensive in Asia, he was put on the defensive in Eastern Europe as well.

Spreading ever higher up in the party hierarchy, the purge began to engulf also officials who had spent World War II under Stalin's watchful eye in Moscow rather than out of his reach in the home underground. Noting that something new was happening, the London *Times* wondered whether he might not have concluded that all he needed anymore was people whose obedience was engaged less by conviction—which can always change—than by greed or fear. If this was indeed the case, then loyalty might appear as suspect as disloyalty, and earlier obedience as the ultimate crime.[100]

The dispatch of additional Soviet police advisers into Eastern Europe showed that Stalin's not-so-invisible hand was getting heavier. Yet the organization and execution of the purge, including the selection of its individual targets, was still being left largely to the locals. Stalin's overall inspiration did not preclude national distinctions in implementation, which could sometimes make the difference between life and death. The search for a conspiracy à la Rajk, pursued diligently in both Czechoslovakia and Poland, as well as in East Germany,[101] produced very different results in each.

By 1951 the Czechoslovak police had arrested for questioning and possible framing-up a greater number of party members than were rounded up in other countries, thus creating a large enough pool to seemingly allow for fishing out the appropriate supertraitor with ease.[102]

Yet the task was actually difficult—not so much because the candidates were all loyal communists (an irrelevant consideration) as because in a party notable for the mediocrity of its officials it was hard to find anyone distinctive enough to impress.

In 1950 the provincial party chief of Moravia, Ota Šling, tentatively identified as the chief conspirator had been made to confess not only to the well-founded charges of self-promotion and favoritism but also to the phony ones of liaison with the U.S. intelligence.[103] (There was such a liaison, however, somewhere else—for the American embassy in Prague was able to get hold of the accusations against him while they were still confidential.)[104] Blown up into a plot of gigantic proportions, ostensibly aimed at overthrowing the country's leadership and exterminating it in the process, in February 1951 the charges were detailed by party chairman Gottwald in a dramatic speech to an emergency session of the central committee, and publicized in 4.5 million copies printed in six languages.[105] Too grand for the wretched Šling to fit in, however, the fabrication subsequently fizzled out, and the search for a more suitable figure went on.

The Polish purge designers proceeded more cleverly. Starting with subsidiary trials, by manipulated interrogations and prearranged testimonies they zeroed in on a person of some stature—Marian Spychalski, the former commander of the wartime communist guerrillas, later made marshal of Poland, who had been a close associate of the fallen party secretary Gomułka. Accused of assorted wrongdoings—from involvement in affairs bearing on the underground party's shady dealings with both German and Soviet secret services during the war to the more preposterous charge of spying for the British—by mid-1950 Spychalski had been sufficiently cowed to confess to anything.[106]

The real target, however, was Gomułka himself. Protected by Stalin in 1948, in July 1951 he was denounced in a message sent from Moscow to Warsaw—hardly without Stalin's knowledge—to alert the authorities to his allegedly imminent flight to the West.[107] Meanwhile, in Prague the resident Soviet advisers and their local helpers instigated the police to disregard in their interrogations the injunction by the central committee prohibiting incrimination of the party's most senior officials.[108] Yet when the investigators finally implicated Gomułka and Slánský—respectively the former and the still acting general secretaries of the Polish and the Czechoslovak parties—even Stalin paused in disbelief.

Informed by telephone of Gomułka's arrest on August 1, the despot is reported as having exclaimed: "I'm not sure it was the right thing to do. I wonder whether they have sufficient grounds to arrest him."[109] Suspecting an enemy provocation, he expressed doubts about the material against Slánský as well. Nevertheless, he had already a week before authorized his demotion—on the ostensibly innocuous charge of "errors in the personnel placement"[110]—and did nothing to discourage preparations for a show trial of Gomułka either.

On July 31, while the Polish party daily *Trybuna Ludu* ran the programmatically entitled article "The Traitors on Trial!"[111] Slánský was still decorated with a medal on his fiftieth birthday. Yet Soviet congratulations on that festive occasion were conspicuous by their absence. Already the police were busily interrogating suspects about his putative treasonable activities, ignoring Gottwald's directive to desist.[112] The machinery Stalin designed to move by itself once prompted functioned perfectly.

The scheme against Gomułka was nevertheless botched. The notorious Col. Józef Światło, the agent who arrested him before himself absconding to the West, later recalled that the arrest, on the top of so many other police atrocities, had "caused tremendous indignation in the Party ranks." Only perfunctorily interrogated and never tortured, Gomułka had the temerity to demand proofs of his guilt rather than confess without asking questions. "No one in the party leadership," according to Światło, "had the courage to cope with him."[113]

Where the ever devious but increasingly remote Stalin fitted into the picture is difficult to judge. His stooge Bierut, the party chairman and state president, was later reported by the former security chief Jakub Berman—a person in a position to know—as having harbored "some sort of a complex about Gomułka" that, despite Soviet pressure for a trial, made him unable to decide.[114] But the testimony of Berman, known to be a shameless liar, hardly carries more weight than that of another insider, politburo member Edward Ochab, who has alleged that Bierut arranged Gomułka's arrest to protect him against being kidnapped by Soviet agents in a staged escape attempt.[115] Of course, suggesting that the man who three years before had engineered Gomułka's downfall now wanted to protect him "out of simple decency" strains belief, too.[116]

In hounding down one another, the Polish communists were more likely to respect certain limits than were those in other countries. Gomułka himself believed that Bierut had something to hide about his own record that could bring Stalin's wrath down upon him.[117] Regardless of whether there was anything to hide, to be sure, the tyrant's wrath could strike him down, as anybody, at any time—yet strike down Poles of any kind it did with a frequency and ferocity that made even the communists among them, if not necessarily less loyal, at least less prone to illusions about him. The resulting collective instinct of self-preservation, if nothing more admirable, helped spare them and their nation the depths of degradation that other communists, who lacked the same sobering experiences, descended into while making themselves more willing accomplices in their own destruction.

This was notably the case with the Czechoslovak communists. In a country exposed to the West because of its geography, they were, admittedly, under particular Soviet pressure to prove their vigilance but, apart from that, were also particularly eager to oblige because of a devotion to Moscow that struck Khrushchev as "extraordinary."[118] Since the purge required, however, that the opposite be proved, more drastic methods had to be resorted to in order to produce the expected results. This ensured that eventually the trial of Slánský and his cohorts would be the most monstrous of all.

The general secretary's slide into the abyss started on September 6 with his demotion to deputy premier, along with the abolition of his party post. Accused at a closed session of the central committee of having allowed "enemy agents" to gain "strong positions," Slánský, unlike Gomułka, tried to save his neck by accepting rather than rejecting the charges against him.[119] Having previously shown no scruples in enthusiastically promoting the purge before it got the better of him, he lacked the strength of character that the Pole derived from his cleaner past.

By then, to be sure, nothing could save Slánský anymore; it still remains intriguing what finally brought him down. The confidential record of the meeting where he had performed the ritual of self-criticism had again reached Washington before its contents were made public,[120] thus lending credibility to the assertion that Western agents had indeed found their way to the party's highest echelons. Moreover, his arrest during the night of November 23 followed the alleged discovery two weeks earlier of evidence of his planned escape abroad, not totally implausible.

Subsequently produced at the Slánský trial, the purported piece of evidence was what was shown as an intercepted written message from the U.S. Counter-Intelligence Corps offering him "secure entry to the West ..., asylum, secure hideout, and later any means of existence, except in a political career."[121] The similarity of the alleged escape scheme and the provocation reportedly prepared by Soviet agents against Gomułka does not in itself exclude the possibility that this time a provocation had been prepared by the Americans against Slánský—as Stalin had feared.

Encouraging high-ranking communists to escape was what the Psychological Strategy Board had decided earlier that year to establish as one of its priorities, intending to use such persons for propaganda interviews on Radio Free Europe besides other purposes.[122] The intercept incriminating Slánský was authenticated by secret codes broadcast by that very station, though perhaps only thanks to some of the many communist "moles" esconced in its Munich headquarters.[123] Whatever the truth may be, the discovery warranted the dispatch to Prague of so senior a Soviet figure as Stalin's troubleshooter Mikoian, before Slánský was arrested.[124]

A scurrilous British book maintains that not only the plight of Slánský but the whole Eastern European purge was the result of a gigantic CIA scheme to obliterate the flower of international communism by incriminating its protagonists through forged evidence systematically planted by the unspeakable Światło and his flunkies.[125] Based on hearsay, this is pure imagination, but the basic idea is credible. For the U.S. secret services, engaged in a deadly struggle with a ruthless enemy, trying to help the communists to the famous rope that Lenin believed them capable of selling to the capitalists to hang themselves on would only have been the most obvious thing to do. On at least one previous occasion, in 1949, Washington is described as having deliberately allowed seemingly plausible but unworkable industrial designs to fall into Soviet hands.[126] But the conclusive evidence is still hidden in the American archives.

Stalin hardly needed any prodding to cultivate his morbid suspicions. With or without Western help, candidates for phony trials had been filling Czechoslovak and other Eastern European jails simply because of their past association with one or another of his failed designs. In November 1951 the arrest in Prague of Mordechai Oren—a functionary of Israel's pro-Moscow Mapai party instrumental in the unsuccessful Soviet attempt three years ago to buy over Israel by supplying it with arms—added the "Zionist" one to the expanding spectrum of "conspiracies."[127]

By then the ramifications had finally reached the Soviet Union itself. The arrest of Slánský coincided with the sacking in Moscow of minister of state security Abakumov, the executor of the 1949 Leningrad purge. The ensuing struggle for the control of the ministry evolved amid signs of incipient crumbling of the position of Beriia.[128] One such sign was the crackdown against his Mingrelian ethnic group in Georgia, contrived by Stalin. Others were the ascendancy in the politburo of the dictator's junior favorite Khrushchev and the rise within the security apparatus of his respective protégés, including Abakumov's successor, Semen D. Ignatev, besides a host of retainers Khrushchev had brought along to Moscow from the Ukraine.[129]

The scene was becoming ominously reminiscent of the time two decades before when Stalin was positioning himself to replace a whole generation of Soviet officials by a younger and more subservient lot. As it had been the case before, the climax was approaching when security personnel who had figured in starting the purge then began to be themselves targeted by it. In February 1952 the Prague party daily *Rudé právo* pointedly linked the great Moscow trials of the 1930s with what it described as an increasingly menacing international situation in the present world.[130]

The mounting U.S. challenge to Stalin's Eastern European empire had not initially discouraged but rather encouraged him to lend a helping hand to facilitate a negotiated end to the Korean War. Once Washington's restraint in prosecuting the war became evident, however, any incentive he had to terminate it evaporated. With limits set to its expansion, its continuation promised to relieve Western pressure on Eastern Europe besides perpetuating Chinese dependence on Moscow.

That was no trifling matter, for the U.S. covert operations, despite all their shortcomings, succeeded only too well in making Stalin more insecure, thus prompting him to let his self-destructive purge mushroom into massive proportions. As a result, the foundations of Soviet power in Eastern Europe became undermined just as the American power in both Western Europe and Japan reached new levels of consolidation. The plunging correlation of forces abroad made Stalin more inclined to look for the resolution of his largely self-inflicted security problems at home.

8

Fits and Starts
March 1952–November 1952

Much like in the prewar Great Purge, in the one that was now climaxing concern about Germany permeated the Soviet mind. Although
Stalin in April 1952 chose to reply to questions asked by U.S. journalists that war was not any closer than it had been two or three years
earlier, he did not say that the threat had passed.[1] He insisted that
peace could be preserved if only the West carried out its purported
obligations and abstained from interfering in other countries' internal
affairs. By this he meant restoration of four-power control in Germany
and acquiescence in Soviet control of Eastern Europe. Unlike its des-
tiny, however, that of Germany was not in Stalin's hands—something
he still refused to acknowledge with a stubbornness making a mockery
of his supposed realism.

Stalin's German Illusion

On March 10, the Soviet government sent its Western counterparts the
later celebrated proposal that went farther than any previous or later
one in envisaging the unification of a nonaligned Germany. Their
failure to act on the proposal gave rise to the protracted and tiresome
German debate about whether an opportunity to end the country's division, if not the Cold War itself, may not have been missed at that
time.[2] In striking contrast to these speculations, retrospective Soviet
assessments never attributed such inflated significance to this long
tantalizing, but now finally explicable, episode.

The origins of the proposal can be traced to the intention, announced
by the Allied foreign ministers at their meeting in Washington on
September 14, 1951, to terminate in the near future West Germany's
occupation status and integrate the country into the projected European

Defense Community.[3] The decision frustrated the quixotic campaign for the convocation of an all-German unification conference that Moscow had been conducting since late 1950 through its East Berlin subsidiary. In the fall of 1951 the Soviet party central committee approved an alternative plan, which envisaged that the East Germans would first sent all four occupation powers a request to accelerate the conclusion of a peace treaty, in response to which the Soviet Union would promptly submit a draft of its own, thus presenting the other powers with a ready-made text they would not be able to ignore.[4]

While the subordinate Soviet officials who, rather than Stalin himself, were looking after the project at this stage were not yet clear in their minds about what the treaty should provide for, at least they took it for granted that its preparation must not entail any weakening, much less a sacrifice, of the East German state. On the contrary, the Soviet occupation authorities there simultaneously took what in their eyes were measures to preventively shore it up. On September 29 their newspaper, *Tägliche Rundschau*, signaled the opening of a campaign to bolster the ideological purity, vigilance, and militancy of the party cadres. Amid much ritual criticism and self-criticism, the resulting purge doomed a good number of provincial officials on charges of "arbitrary behavior" and assorted other misdeeds. Unlike in other countries, however, the cleansing did not go so far as to jeopardize the integrity of the top leadership.[5]

Although the East German leaders were made privy to and accepted the Soviet plan, they had reasons to be worried about its possible consequences. To a sympathetic Italian socialist, Pietro Nenni, President Pieck and Premier Grotewohl later expressed their view that "Stalin wanted to put us into a new situation, and we do not know how we would have come out of it."[6] In vain did East German officials seek Soviet reassurance that the proposed conclusion of a peace treaty was merely a diversion, intended to complicate for the West the planned transfer of power to the Bonn government.[7] Even if they did not outwardly panic,[8] the future of their own power would have been put in doubt if Moscow were to produce—as it intended to—a draft of the peace treaty acceptable enough for the West to serve as the basis for an open-ended discussion.

Yet the draft that the Soviet foreign ministry was preparing under Gromyko's supervision was hardly of the kind. It was to be made public rather than sent through diplomatic channels, as would have been appropriate to ensure the necessary confidentiality for serious discussion. It envisaged holding a conference of foreign ministers without specifying an agenda. It made its later preparation contingent upon the West's response to the initiative described by Gromyko in his cover letter to Stalin not as an overture for negotiations but something that could have "great political significance for the strengthening of the struggle for peace against the remilitarization of West Germany, thus helping supporters of the unity of Germany . . . to expose the aggressive intentions of the three Western powers."[9]

The draft Gromyko forwarded to Stalin for his approval on February 18—five days after the East Berlin government had duly sent its public request for a peace treaty[10]—ignored blissfully the existence of the West German state. Instead it expatiated upon the supposed virtues and accomplishments of the East German one as the sole suitable foundation of a reunited Germany, describing its future government as one that would "strengthen and continue the democratic changes achieved in [East] Germany, including those pertaining to the organs of administration, economy, labor legislation, justice, education, public health, etc."[11]

Specifically, the projected peace treaty envisaged the repatriation of all citizens of the victor nations currently residing in Germany in consequence of the war—that is, the extradition by the West of all anti-Soviet refugees—and the return of the Ruhr area to Germany—which would have the effect of undoing its integration into the West's European Coal and Steel Community. While pre-1945 German debts—all owed to other than Soviet creditors—would be canceled, Moscow would claim for itself $6,829 million worth of reparations. In this original version of the proposal, Germany was not to be allowed either its own army or any military production. The foreign occupation forces were to leave its territory within a year, but a four-power control commission was to remain there for two more years to supervise the fulfillment of all the above obligations.

So preposterous was the list that it had to be discarded, all but certainly at Stalin's orders, after which Gromyko submitted to him a revised shorter version on February 23.[12] It was still not good enough to be sent out five days later, as had originally been planned. Indeed, no draft with any specifics of a peace treaty was ever sent; the note eventually handed to the Western diplomatic representatives on March 10 merely included its general principles as a putative basis for negotiations. Made public at the same time, in its final version the note pointedly omitted the most outrageous demands from the previous drafts while including two items that had not been there before, calculated to appeal not so much to the Western governments as to the Germans.[13]

The first was the provision allowing the future nonaligned Germany to have its own armed forces and military production within limits to be specified. The second was an assurance to all former Nazis, except those already sentenced for specific crimes, that they would enjoy full civil and political rights. Stalin's way to German unity was thus based on the dubious assumption that Germans were so infatuated with soldiers and so eager to whitewash Nazis that by pandering to their vices they could be won over to the Soviet side. Since, according to his reasoning, "whoever supports German unity will also win the German people," by his own lights Stalin was making a "realistic" offer.[14]

The expected outcome was described in the note as a "united, independent, democratic, and peaceloving Germany in conformity with the Potsdam agreements." In the Soviet vocabulary "united" meant a single state within the boundaries provisionally set at the 1945 Potsdam conference—hence not only their confirmation as permanent but also the obliteration of the subsequently constituted Federal Republic. "Independent" referred to the reunited nation's obligation to shun any "coalition or military pacts" directed against any state that had taken part in the war against Hitler—that is, prohibition not only of its membership in the European Defense Community or NATO but also of its participation in any nonmilitary grouping that Moscow might find objectionable. Finally, the country was to be kept "peaceloving" through those "democratic" organizations of Soviet choice that alone would be permitted to function there.

Stalin had before his eyes the model of the interwar Weimar Republic, with which the Soviet Union had once colluded in a mutually convenient partnership, although his proposals, if implemented, would have actually resulted in anything but a partnership. In an allusion to the Germans' predicament after World War I, the note alleged that without a peace treaty the divided nation would remain the pariah of Europe, identifying as the remedy a treaty to be negotiated with an all-German government. Unlike earlier Soviet proposals, however, this one was not about electing such a government. Not offering a procedure for reunification, it dwelt instead upon the peace treaty, proposed in terms general and flexible enough so that they would not preclude raising Moscow's unstated specific goals anytime later on, depending upon the outcome of the discussion that the note was calculated to encourage.

The discussion to be encouraged, however, was not one between the Soviet Union and the other occupation powers responsible for the German settlement but that which was climaxing in West Germany as the time was approaching for the conclusion of the crucial agreements that would anchor it in the Western alliance system as its main European bulwark against Soviet expansionism. These were the General Treaty that would end West Germany's occupation regime and establish its sovereignty, scheduled to be signed on May 26, and the treaty on the European Defense Community with West German membership, which was to be concluded on the following day.

In confidentially explaining the purpose of the Soviet note to party activists, the East Berlin politburo made no qualms about its being intended to derail the obnoxious agreements.[15] In trying to enlighten East German foreign minister Dertinger, however, Soviet envoy Georgii M. Pushkin conceded that, since West Germany's eventual integration into the Western defense system was a foregone conclusion, the note's only practical goal was bringing down the present Bonn government.[16]

If this was the goal—and the Soviet foreign ministry preparatory materials leave no doubt that Stalin expected the note to be effective— the Western response showed the extent of his genuine, though unjustifiable, miscalculation. He surely could have expected that the Bonn government, unwilling to collaborate on its own demise, would do what it could to convince its Western patrons of the necessity to pre-empt any divisive public discussion by rejecting the Soviet proposals out of hand and that the Allies, loath to see their policies ruined, would need little convincing. So on March 25 they predictably turned down the proposals, restating their own demand for free elections as the only way to German reunification—for the record, if not as a policy they believed to be achievable any time soon.[17]

Only afterward did Stalin begin to act as if he had finally concluded that nothing could be done to prevent the inevitable. Parting with his long-cherished illusion of a united and pro-Soviet, though not neces-sarily communist Germany, he reluctantly embraced the second-best, but real, alternative of a Germany run along the Soviet model by his stooges in the part of the country he could control. Eager to run it for him, this was the solution that the East German communists had been so far unsuccessfully maneuvering to impress upon him as being not only in their but also in his best interest. On April 1 and 7 he discussed with them their new assignments during important meetings from which Pieck left his candid notes on the topics covered there.[18]

According to those notes, Stalin did not take seriously his pretense that the European Defense Community was intended to be a tool of military aggression. What he rather feared was that, once established, it would enable the Americans to project their power on the Continent more effectively by other means, with potentially calamitous conse-quences especially for East Germany. Cautioning Pieck that the "pacific period" was over, he warned that "we must expect terrorist acts"—the reason why not only a militia but also a regular army needed to be built up—"without much noise, but with perseverance."[19] He regarded the police as the first line of defense, and when Pieck inquired what to do about the already overcrowded prisons, Stalin had the solution: "Stage trials. Fill yourselves with a fighting spirit, and we shall help you."

That Stalin by now had discarded any idea of trading communist East Germany for a merely subservient whole one followed from his further directives to Pieck. He wanted the country to be set on the "road to socialism" no later than the fall, but "cleverly." This meant, for example, that in collectivizing its agriculture the communists should not "cry 'kolkhozes' and 'socialism,' but create facts." He left no doubt anymore that the goal now was a "socialist" economy Soviet-style rather than any mixture of the sort he used to call "democratic." According to the Soviet record of the same conversation, he finally came to the conclusion that the West would not agree to any of his proposals on Germany, and hence told the East Germans: "You must organize your own state."[20]

Accordingly, not much could be expected from the second Soviet note on the German question, on April 9, which was a reply to the

Western demand for free elections.[21] Moscow countered it by proposing to first create a four-power commission that would supervise the elections, but avoided any commitment to hold them before the peace treaty has been agreed upon. Otherwise it truculently restated the terms outlined in the previous note. In presenting the new one to British chargé d'affaires Grey, Vyshinskii was evasive about the proposed commission—its only new feature. When asked whether this was a formal proposal, he hedged that "it might be."[22] He was keeping the door open for discarding it should the Western governments unexpectedly agree to creating the commission, and then try to turn it around against the Soviet Union by using it to press for free elections.

Taking no chances, the West rejected the idea, proceeding to finalize, as planned, both the General Treaty and the European Defense Community. The respective documents were signed on May 26 and 27. In assessing the outcome for Nenni during his Moscow visit two months later, high Soviet officials acknowledged that by then their government had regarded any further exchange of notes "completely phony and futile."[23] This confirmed Kennan's estimate of the third note Moscow still bothered to send on May 24 as a statement "remarkable for its weakness, its mild discursiveness, its lack of enthusiasm, its failure to add to discussion any important new element.... The bid for oral negotiations is perfunctory.... [It] is not the authentic, terse, collected, menacing voice of Stalin's Kremlin when functioning in high gear and pursuing an important Soviet initiative."[24]

The character of the note tallied with the notable lack of enthusiasm shown by Moscow in letting the East Germans themselves implement the directives they had received from Stalin in April. Ulbricht and his companions jumped on them as giving them the green light they had been waiting for to proceed with the "accelerated construction of socialism" they wanted.[25] Invoking Stalin's dictum that its approaching victory makes more class struggle necessary, they immediately cracked down on farmers, churches, and the remaining private businesses. They conducted a new round of show trials of "terrorists and saboteurs"— some of whom had to be kidnapped from West Berlin so that they could be shown—which for the first time resulted in death sentences.[26] They also speeded up the party purge, the ax hovering ominously around politburo members Franz Dahlem and Paul Merker as candidates for a supertrial.[27] Yet not until the day before the party conference that was to formally inaugurate the new course opened on July 9 did Moscow signify its final approval.[28]

Whatever the opportunity for a German settlement in 1952, it had been lost before it could even be created. If it was never more than an illusion, this was ultimately because Stalin's wishful thinking that he could divert the Western powers from their course in Germany and achieve there a settlement on his terms. Amid the amply justified mutual suspicions, which he had so diligently cultivated, there was no longer enough room for accommodation even on less contentious matters than the intractable German question. This was a distressing fact of life that the rapidly declining Soviet dictator found much more difficult to bear than his ascendant Western adversaries.

The Despot Loses His Grip

In the summer of 1952 Kennan thought Stalin's personality could be felt only "from time to time," his participation in public affairs being "sporadic and relatively superficial," not giving anymore an "impression of one-man dominance."[29] His army chief of staff Gen. Sergei M. Shtemenko later recalled that the generalissimo had often left even urgent military business unanswered.[30] According to Molotov, Stalin was no longer in full control of himself.[31] Khrushchev noticed lapses of mind and losses of memory, which drove the despot crazy.[32]

These casual impressions concurred with the expert opinion of Professor Ia. L. Miasnikov, later present at the dissection of Stalin's body, who diagnosed him as having had a brain disease that leads to the loss of orientation. According to the physician, he had difficulty deciding what was right and what was wrong, what was useful and what harmful, who was a friend and who a foe. Making matters worse, the ever-suspicious Stalin refused any medication.[33] Still, the illness—which merely magnified his obvious predispositions—need not have made that much difference if the system he had fathered had not influenced his policy in much more fundamental ways.

The Soviet totalitarian model came the closest to perfection precisely during Stalin's sickly last years. After all conceivable opposition had been crushed or cowed into submission, even the ruling politburo—a mere semblance of an oligarchy—was formally convened only four times in 1952—the lowest number of meetings in any year.[34] By then, whatever little government business was being transacted at that level ran through the autocrat's small chancery, managed by the sinister but otherwise unremarkable Aleksandr N. Poskrebyshev.[35] As a result, the totalitarian system's capacity to function as such was paradoxically diminished. Stalin was still quite adept at devising his schemes but no longer as capable as he used to be at following them all through.

The developing mismatch between the tyrant and the tyranny meant that, as Khrushchev remembered it, the government "virtually ceased to exist. . . . Everyone in the orchestra was playing on his own instrument anytime he felt like it, and there was no direction from the conductor."[36] The ensuing cacophony, of which little could be heard outside of the Kremlin's sound-proofed walls, did not necessarily mean that anything was being played against Stalin's will, only that he could not be as certain as before that nothing was—an uncertainty adding to his already monumental sense of insecurity. Hence it became difficult, and still is, to determine beyond doubt which of the contradictory policies during his last years were actually his and which were not. This was particularly the case in Eastern Europe.

In Romania—the most thoroughly penetrated and subjugated country of the area—the purge that climaxed in 1952 surprisingly showed substantially more of local input than was seen in any of the other countries. When the party chief Gheorghiu-Dej had gone to Moscow in the year before to complain about intrigues by his foreign minister Pauker, Stalin refused to be bothered.[37] Only when he came again, in April 1952, this time accompanied by Beriia's man in Bucharest Iosif Chișinevschi, did Moscow give Gheorghiu-Dej the green light he wanted.[38]

In May, Pauker and two other protagonists of what was Eastern Europe's most byzantine communist hierarchy—the minister of finance Vasile Luca and that of the interior Teohari Georgescu—fell into public disgrace. Yet nothing more serious was held against them than their having presided over wrong agricultural, industrial, and banking policies, besides having flaws of character supposedly conducive to, of all things, "aristocratization." Conspicuously missing among these odd charges was anything anti-Soviet. Nor did they cater—despite Pauker's being Jewish and the other two not ethnically Romanian—to either anti-Semitism or nationalism.[39]

None of the trio was ever tried in public or executed. They all survived Stalin—though not Gheorghiu-Dej, who had Luca's and Georgescu's lives shortened by imprisonment without any additional publicity. So was shortened, more abruptly, the life of his already fallen rival, Pătrășcanu, who had been languishing in prison even before the Stalinist purges had started—and would be surreptitiously put to death after they had already ended when Stalin himself was no more alive.[40] By then the inimitable Gheorghiu-Dej would already be getting ready to beat the drum of Romanian nationalism, of which he had shown no traces while the master was still living.

On July 7 the arrival in Bucharest of a new Soviet ambassador heralded Moscow's closer attention to the local purge—but of different people. He was the Lavrentev who in 1948 in Belgrade had set into motion the purge of Tito, now fresh from an assignment in Prague, where he had supervised the unfolding of the purge of Slánský and other persons whose security jobs linked them with Beriia.[41] It was under Beriia's auspices that the purge in Romania had started, and soon similar sort of people came to grief there, too, replaced in conspicuous numbers by successors connected with the neighboring Ukraine.

The reshuffle was directed from Moscow by Aleksei A. Epishev, recently given there the second-highest post at the contested ministry of state security as one of those whom Khrushchev had originally brought along from his Ukrainian fiefdom.[42] Three other deputy ministers were also said to be his retainers.[43] An intriguer in his own right, Khrushchev's star as the fastest-advancing member of Stalin's entourage was rising as that of Beriia was falling. Yet as long as Stalin was living, the shadowy struggle that was progressing was not so much for power as for positions in its growing vacuum.

While in Eastern Europe Soviet control remained unimpaired, the vacuum impaired Soviet relations with Western European communists. Although Stalin had given up on revitalizing the Cominform, he did not on the assistance they could provide him if properly manipulated by other means. This put into the limelight the French party which, after his proposals on Germany had fallen flat, was crucial to the success of continued Soviet efforts to dismantle the security system that Washington was constructing in Europe. Yet despite the French communists' famous servility to Moscow, the manipulation of a party whose propensity for infighting rivaled the Romanian one proved difficult.

In April 1952 the Paris politburo member François Billoux received in Moscow new directives which, once published, caused consternation in the party's ranks.[44] As in 1947, when Zhdanov had humiliated its leaders by making them to reverse a policy they had been conducting at Soviet prodding before, the directives mandated the same kind of abrupt turnabout—this time of the policy of keeping the party and its subsidiary "peace" movement formally separated, so that it could attract a wider following. Resurrecting Zhdanov's injunction against coalition-building, the "Billoux theses" demanded that the separation be ended and the masses, comandeered by the party, be brought into the streets under its banners, to intimidate the government by rioting and strikes. It was the same turn toward violence as that which Moscow had previously enforced in Western Europe in September 1947, in South East Asia in February 1948, in Japan in January 1951—and the outcome was to be similarly disappointing, too.

Simultaneously, the party central committee charged two of its outstanding militants with impeccable revolutionary credentials, André Marty and Charles Tillon, with having already the year before conspired to erroneously separate the peace movement from the party.[45] They were even more preposterously accused of having also balked at Zhdanov's demand for the change of course in 1947. More to the point, they were taken to task for drawing the wrong distinction between Germany's "militarists" and its people. This, of course, used to be Stalin's own favorite distinction—until it had recently lost its appeal when the results of his note designed to sow dissension among Germans had fallen short of expectations.

In the only westward extension of the Eastern European purge, before losing their party posts Marty and Tillon were put on a sort of show trial by the central committee—the former (once an agent of the Soviet secret police) as a supposed agent of the French one, the latter (the firebrand former chief of the communist guerrillas in World War II) as a flabby compromiser.[46] If all of this was vintage Stalin, "the contradictions of the French [communist] policy thus reflecting the contradictions of the Soviet policy,"[47] the profusion of indigenous factors in the infighting that increasingly rocked the party showed how much his control over it was slipping. While its sick general secretary Maurice Thorez was kept away in Moscow, within the rest of its leadership an internally generated purge was raging, allowing especially his wife, Jeannette Vermeersch, to indulge her many enmities.[48]

Under such circumstances, the Soviet-inspired campaign against the German General Treaty and the one on the European Defense Community was bound to end badly. On May 28, the day after both treaties had already been signed, the communists mounted in Paris what they had conceived as violent mass demonstrations against the visiting U.S. commander in Korea, General Ridgway. But not enough people came, and the violence they provoked only helped the police to round up the ringleaders, on one of whom, Duclos, was found a copy of Billoux's notes about his Moscow conversations that had started all the trouble.[49]

True to form, Moscow promptly disowned the rioters and reinstated the previous line, insisting that it was necessary, after all, to keep the struggle for "peace" and for "socialism" separate, while blaming poor Billoux for having deviated from the line. After this new turnabout, the Paris politburo could be excused for behaving as if it had no clear signals from the Soviet headquarters.[50] Downplaying the significance of the continuing disagreements about strategy and tactics among the French communists, Moscow was conspicuously reticent to publicize the issue anymore. It all looked as if Stalin had lost for good his ability, desire, or both, to look after the directions that international communism should take.

As a postscript, there followed yet another *tournant* in December, already the third that year, which may or may not have been related to the shifting of the political sands in Moscow, particularly the increasingly wobbly position of Beriia. The French politburo entrusted to the same Billoux whom it had previously disciplined for wanting to merge the party and the peace movement the restoration of that very policy. As the dust was settling, the loser in the game appeared to be Duclos, reputedly Beriia's man, made into the whipping boy of the party's "opportunism" during the period when Billoux had been in charge.[51] Yet when Yves Farge, another resistance figure like Tillon, the next month traveled to the Soviet Union to more plausibly protest against the high-handed manner in which Duclos had been running the peace movement (while also taking the much greater risk of carrying with him a written protest against the Eastern European purges by the purged Marty), he met his death in one of those mysterious automobile accidents that Beriia's men were so expert in staging.[52]

It is difficult to determine how much the growing incidence of such imperspicuous events was indicative of Stalin's old predilection for them rather than his new inability to control them. The man who in June slipped past the police guards into the U.S. embassy in Moscow with an offer to the dumbfounded ambassador Kennan to do away with the Kremlin leadership in return for money and arms—before being speedily ushered out to be taken away by the same guards that had let him in—may have been, as Kennan thought, Stalin's agent provocateur. If he was, this was an uncharacteristically clumsy way of creating a pretext under which the American diplomat could be expelled as a persona non grata, as in fact Kennan soon was under a different pretext.[53] But there are other plausible explanations of this remarkable Cold War episode.

The untrustworthy but otherwise excellently informed Soviet insider Georgii Arbatov has reported from secondhand knowledge that the intruder was positively one N. N. Iakovlev, identified as the son of a recently arrested Soviet marshal of artillery and himself probably both a victim and an agent of the security services.[54] Since, however, the man introduced himself to Kennan as the son of the recently dismissed minister of state security—who was Abakumov, in turn regarded as either a victim or an agent of Beriia—he could just as well have been on an independent errand for him, the person who as the overall security boss was in a position to control both the messenger and the guards. Whatever happened, and something important did happen, the incident added to the multiplying signs of Stalin's slipping.

Nenni, who saw Stalin in the Kremlin on July 17, reported that he was well and in full control. Yet the admiring and ambitious Italian—who had received the Stalin Peace Prize and was proud of it—was no reliable witness. He treated Western diplomats in Moscow to the self-promoting story that the generalissimo had confided to him such important proposals as that for a nonaggression pact between their two countries, which could be concluded as soon as the current right-wing government in Rome has been replaced by a left-wing one, bringing into power such people as Nenni.[55] To his diary, however, he confided the more candid description of a conversation in which Stalin had said nothing substantive, blabbering about the nefarious influence of the Vatican and other nonsense.[56]

A month later, French ambassador Louis Joxe, astonished to have been invited to the Kremlin after the increasingly reclusive dictator had all but ceased to see any foreigners, found Stalin looking "old and tired ... , sickly."[57] Offering no clue as to why he had wanted to meet in the first place, he was "beating around the bush," complaining about German rearmament but not making any proposals. He was asking stupid questions: whether De Gaulle thought that the French-Soviet pact had been good, whether Iceland was aggressive, whether the Soviet Union might join NATO.[58]

In this sickly and thickening atmosphere, the announcement on August 20 of the convocation of the Soviet communist party's nineteenth congress came as something of a bolt of lightning. Holding such a gathering after an interval of more than thirteen years was no routine matter. They were the years during which the Soviet Union had fought and won a major war, only to start another, the Cold War, not to speak of the one in Korea. Even if the accumulation of internal tension was the main reason for no longer delaying the congress, its outcome was bound to have important repercussions for foreign policy as well. The early October date chosen for its opening was significant—a month before the forthcoming U.S. presidential elections, whose result could make a critical difference in how the Cold War would be subsequently waged.

Win the War at Home

In one of the Cold War's many ironies, Western fear of Soviet threat increased just as Stalin's rule was entering the stage of its terminal paralysis. The slowing down of hostilities in Korea had the paradoxical effect of generating the concern that Moscow was becoming more dangerous because of its presumed ability to convince enough people in the West that it was becoming less so. NATO suspected its design to expand its power by political means "whenever an opportunity is offered to do so without serious risks to its interests."[59] Even the perspicacious Kennan, while allowing for divisions of opinion in Moscow and ups and downs in Soviet estimates of the correlation of forces, tended to overstate the confidence with which the adversary was presumably acting. In his penetrating analysis of Soviet perceptions of the Western alliance, he correctly identified Stalin's 1949 belief that capitalism was preparing for war but was politically in crisis,[60] yet more questionably presumed that he was therefore ready to act upon the premise of the Soviet Union's own ascendancy.[61]

In 1952 the Russians had growing reasons to be more concerned about the capabilities and intentions of the Americans than vice versa. In the spring, the first U.S. tactical nuclear arms were sent to Europe with the intention of being used if needed.[62] In May Washington, exasperated by the communists' procrastination in Korea and what Truman detested as their total "lack of honor and . . . moral code,"[63] was again considering the deployment of atomic bombs in Asia. A war scare reappeared in Europe, although in its Eastern part, where many people regarded war as their only way to freedom, there was more excitement than fear about it. Rumors spread through the city of Prague, passing the word that the Americans were coming, maybe even "within the next two months."[64] The regime there became so edgy that it even put on public trial and sentenced to long prison terms a group of Boy Scout leaders for "hoping for war."[65]

During that chilliest month of the Cold War, when East Germany was finally set on the road of Soviet "socialism,"[66] orchestrated anti-American propaganda reached a level without "equal in viciousness, shamelessness, mendacity, and intensity."[67] This applied particularly to the charges that Washington had been conducting "bacteriological warfare" in Korea. It was not, although there had been enough secret U.S. research and experiments with biological weapons[68] to make Stalin— with his own predilection for surreptitious methods of killing—hysterical if he was able to find out anything about what was going on.

New themes, alarming to Moscow, permeated the increasingly acrimonious U.S. presidential campaign. Even aside from the antics of the ludicrous Sen. Joseph McCarthy of the House Un-American Activities Committee, searching for such "traitors" as Acheson, anti-Soviet demagogy flourished. The Republicans demanded the repeal of what they denounced as the treasonable Yalta agreement, vowing to do away with what they lambasted as the Truman administration's intolerably timid and bankrupt policy of Moscow's mere containment and provide a more assertive substitute.

On August 29 *Pravda* accused the Republican presidential candidate, Gen. Eisenhower, of wanting to dictate the Soviet Union a "military decision."[69] The reference was to his speech four days earlier at the New York convention of the American Legion, the staunchly anticommunist association of U. S. veterans, in which he had sounded more militant than in the printed version distributed beforehand. Speaking as if peace depended upon the liberation of Eastern Europe's "captive peoples," in alluding to the means by which this was to be accomplished, he had omitted the word "peaceful" that was in the written text. Nor did Dulles, the Republican party's chief foreign policy expert and its designated choice for the secretary of state, use this crucial qualifier during his still more rousing press conference two days afterward.[70]

Having as early as 1950 described "rollback" as the desirable substitute for containment in his suggestively entitled book, *War or Peace*, in May 1952 Dulles elaborated the strategy he advocated in a programmatic article in *Life* magazine,[71] the main themes of which were subsequently incorporated into the party platform. Using the threat of "massive retaliation" by all nuclear and other weapons at America's disposal as an effective deterrent to any Soviet aggression, he insisted that the communist empire could be broken up, and proceeded to explain how: "What we should do is try to split the satellite states away from the control of a few men in Moscow. The only way to stop a head-on collision with the Soviet Union is to break it up from within ... [by encouraging] passive resistance, slow-downs, and non-cooperation."[72]

Although Dulles, in response to public concern about his rhetoric, toned it down, taking particular care to insert the key word "peaceful" where appropriate, enough remained to make Moscow worried should his recommendations be ever put into effect. Talking about specifics, he made no secret of his desire to stir up resistance in Eastern Europe by inflammatory broadcasting and covert action, whereupon ostensibly private organizations like the Free Europe Committee could then fan the resistance by supplying freedom-fighters by air and by other means.

Although arms were not specified, neither were they a priori excluded in supporting Dulles's argument that the Soviet Union, when faced with the unraveling of its empire, would throw in the towel rather than risk war.[73]

The projected strategy was not so much utopian as too realistic. Anticipating correctly what would in fact happen in 1989, Dulles grasped better than Kennan Stalin's perception of his vulnerability, proposing to exploit it. The perception was palpable in his being so much alarmed about the threat of war, yet unwilling or unable to counter it by any substantive new initiative in foreign policy. Despite Moscow's thundering about U.S. imperialism, when it came to action not much was happening anymore. When in June the Americans attempted to break the Korean deadlock by renewed massive bombing of North Korea, they scarcely elicited a Soviet whisper.[74] On July 1 Soviet ambassador to Beijing Roshchin left for home, never to return, after which several of his colleagues in other Asian capitals were replaced, too, nourishing speculation that something important would follow, but nothing did.[75]

Spreading panic, the systematic destruction of North Korea finally moved Kim Il Sung to ask for, and receive, Chinese backing for a secret trip to Moscow to persuade Stalin to help bring the war to a negotiated end.[76] It is not clear from the documents whether he actually went, but Zhou Enlai did. He arrived in mid-August to discuss with Stalin how the armistice could be achieved and, pending its achievement, maximize Soviet economic and military assistance to China. Zhou stayed five weeks. The record of their conversations shows how much, despite the veneer of amity and mutual respect, the Sino-Soviet relationship had deteriorated as the war continued without Stalin's doing anything to end it.[77]

The talks were businesslike. Bargaining took place, though within the limits of propriety, and conflict of interest was present, even if not allowed to come into the open. Considering Stalin's rapidly declining physical and mental condition, he still showed an impressive command of economic and military facts; only in the later sessions did his reasoning get muddled when he tackled the larger questions of diplomacy and war. Living up to his reputation as a cool and deft negotiator, Zhou never stated his goals clearly and sometimes even seemed to be contradicting himself. He affirmed China's refusal to entertain any concessions to the United States as he and Stalin were outdoing each other in professions of intransigence toward the "imperialists."

Playing a weak hand as a *demandeur*, Zhou had the difficult task of convincing the Soviet dictator to provide enough material help for both the prosecution of the war and China's economic development, while dissuading him from blocking a compromise that alone could lead to an armistice. Vowing China's determination to fight on for several more years, if necessary, rather than to make concessions, Zhou secured

Stalin's promise of huge military and economic assistance. As a deterrent to any American attempt to expand the war into China, Moscow also complied with Beijing's earlier request to keep Soviet forces in Port Arthur beyond the previously agreed time limit.[78]

Yet Stalin was not helpful in advancing any practical proposals conducive to the cessation of hostilities, insisting on demands known to be unacceptable to the Americans.[79] For his part, Zhou Enlai tried to call his attention to different possibilities of expediting the armistice talks, deadlocked by the dispute about the disposition of the Chinese and North Korean prisoners of war unwilling to be repatriated. While paying lip service to the principle of unconditional repatriation of all prisoners, Zhou outlined his plan for the transfer of the unwilling ones to a neutral country, such as India; noting the inconsistency, Stalin demurred. Nor did Zhou fare any better with his alternative proposal that a cease-fire should come first and the question of the captives be settled only later.

The inconclusive outcome of the discussion about this key issue was a victory for Stalin, made possible by the vast disparity of power between China and its Soviet protector, which no amount of Chinese dexterity could bridge. Nor did a three-month stay in Moscow by Mao's other top aide, Liu Shaoqi, lead to any progress toward armistice.[80] Despite the appearance of Beijing's agreeing with Moscow in taking an intransigent position against it, Stalin was the one who stalled, while the Chinese were seeking ways to achieve it.[81]

No sooner did Zhou return home than the omission of the obligatory references to the necessity of armed struggle in statements emanating from the Chinese-sponsored Asian and Pacific "peace" conference in Beijing signal its more accommodating attitude.[82] In contrast, Moscow signaled the opposite by issuing a public denial of rumors about U.S.-Soviet discussions about peace in Korea.[83] In November, the Chinese showed interest in the Indian plan for an exchange of prisoners of war that might unblock the way to peace—the idea Zhou had commended to Stalin.

The plan was rebuffed by Vyshinskii's preventively turning it down at the UN, leaving the Chinese no choice but to follow suit.[84] Early the next month, the arrival in Beijing of Moscow's new ambassador—after the post had been vacant for almost half a year—seemed to be indicative of its desire to tighten the rein on them. He was the former envoy to Washington, Paniushkin, who in his previous incarnation as the head of the international department of the Soviet party central committee had been responsible for supervising foreign communists. Yet during the mere three months that he stayed on the new job, he gave no signs of doing anything important.[85]

Instead of taking charge to cope with the looming foreign threats he was invoking, Stalin turned to theorizing about them. Having already in February finished writing a lengthy reply to the "request" by Soviet academics for an authoritative opinion on the prospects of war and peace, he waited another seven months to make it public, just in time for the opening of the party congress, under the misleading title "Economic Problems of Socialism in the USSR."[86]

Meanwhile, according to the amazing anecdote recounted by Stalin's German hand Semenov—which, if not true, is certainly well invented— the great dictator for three months had been drafting his speech for the congress, sticking the drafts into his shoes, reading them in the morning, then burning them, raking the ashes with bare hands, and then emptying what was left into a trash container.[87] In any case, the finished product turned out to be a meager seven-minute lecture about how Western communists should help prevent war by making their capitalist governments incapable of waging it.[88]

In the passages about war in his essay on economics, Stalin added a seemingly reassuring twist to Lenin's theory by positing that the contradictions among capitalist states were even greater than those between them and the "socialist" camp, thus suggesting that its enemies would destroy themselves in a fratricidal war before they would be ready to fight against the Soviet Union. He considered particularly the recovery of Germany and Japan to be conducive, sooner or later, to such a war, as they would turn against the United States.[89] Yet he did not unambiguously spell out the implications for Soviet policy, making it possible to invoke the kind of war that was impending as a justification for both restraint and its opposite. While the inevitable self-destruction of the capitalists could be construed as a reason for not taking any risks in waiting to pick up the pieces, the irrepressible discord among them could also be seen as making it safer to take the risks in order to give them a push.

Stalin did not make his preference for either course any clearer by insisting that particular wars could be prevented, thus implying— though never stating explicitly—that an East-West Armageddon could be avoided. In his contribution to the party congress, he expressed confidence in the ability of the people in the capitalist countries to nullify the aggressive propensities of their rulers. Yet the belief in their succeeding was difficult to reconcile with the notion that war was inevitably arising from contradictions inherent in the capitalist system.

There was more than a touch of desperation in Stalin's sloppy reasoning amounting to the acknowledgment that capitalism's collapse —which alone could safely prevent the inevitable war—could not really be accomplished by anything within the realm of foreign policy but only by the attainment of "communism" in the Soviet Union. This was, in a return to the beginning, a variation on the "socialism in one country" formula that had originally helped Stalin to grab power, refurbished to save him from losing it against the seemingly hopeless odds his country was facing in an implacably hostile world.

Such was the dismal message of Stalin's last written work, eulogized as something of his last will and testament, which permeated its dreary text, masquerading as an obtuse theoretical discussion about issues supposedly relevant to domestic policy. In trying to shore up his slip- ping power, Stalin professed having discovered the "laws" leading inex- orably to "communism," once again taking shots at the "voluntaristic" idea of the already executed Voznesenskii that its attainment was merely a matter of proper economic planning. Instead he proclaimed the "law of ever-increasing socialist production" which, having already brought that happy state of totalitarian perfection within reach, man- dated the methods that were to be used to make the final stretch.[90]

According to this providential law, still more power needed to be wielded by the state, that is, by Stalin himself. Clothed as an "objec- tive" necessity, here was the rationalization of his insatiable demand not only for more sacrifices in fulfilling the simultaneously enunciated ambitious new Five-Year Plan, with its further buildup of the milita- rized heavy industries at the expense of peaceful consumption, but also for more human sacrifices in the ever-expanding purge. To its consum- mation, Stalin devoted most of his remaining energy.[91]

Attesting to Khrushchev's ascendancy in the purge, the program- matic "theses" of the congress were published over his signature rather than Stalin's, as would have normally been required because of his posi- tion as the party's general secretary. Stalin also abstained from deliv- ering the keynote address at the plenary, leaving the honor to his anointed successor, Malenkov—possibly because of the writer's block described by Semenov if it is true, as Khrushchev says, that the assign- ment was a last-minute decision.[92] Subsequently the gathering per- formed like clockwork, singing paeans to the leader and lavishing praise on his last scholarly masterpiece as an inexhaustible fountain of wisdom.[93]

In this orgy of adoration, Beriia alone struck a discordant note in his speech on nationality questions—Stalin's particular specialty—which expatiated on the rights of Soviet nationalities while condemning "Great Russian chauvinism"—Stalin's particular affliction.[94] The least Russianized of his closest associates, Beriia was also the only one among them to possess in his native Georgia an independent power base. It is impossible to tell whether the challenge implied in his speech was deliberate or, if it was, whether Stalin noticed. In any case, the "objective" circumstances being for him more important than intentions, taking precautions was in order.

The most notable outcome of the congress was the substitution of the party's small politburo with a large presidium, thus making it more difficult for any power but Stalin's own to coalesce.[95] If he was thus both asserting his power and losing his grip on it,[96] this was still not because anybody else was grasping it. Malenkov's promotion as his heir apparent and Khrushchev's rise as his favorite made both more depen- dent on the despot's good graces, while his simultaneously denying those graces to others was his own choice. This applied particularly to Molotov and Malenkov, left out of the smaller bureau that reconstituted the former politburo within the new presidium. Other losers, despite their remaining in the inner circle, included Voroshilov and especially Beriia, demoted from number four to number six in the Kremlin pecking order.

At a still mysterious closed meeting of the central committee convened secretly after the main congress was already over, Stalin on October 16 reportedly delivered an hour-and-a-half speech, of which no written record was taken (perhaps because he could not write the speech in the first place).[97] He is said to have stunned his audience by offering to resign as general secretary and remain only the head of the government.[98] Even the devout Molotov did not believe he meant what he was saying, although he may have been quite serious in intending to effectively destroy the party in the purge and replace it with the machinery of the state, whose power he had exalted in his essay. According to Malenkov, however, Stalin merely wanted to find out who his hidden enemies were.[99] If he did and there were any, they remained in hiding as the meeting anxiously implored the helmsman to stay in his place, which he graciously deigned to do.

Cultivating the notion of his indispensability, Stalin, according to Khrushchev's colorful account, berated his aides: "You are blind like young kittens; what will happen without me? The country will perish because you do not know how to recognize enemies. . . . When I am gone the imperialistic powers will wring your necks like chickens'."[100] By the time Stalin supposedly offered to resign the worst that he could expect to happen in Washington did happen—the electoral victory of the Republicans—estimated in Moscow as having measurably increased the danger of a military collision with the United States.[101] According to the same Khrushchev, Stalin "trembled at this prospect. How he quivered! He was afraid of war. He knew that we were weaker than the United States."[102]

The Soviet Washington-watchers, having, for no good reasons, expected a softening of U.S. policy from the reelection of Truman in 1948, this time had better reasons to expect its hardening. Even Churchill thought Eisenhower's victory made "war much more probable."[103] Not only had the United States shortly beforehand exploded a prototype of its first hydrogen superbomb, but between Eisenhower's election and inauguration bellicose signals also emanated from Washington.[104] General MacArthur, out of command but for that not any less ardent a supporter of the administration which was about to replace the one that had sacked him, publicly clamored for a "clear and definite solution to the Korean conflict." Privately he proposed to the president-elect to send the Russians an ultimatum, demanding that they had better agree quickly to the unification of both Korea and Germany by free elections, and then conclude a pact outlawing war with the United

States—or else nuclear bombs would start raining on North Korea and on China, too.[105]

Although Eisenhower ignored the proposal, after his demonstrative December visit to the Korean battlefield he proclaimed that the enemy could not be "impressed by words ... but only by deeds."[106] His words certainly impressed Gen. Mark Clark, the chairman of the Chiefs of Staff, making him to expect from the incoming administration approval of the extension of hostilities into North Korea.[107] As far as Moscow was concerned, it could not yet judge any better than the general whether or not the military president would eventually live up to his party's campaign rhetoric, although the mere possibility that he might was apt to send shivers down the Soviet spines.[108]

After Stalin's ineffectual German notes had sung the swan song of his Cold War diplomacy, he gave abundant other proofs of how much his overall deterioration impaired his ability to act effectively, generating drift and incipient paralysis. Having in all but name "broken diplomatic relations with the Western world,"[109] reducing them to little more than exchanges of insults, he turned increasingly inward, trying to find at home the solution of all his real and imaginary security problems. Yet in concentrating his mind on the purge as the panacea, the despot's diminishing capacity to wield his despotic power threatened to frustrate his domestic policies as well. The spreading paralysis heralded the pathetic dead end of any policy as Stalin's own life was coming to an end, too.

9

The Dead End
November 1952–March 1953

The last three months of Stalin's life were the darkest of his dark ages. Not only did dark deeds proliferate, but also all their motives may forever remain shrouded in darkness. Yet they are indispensable for the explanation of Soviet policy, or the lack of it, at this critical juncture of the Cold War. Did Stalin's terror make his empire, as American officials believed, "well nigh invulnerable" or, on the contrary, all the more vulnerable because of the paralysis it spread? What did Stalin himself think and what were the implications of his adversaries' perceptions of the empire's unnatural condition for Western policy—or the lack of that?

The Nadir of Stalinism

In one of its less perceptive appraisals of the situation, the U.S. Board of National Estimates on November 21, 1952, estimated that on the Soviet side "internal stresses and strains appear less serious now than ever before."[1] This was the day after Eastern Europe's most gigantic political trial of the deposed Czechoslovak party secretary Slánský and his accused confederates opened in Prague, bringing the stress and strain to an almost unbearable level. It was a time when no one, not the least those in the highest places, could be certain when, whence, and why the next blow might come. The conspicuous reluctance of the controlled Soviet media to comment about the court proceedings despite their extensive factual coverage added to the uncertainty about what, if anything, the future might hold.[2]

In Czechoslovakia, the managers of the trial deliberately whipped up a mass hysteria that, once incited, kept spreading by itself. Gatherings in factories and offices, goaded by party activists, bombarded the court with demands to be merciless, urging that the "monsters" be put to death," preferably after torture. Much though the bulk of the populace

153

may have secretly rejoiced at the communists' slaughtering each other, a significant minority reacted the way the authorities wanted. The police noted a marked increase of people's spontaneous denunciations of their fellow citizens.[3] Most of the party faithful believed that the charges raised against the accused were true. Even the French wife of one of them, having been brought up in a prominent communist family, became convinced that if the party said so, her husband must be guilty, and urged the court in writing to punish him accordingly. Nor was the open letter by the teenage son of another of the defendants, which shocked the world by demanding the death sentence for his father—described as a "creature, unworthy to be called a human being"—extracted from an unwilling writer.[4]

While touching a responsive local chord, indicating how far Eastern Europe's moral fiber had deteriorated, the trial was meticulously managed from Moscow. The indictment read as if it had been translated from Russian word by word. In identifying the defendants by their last names first, as well as by nationality, it followed the Soviet rather than the Czechoslovak practice. This mattered particularly in the listing of the nationality as Jewish for eleven out of fourteen of them.[5] In a country which could justly pride itself on having previously avoided the excesses of anti-Semitism for which the region was notorious, the "Zionist conspiracy" clearly stood out as a Soviet import.

As the crowning piece of evidence of the bogus conspiracy, the prosecution invented a "Morgenthau plan," said to have been hatched in 1947 at a secret Washington meeting of the Jewish former U.S. secretary of the treasury Henry Morgenthau Jr., Truman, Acheson, Israel's later premier David Ben Gurion, and its future foreign minister Moshe Sharet. At their conspiratorial gathering, these prominent Zionists had supposedly promised to deliver Eastern Europe into American clutches in return for the U.S. promise of support to their state.[6] Not to be confused with the real, though hardly less improbable, Morgenthau plan of 1944, which envisaged the pastoralization of Germany as a cure against its aggressiveness, its imaginary Soviet namesake of 1952 was calculated to offer a ready-made explanation of the recent upsurge in U.S. efforts to subvert Eastern Europe.

Yet the anti-Semitic slant, while important, was secondary to other purposes of the trial. Its selected protagonists, who included both prominent party figures and lesser fry, had been chosen not so much for what they were—even less for what they had done—as for what they could be made into because of their particular association with a cause or institution that was to be highlighted. Accordingly, their purported transgressions did not matter any more than those of the scores of additional individuals brought into the courtroom as witnesses, whose own turn in the dock was to come later. The resulting spectacular amounted to the most comprehensive review so far of the

Soviet policies that had gone wrong or could have gone wrong during the preceding two decades.[7]

The already familiar broader topics—from Moscow's unsuccessful intervention in the Spanish civil war to its troubles with the communist undergrounds during World War II, its ill-fated collaboration at that time with the Western powers and the disappointing postwar maneuvers with coalition governments as well as the disillusioning courtship of the noncommunist Left,[8] besides Stalin's miscalculations about Tito and Israel—were supplemented by an array of local topics. These ranged from the ambivalent roles communists had played in the Slovak and Prague uprisings of 1944–45, which had cast a doubt on the monopoly the Soviet Union claimed for Czechoslovakia's liberation,[9] to their later controversial performance in what had been Stalin's longest-lasting coalition experiment in the region before he had decided to terminate it as an error.

During the three years that the experiment lasted, the Czechoslovak communists had inevitably become involved more extensively than their comrades elsewhere in both the subsequently discredited collaboration with other parties and Moscow's various international schemes which they had been particularly well-qualified to assist, but which later misfired. Having consequently become that much more vulnerable to being blamed for anything that had gone wrong, their special vulnerability ensured that the Prague trial would reach higher, sweep wider, and prove more deadly than any of the other Eastern European trials. In the end, of the fourteen men indicted, as many as eleven were hanged.

In attempting to connect the disconnected topics, the managers of the trial highlighted covert operations indirectly incriminating Beriia and his international staff as inefficient or worse. Several of the defendants belonged to that staff or could be associated with it. Apart from Slánský himself, who had been overseeing the purge in his country until it had caught up with him, they were Bedřich Geminder, formerly the head of the Czechoslovak party's international department and as such the supervisor of the important Prague organization of anti-Tito Yugoslav exiles;[10] Bedřich Reicin, once the chief of Czechoslovakia's military counterintelligence and the purger of its armed forces; Karel Šváb, Slánský's appointee in the security department of the party central committee; and Otto Katz, under the pen name of André Simone previously employed as the editor-in-chief of the party daily *Rudé právo*, but more importantly as an agent on particularly shady assignments in Western Europe, especially France.[11]

It was not so important that these men and their codefendants could already by contemporaries be recognized as being "guilty of [the] basest crimes" but "innocent of the majority of the crimes for which they will be sentenced."[12] More to the point, they were accused precisely of those misdeeds that they had been the farthest from committing, thus bringing still closer the sword that had been dangling above the heads of those who, while still at large, could perhaps more appropriately be charged with the same wrongdoings. This made the position of anyone in any position absolutely insecure.

Slánský had been a preeminent lackey of Moscow, whereas the person who could have been more plausibly—if unfairly, too—accused of faltering in his obedience to it was Gottwald, the party chief during the controversial coalition period. As the current head of state, he was the person to whom Stalin insidiously deferred the decision about which of Gottwald's former associates found guilty of treason would die and which would be allowed to live. By then thoroughly terrified of Stalin, lest he himself be not allowed to die his "natural death" by alcohol poisoning, Gottwald chose to exercise no mercy.

By the end of 1952, Stalinism's indiscriminate corruption of both its supposed beneficiaries and its victims had brought the system to its nadir. Yet the Prague trial was not intended to be the end of the terror, but merely the prelude to its grand finale in other parts of Eastern Europe and ultimately the Soviet Union itself. No sooner did the court adjourn than Gottwald's Polish counterpart, Bierut, prepared to dispatch a team to the Czech capital to explore there the possible ramifications of the Slánský case inculpating Gomułka.[13]

Having so far hesitated to move against the former party secretary, Bierut now had a powerful incentive to do so, for he himself was the foremost candidate for the role of a Polish Slánský. As the highest-ranking Warsaw confidant of the Soviet security services, he had worked closely with both Stalin and Beriia. Recently, the trial of Gen. Wacław Komar and other Polish veterans of the Spanish civil war had implicated Bierut as well as his Jewish security chief Berman, a creature of Beriia.[14] Beyond this, however, it is all but impossible to tell who was moving whom and why.

Like many of the Prague defendants, Komar was a former purger who got caught in the net he had helped to weave. His arrest, organized by officers from Moscow, was described by the knowledgeable politburo member Ochab as having been Beriia's doing. But it could just as well have been a part of the scheme aimed at his undoing, for once made to talk, Komar incriminated people linked with Beriia. If so, the timing of the public exposure in December 1952 of the underground organization Freedom and Independence, which had been directed from Washington but infiltrated and controlled by the police for years, would have been right to enable the embattled Beriia to fight back. In any case, the publicity of this feat advertised the prowess of his personnel in Poland as both their competence and their loyalty were increasingly under attack elsewhere.

While the Polish communists' penchant for procrastination hampered the purge's progress in their country, a greater native fervor was advancing it in East Germany. On December 20, the party central committee publicized in a special declaration its readiness to draw proper "lessons" from the Slánský trial.[15] The document cast into the role of a German Slánský former politburo member Merker, vulnerable because of his having spent the war out of Soviet reach, running the party's organization in exile in Mexico. He was further conspicuous as the rare East German communist of any stature who, while himself not Jewish, was decent enough to believe that Germans owed Jews a special debt because of the Holocaust.[16] Yet he, too, was gaining valuable time as the case against him, difficult to build because of his relatively low rank, amounted to little more than the fairly innocuous accusation of having colluded with Zionists to help channel compensation payments for Nazi victims to Israel.

As the epicenter of the purge was finally shifting to the Soviet Union, Stalin's hand could be detected in its increasingly anti-Jewish thrust. This had already permeated the secret trial in Moscow in July 1952 of over twenty prominent Jews, most of whom were subsequently executed. Besides cultural figures, they included former deputy foreign minister Lozovskii and other individuals once associated with the dissolved Anti-Fascist Jewish Committee, supervised by Beriia.[17] In November there followed the public trial in Kiev of a group of "saboteurs," which foreshadowed the Slánský case by featuring mostly Jews and prominently identifying them as such. Allusions to a wider conspiracy about to be uncovered were evocative of the Great Purge; what differed from it was holding the trial before a military rather than a civilian court.

Whether this was Stalin's way of courting the armed forces as the purge was approaching its peak is uncertain. What is certain is that, in contrast to the nineteen-thirties, when their decimation had catastrophic consequences for the later war, this time they were spared, while the security establishment, as before, was not. As the year turned, Jewish members of Beriia's security apparatus had been losing jobs in disproportionate numbers,[18] adding to the growing evidence of Stalin's rampant, if inconsistent, anti-Semitism. While bent on destroying Jewish culture—as any culture he deemed subversive—he kept in his service Lazar M. Kaganovich, the only Jew on the party presidium, besides lesser thugs of Jewish ancestry.

Nor was the thrust of Stalin's scheme against Beriia—who, too, had been promoting some Jews and demoting others[19]—unequivocally anti-Semitic. Despite the multiplying signs that it was under way, its exact nature was, and remains, elusive. It could be misleading trying to read consistency and logic into what was happening when inconsistency and illogic suited better Stalin's purpose of perpetuating uncertainty about his goals and means.

The publication on December 24 of a *Pravda* article by the ideological watchdog Suslov was obviously important, but not for any immediately obvious reasons. It lambasted an obscure economist, Petr

Fedoseev, for having praised Stalin's essay on economics but neglected to specifically acknowledge that three years earlier he had himself subscribed to some of those erroneous Voznessenskii views that were being attacked there. Since, however, a mere public apology sufficed for the culprit to be forgiven, the real purpose of the mighty Suslov firing at the lowly target was to justify the unprecedented publication at the same time of a secret politburo decree.

The decree was the one that in 1949 had precipitated not only the sacking of the editors of the journal *Bolshevik*, of whom Fedoseev had been one, but also, more importantly, the unleashing of the murderous Leningrad purge of Voznesenskii and other former followers of Zhdanov.[20] Supervised on Stalin's behalf by Malenkov and carried out by agents of Beriia, at that time the purge had not cast any pale over the already dead Zhdanov himself; what it did cast a pale over now, by rehashing the three-year old story, was the two purgers.

As far as Zhdanov is concerned, his merits as a champion of conformism were again conspicuously emphasized and his son Iurii was given the honor of putting his name under an article inaugurating a campaign against politically deviant scientists. But the simultaneous resumption of the campaign against cosmopolitanism—which figured prominently in the charges against Beriia's foreign operatives at the Slánský trial—could also be construed as a repudiation of Zhdanov's internationalist strategy in which they had been instrumental, thus threatening anybody who could be associated with it.

All this was very complicated and possibly misleading, but then this is what it was most probably intended to be. Just how and against whom these elaborate traps were subsequently to be used need not have been decided in advance, nor was Stalin necessarily the only one who set them up. On Christmas Eve, responding to the request by James Reston of the *New York Times*, he made what was to be his last public pronouncement on the subject of war.[21] Insisting that it was not inevitable after all, he tried to reassure the world about his intentions abroad just as he was about to take the final steps to win the war at home. Rather than any policy worthy that name, however, these merely entailed an intrigue—so intricate that in setting it the master could no longer consider even himself safe from its workings.

Stalin's Last Plot

When Stalin on January 12 made his last public appearance in a Moscow theater, he gave no inkling of the bombshell that would explode the next morning—the news of the "doctors' plot."[22] This was the incredible story that a group of Kremlin physicians, most of them Jewish, had conspired to shorten the lives of their distinguished patients, including Zhdanov, and that only the exemplary vigilance of a junior woman doctor, a Lidia Timashuk, had prevented them from succeeding. That the purported conspiracy had been set up is more easily determined than the reasons why.

Incredible as it seemed, the story was not entirely without foundation. Their professional eminence notwithstanding, the accused physicians were no paragons of Hippocratic integrity. Some of them had been Stalin's flunkies in denouncing their colleagues in the 1930s for the same kind of malpractice of which they themselves now stood accused.[23] Since the tyrant himself was prone to occasionally employing doctors to expedite the deaths of those he wanted to be rid of—such as the Red Army commander Mikhail Frunze and probably the writer Maxim Gorkii—he had reasons to be wary of the profession. Toward the end of his life he refused to submit to any serious medical care.[24] And since the person most knowledgeable about his tricks, including those with doctors, was their principal executor, Beriia, Stalin had particularly good reasons to be wary of him.

According to Beriia's former subordinate Sudoplatov, the agent for "special tasks" who as supervisor of the poison laboratory in Moscow secret police headquarters was in an excellent position to know, the vigilant Dr. Timashuk had first filed her report in 1948, shortly after Zhdanov's death, but it had inexplicably not been acted upon.[25] This time the untrustworthy Sudoplatov may be trusted, for the piece of information he provides neither harms nor helps his own terrible reputation. According to another disreputable source, Stalin's minister of public health and drinking companion Efim Smirnov, it was only in the summer of 1952 that the dictator expressed interest in the name of the physician who had treated Zhdanov, insinuating his responsibility for the patient's death.[26]

It is possible but not probable that Stalin was trying to recall the name of the expert he himself had commissioned to do the job; more likely he was simply getting worried about the harm doctors might do to him. By the time of the October 1952 party congress, he had circulated rumors about their putative conspiracy among his entourage.[27] The topic also came up at the Slánský trial in the form of fabricated testimonies incriminating Gottwald's personal physician for having supposedly prescribed him a treatment that could harm him.[28]

On December 4 the Soviet party central committee passed a secret resolution ominously calling attention to "The Conditions in the Ministry of State Security and Medical Sabotage."[29] Within the high party circles Zhdanov was now being mentioned as one of the early victims of murderous doctors, as was the wartime chief of the Red Army's political directorate, Gen. Aleksandr S. Shcherbakov, a particularly repulsive character described by Khrushchev as "a poisonous snake."[30] In the same month, Smirnov lost his job—but nothing more.[31] He must have had more to hide than he was later willing to reveal.

So did Khrushchev, who later on could have added the responsibility for the doctors' plot to the tall list of misdeeds attributed to Beriia but did not, although he allowed the idea to be briefly aired in *Pravda* in 1956. Nor did he blame Stalin for arranging this particular ploy, as so many others. He oddly depicted the dictator as the victim of the woman who had reported the conspiracy, although he did not go so far as to suggest that she had invented it all by herself. Khrushchev never volunteered an opinion about who else may have invented it, nor did he link the doctors' plot with Stalin's scheme to liquidate under some pretext most, if not all, of his closest associates.[32] It is difficult to avoid the conclusion that among those few to be spared Khrushchev was Stalin's willing or unwilling accessory in precisely such a scheme, directed mainly, though not necessarily only, against Beriia—a role that, though no guarantee against Khrushchev's being himself slated for liquidation at the time, was certainly not something to be later advertised.

By 1953 the despot had acquired compelling reasons to get rid of Beriia. Not only was he the person most familiar with the methods Stalin had used in committing his crimes, but as the security chief was also best positioned to apply them against him if he got the incentive. And this is what Beriia had been getting ever since starting to fall out of favor in 1951, if not before. As Khrushchev succinctly put it, since "the practical means for achieving Stalin's goals were all in Beriia's hands," he could also use those means to further his own goals, thus giving Stalin enough reason to fear "that he would be the first person Beriia might choose."[33] None of this necessarily meant, however, that Beriia had actually decided to act against Stalin before prompting Stalin to decide to preempt him.

The key figure on the stage was Mikhail Riumin, one of the deputies in the ministry of state security built up by Stalin with Khrushchev's help in trying to dilute Beriia's control of the security establishment. Riumin and his men had been instrumental in organizing both the Slánský trial and the crackdown on alleged followers of Voznesenskii among Soviet economists, which paralleled the unfolding of the doctors' plot.[34] Although never accused of actually concocting the plot by anyone but the author Aleksander I. Solzhenitsyn, writing from hearsay evidence,[35] soon after Stalin's death Riumin was arrested and interrogated on orders from Beriia.[36]

Afterward Beriia allowed selected members of the central committee to read some of the documents gathered in connection with the interrogation, particularly the transcripts of Riumin's conversations with Stalin. One of those privileged readers, the writer Konstantin Simonov, later remembered that the papers showed Stalin as being pathologically suspicious—nothing more startling.[37] After a week, they were in any case taken away, never to be seen again. Eventually Beriia's enemies, led by Khrushchev, had Riumin executed after a secret trial without specifying his guilt.

The point of the convoluted story is that several members of Stalin's entourage who had reasons to feel threatened by him during the last weeks of his life subsequently wanted to get rid of Riumin as the person suspected to have been his chosen instrument to destroy them. But only Beriia—the main target—acted as if he tried, though apparently without success, to find out the truth about Stalin's scheme and wanted some, though not all, of his findings to become more widely known. And only Khrushchev—Stalin's likely accomplice—acted as if he had been sufficiently implicated in the scheme to try—and succeed—to keep its sordid particulars out of anybody's sight.

Khrushchev protected Riumin's nominal superior Ignatev, the minister of state security, who as the person appointed under Stalin to investigate the bogus doctors' plot, was understandably sacked by Beriia promptly after his death, a month before the arrest of Riumin.[38] But unlike Riumin, Ignatev—who was in a position to know more about the real plot than any of the top leadership with the probable exception of Khrushchev—survived. Indeed, at the twentieth party congress in 1956 he was treated by Khrushchev as a respected witness of the time—though never called upon to tell what he actually knew.[39]

In whatever Stalin was contriving, the military were bound to play a critical role, since the police, whose chief was to be the target, could not be trusted. Not only were the armed forces the only segment of the Soviet establishment spared any significant damage by the purge, but the intention to damage them was also prominently attributed to the "assassins in the white coats." Besides the serpentine General Shcherbakov, whom they had supposedly slain already in 1947, other high-ranking officers were said to have been on their hit list, and barely escaped with their lives.[40]

The list included Marshals Aleksandr M. Vasilevskii, Ivan S. Konev, and Leonid A. Govorov, besides Chief of Staff General Shtemenko and Adm. G. I. Levchenko. The investigators produced a letter by Konev, in which he complained of having been subjected to the same kind of medical treatment that Zhdanov had failed to survive. As far as Govorov is concerned, he was one of the many generals with whom Stalin had packed the central committee at the October party congress. But his appointment was special in having been announced after a two-week delay, with the improbable explanation that his name had previously been omitted "by mistake."[41]

What could be no mistake was Stalin's intention to exploit the doctors' plot in accelerating the purge both in the Soviet Union and in Eastern Europe. The same January 13 issue of *Pravda* that announced the scary news also carried a keynote article by Frol R. Kozlov, then an undistinguished party secretary from Leningrad but later distinguished as a prominent supporter of Khrushchev, which pleaded for trying still harder in unmasking ubiquitous enemy conspiracies.[42] On the same day in Prague, the decision was made to follow the Slánský trial with successor trials to process the many designated conspirators already collected in jails.[43]

The Kozlov article singled out Jews as outstanding perpetrators of heinous crimes, now regularly reported in the Soviet media in prurient detail.[44] Yet the extent of the ensuing anti-Jewish persecution was both more or less than what has usually been attributed to Stalin's raging anti-Semintism. It was more in Eastern Europe where, in taking the cue from Moscow the local regime sometimes overdid what they were commissioned to do, while in the Soviet Union less actually happened than it has retrospectively been seen as happening.

In East Germany, offices of the few remaining Jewish communities were raided by the police for the first time since Hitler's era, prompting Jews to flee to the West by the hundreds. The Association of the Victims of the Nazi Regime, the communist front organization set up to rally West Germans under the antifascist banner, was dissolved on the grounds of having been infiltrated by Zionist agents.[45] In Hungary, mimicking the doctors' plot, thirty prominent physicians were arrested besides the Jewish community leader Lájos Stöckler, amid rumors of an impending pogrom.[46]

Péter, the Jewish chief of the Hungarian secret police, had already fallen, together with Belkin, his Soviet supervisor in organizing the Rajk trial, amid insinuations that the trial had really been designed by these two Beriia creatures to benefit the British and Israeli intelligence.[47] Israeli diplomats were expelled from Prague and Warsaw, while the explosion on February 9 of a bomb at the Soviet embassy in Tel Aviv gave Moscow a pretext to break diplomatic relations between the two countries. Yet, contrary to what might be suspected, the bomb had not been planted there by Soviet agents but rather by a local terrorist group, linked with the later Israeli Prime Minister Menachem Begin.[48]

In the Soviet Union, the picture was also mixed. On February 14 the death of the Jewish predecessor of Shcherbakov as political commissar of the armed forces, Gen. Lev Z. Mekhlis, added to the lengthening list of controversial figures passing away at convenient times. Associated with Beriia as the minister of state security before Abakumov and subsequently as the minister of state control, Mekhlis was rumored to have "taken mysteriously ill" after having taken refuge in the provinces—before expiring in Moscow at a police infirmary. His burial with full honors did not dispel the growing fears that a pogrom was on its way.[49]

What was really on its way was being exaggerated by the officially encouraged hysteria. Khrushchev later recalled Stalin's order to "organize armed men" for a pogrom and other dirty assignments.[50] Another of the dictator's former associates, Marshal Nikolai A. Bulganin, even professed remembering a plan, supposedly designed by Stalin together with Malenkov and Suslov, to publicly hang each one of the doctor plotters in the main square of a different Soviet city.[51] Stalin is said to have entrusted to his "court Jew" Kaganovich, reputed as an organizational wizard, the preparation of mass deportation of other Jews from cities to the countryside, ostensibly to protect them from people's wrath.

None of these retrospective allegations is supported by sufficient evidence. Soviet Jews have later reported having heard of a petition that the more prominent among them were allegedly forced to sign to demand the deportations for their own protection, although the lack of any extant copy of this presumably widely circulated document casts doubt on its existence.[52] According to another improbable version, the petition was merely being prepared to be sent as a letter to *Pravda*, but then was not sent, after a mild protest to Stalin by his acolyte, the writer Ilia Erenburg, had sufficed to persuade him to drop the projected deportations.[53] What has been said, but not documented, gives credence to Sudoplatov's assertion that they were mere rumors, not substantiated by actual preparations.[54]

Whatever new atrocities Stalin may or may not have been contemplating against his people, more important was the threat he posed to his cronies. The programmatic statements which, alluding to the doctors' plot, had been appearing in the media over the names of largely unknown authors, were increasingly suggestive of his intent to replace the existing ruling elite with new faces, as he had done in the 1930s.[55] Yet while he retained enough of his propensity to spin evil thoughts and entertain sinister designs he was no longer as capable as he used to be of devising, much less implementing, a coherent plan to put them into effect.

The despot's physical and mental condition had been plummeting. Khrushchev later recalled that "in the last time comrade Stalin read no papers and received no people"[56]—a situation suitable for being exploited to prevent him from making decisions or to manipulate him. Beriia had both the best reasons to do so and the best opportunity to try. The circumstantial evidence of his possible counterplot against Stalin, perhaps initiated as early as December 1952, includes the removal at the critical time of at least three persons controlling access to the dictator. The terse announcement on February 15 of the sudden death of the commander of the Kremlin guards, Gen. Petr Kosynkin, followed the unannounced departures from their posts at the end of the preceding year of Stalin's longtime personal secretary Poskrebyshev and chief bodyguard Gen. N. S. Vlasik, both trusted members of his household.[57]

The former had been responsible for the leader's appointments, the latter for his personal security. The arrest of Vlasik on December 16 convinced his daughter that "Stalin's days have been counted. He has not much life left."[58] Kosynkin, according to a high-ranking defector from the Soviet security services, was killed on orders from Beriia, acting in league with Bulganin, Malenkov, and Ignatev.[59] Malenkov had been in eclipse since late 1952, while Ignatev's future was made precarious by his assignment to supervise the investigation of the doctors' plot.

Whether those key personnel had been disposed of by Beriia directly or indirectly—by his sowing nagging suspicions about their loyalty in the despot's receptive mind—[60] intrigues involving the top military complement the circumstantial evidence of a counterplot. They concerned the ambivalent role of Stalin's sycophantic former army chief of staff Shtemenko, one of the listed targets of the killer-doctors, whose recent return and brief ascendancy could have been indicative of a maneuvering for support from the armed forces.[61] On February 22 the vigilance campaign in the media, which had surged so rapidly after the announcement of the doctors' plot five weeks earlier, subsided just as fast.[62] Could it be true, as Washington still believed, that "the Soviet regime is securely entrenched in power, and there is no apparent prospect of its control being threatened or shaken"?[63] Or was the regime rather critically undermined?

The Unopened Window of Opportunity

In his first State of the Union message on February 2, Eisenhower proclaimed that "there is but one sure way to avoid total war, and that is to win the Cold War"—a statement he believed was as apt to "scare the Russians as hell."[64] He also announced the withdrawal of the U.S. Seventh Fleet from the Taiwan Strait, thus no longer "shielding" mainland China from a possible attack by Chiang Kai-Shek,[65] and authorized the heaviest bombing of the enemy lines in Korea in a year.[66] In trying to force the end of the war, Washington was seriously considering the use of nuclear weapons, and the Soviet intelligence knew about it.[67]

At the February 11 meeting of the National Security Council, Dulles pleaded for discarding the "false distinction" between nuclear and other weapons, which he argued the Russians had been making to suit their own purposes.[68] He served them notice that "power never achieves its maximum possibility as a deterrent of crime unless those of criminal instincts have reason to fear that [it] will actually be used against them."[69] This was the closest the highest-ranking officials in Washington came to contemplating the use, if not of war, then at least of the threat of war as a tool of American policy in dealing with the Soviet Union.

In trying to end the Korean War, the new administration realized how much that "wrong war, at the wrong place, at the wrong time" distracted the United States from its goal of winning the Cold War. What it did not realize was how weak the real enemy—the Stalinist tyranny—had become as a result of its self-inflicted paralysis. Accordingly, if there ever was the right time to wage the right war (to dispose of the tyrannical system once and for all) in the right place (Europe), this would have been the time. The war to be waged was not a fighting war—the kind described by Carl von Clausewitz as a desperate affair in which anything that could go wrong is likely to go wrong—but rather the one that his Chinese predecessor, Sun Tzu, had in mind in describing the consummate strategy as that of "forcing the other party to resign to our will without fighting a battle."[70] In short, the situation called for a vigorous diplomatic offensive from the position of strength that Dulles insisted so much the United States was in, backed by the implied threat of force.

Instead of rattling nuclear weapons at the Chinese, who rightly refused to be scared,[71] it would have been more appropriate to do some rattling at the Russians in Europe, using the conventional balance there as the starting point. In Asia, what Washington should have been doing was reassuring the Chinese. It could have tried secretly to send a messenger to Beijing, best one of those experts on China whom Sen. McCarthy had hounded down for having "lost" it, to see not so much Mao Zedong as Zhou Enlai for a very private talk and to make sure that the Soviet ambassador Paniushkin would find out about the trip, though not necessarily about the main subject of the talk. That subject would have been the separation of the prisoners-of-war issue from the conclusion of a cease-fire that by then both Washington and Zhou, though not necessarily yet Mao, understood as being the only way to the armistice they wanted. If anything could have swayed Mao to go along, this would have been it, while Paniushkin's Moscow bosses, including or excluding Stalin, would have hardly been in a position to do anything but go along, too.

In Europe, meanwhile, a new and comprehensive Western proposal for the unification of Germany by free elections, perhaps even making use of Stalin's rejected March proposal now more usable to the West than to him, would have been in order. If the preparations for the creation of a West German army were accelerated at the same time, this could have done no harm. In addition, Washington could have as well signalled its intention to bring Austria, at least the bulk of it which was under Western control, into the alliance system it was building. Eisenhower could have even followed on his trip to Korea by bringing into being, for the little it would have been worth, that force of Eastern European volunteers whose authorization by Congress more than a year before had rankled the communists so much.

None of this would have necessarily brought quick results, if any; for such results, the time was, in any case, too short. But precisely because it was so short Washington should have been buzzing, with officials mapping out strategy and getting proposals ready, rather than just talking. Even if they would not have gotten any serious response from Stalin before he would die, it would have been to the West's advantage to have his successors faced with comprehensive demands, backed by its superior power which they could not ignore, rather than to leave the initiative in their hands.

This would have created the optimal preconditions for the grand bargaining that should have taken place after World War II but did not, leading instead to the Cold War, as well as for the termination of the Korean War and perhaps even normalization of U.S.-Chinese relations. Moreover, once such bargaining on Western terms got under way, it would have been exceedingly difficult for the Soviet leaders to keep their Stalinist system intact, and their control of Eastern Europe, too. They would have been under great pressure to replace that system, or allow it to be replaced, by a more desirable one, thus not only making their and the Eastern European people happier but also a settlement of the Cold War that much easier.

Instead, the window of opportunity remained closed. While nothing important was happening in Washington and the important things that were happening in Moscow behind the walls of the Kremlin were proceeding out of public sight, Stalin could afford to keep ignoring foreign policy and to drag his feet about Korea. Three days after having told the *New York Times* that war was not inevitable, he secretly gave Mao some singularly unhelpful advice about what to do. To Mao's query about whether the new Washington administration might follow the plans of the old by launching a major attack, Stalin replied that it might but also it might not, so that he should brace himself for the worst.[72] Mao took the advice, ordering a costly program of massive construction of fortifications that proved totally useless.[73]

Stalin being of no help, the celebration in Moscow of the third anniversary of the Sino-Soviet alliance on February 16 was a painful affair. Neither Molotov nor Mikoian—not to mention Stalin—showed up, leaving it to the dullest member of the inner circle, Bulganin, to deliver an uninspired speech before the uneasy Chinese.[74] Nor did Beijing's dispatch a week later of its leading nuclear scientist to the Soviet capital—perhaps to seek reassurance against possible U.S. nuclear attack—stir Moscow into action.[75]

The day after the lackluster celebration, Stalin received separately two visitors from India: its ambassador, Krishna Menon, and Saffrudin Kitchlew, the chairman of its pro-Soviet Peace Council. Showing no interest in their government's attempts to mediate in Korea, the despot was described by Menon in his memoirs as behaving ominously. He read into Stalin's doodling of pictures of wolves on a piece of paper and his blabbering about the peasants' need to kill them his supposed readiness to strike against the American predators, presumably to save the Korean and Chinese sheep, thus unleashing a world war.[76]

Nothing in contemporary evidence supports Menon's conjecture. The substantial reduction of Soviet military spending and the relative neglect of the management of defense industries precisely during his last year in power[77] are the best evidence that any martial adventure was far away from his admittedly declining mind. Immediately after the interview, Menon himself gave U.S. chargé d'affaires Jacob Beam a considerably more reassuring account than the one he later included in his memoirs. He reported that despite complaints about the movement of the Seventh Fleet Stalin showed no particular alarm. Evidently, he was not as "scared as hell" as Eisenhower thought; in fact, in his talk with Kitchlew, Stalin paid his former companion-in-arms in World War II the tribute of crediting the general with greater realism than the civilian Truman.[78] Although Stalin bemoaned the omnipresent capitalist conspiracy holding sway of the president, he ended his conversation with Menon on an upbeat note, expressing a desire to meet other foreign envoys as well.[79]

He never did. Menon and Kitchlew were the last foreigners known to have seen Stalin alive. Instead, if Sudoplatov has it right, the dictator received him—the man for "special tasks" (such as had been the murder of Trotskii)—to discuss a new assignment: the murder of Tito.[80] If this was so, there could have hardly been a better proof that Stalin neither perceived a military threat nor was preparing a military act, but still felt so hard pressed for other reasons that he became desperate. According to Sudoplatov, he also ordered a radical reorganization of Soviet intelligence and covert operations in trying to make them more effective.[81] This was his last recorded act of state; in a few days, Stalin would lie dying, amid indications that the fate he may have been preparing for Tito, as well as for Beriia, befell him.

On February 27, Western observers noted that Soviet troops in East Germany went on alert.[82] This was the day before the four more or less permanent members of Stalin's inner circle—Beriia, Malenkov, Khrushchev, and Bulganin—last saw him together in good health. They were the same people who, together with Molotov, would become his successors. Khrushchev (who was present) and Molotov (who was not) subsequently offered strikingly different accounts of what had happened on February 28, the former more dubious than the latter.

In 1957 Khrushchev let his protégé Panteleimon Ponomarenko leak into the Western press the story of a party presidium meeting held on or about that day. At the meeting the normally docile Voroshilov—who had a Jewish wife—allegedly protested Stalin's plan for a pogrom so vehemently that the old man got a stroke on the spot, never recovering.[83] In his later memoirs, however, Khrushchev had Stalin leave his companions after an uneventful dinner-and-drinking party in good enough health to shock them by collapsing only the next day, March 1, at his country dacha.[84] Whether because of the shock or in spite of it, he was then left unattended for a good twenty-four hours, except by bodyguards posted outside his room.[85]

Molotov provided a less evasive but more contentious account than the inconsistent Khrushchev. When asked in 1984 by his admiring

interviewer Feliks Chuev whether it "could be that they poisoned
Stalin when they were drinking with him on the last day before his
illness," the despot's former intimate immediately responded that "it
could have been." Speculating about suspects, he noted that "Beriia and
Malenkov were close to each other. Khrushchev joined them having his
own goals." Summing up, Molotov concluded that Stalin could have
been "killed by Beriia, or by a chekist or a physician on his orders. . . .
He did not die a natural death. He had no particular illness. He worked
all the time. . . . He was alive, and how."[86] Molotov further remembered
a conversation with Beriia during which Beriia boasted: "I put him out
of the way. . . . I saved you all."[87]

With the passage of time, the murder theory has received more
credit than discredit. Initially, it was expounded mainly by Stalin's dis-
solute son Vasilii, who caused so much embarrassment by spreading it
in public that he was locked up in prison to be silenced.[88] Among
others who disappeared, according to the dictator's similarly unbalanced
daughter, were two of his bodyguards reported as suicides soon after his
death; in her opinion, however, there had been no foul play.[89] Assuming
that there was some has been more typical of Soviet observers, looking
at closer range and with more empathy,[90] than of Westerners, taking the
view from a distance tempered with detachment. Although the suspicion of murder is unlikely ever to be substantiated by evidence that
could stand in a court of law, in the court of history, where plausibility
is what counts, it makes the ensuing developments more rather than
less explicable.

What for Molotov was a gang of pygmies plotting against a man of
genius was for Khrushchev a group of leaders formed within the party
toward the end of Stalin's life to mitigate the tyrant's excesses. No
contemporary sources, however, support the claim that such a fine
group coalesced during his lifetime. Anton Antonov-Ovseenko, the
scion of a prominent Bolshevik family who writes without footnotes
but with an insider's feel for the subject, sees Stalin as detaching
himself from the loyal servants Molotov and Voroshilov, thus unwittingly opening the door for the disloyal ones—Beriia and his buddy
Malenkov—while vainly trying to balance off the pair with his favorite
Khrushchev and the harmless Bulganin.[91]

Malenkov's son Andrei, however, considers the man closest to Beriia
to have been not his father but Khrushchev, whom he describes as
enough of a match for the police chief as an intriguer.[92] If Khrushchev
was indeed so adept at scheming, and his later record confirms that he
was, he may even have been capable of outwitting both Stalin and
Beriia by being privy to their respective machinations against each
other yet not arousing the suspicion of either. In view of the enormous
risk involved, however, any conspirator against Stalin would have
preferred to act alone, without relying for assistance on anyone else
except, perhaps, those bodyguards under Beriia's command who alone

kept the watch over the paralyzed dictator during the critical interval before the doctors were called in; this applies particularly to those two officers later reported as suicides.

Whether Beriia had actually acted to shorten Stalin's life or, more probably, was merely preparing to do so while waiting for the right opportunity, once the despot lay dying, he jumped into action, taking charge in putting into effect the particular emergency measures introduced to cope with the situation. In handling publicity about the supreme leader's illness and the organization of his medical treatment, the measures were suggestive of an attempted cover-up, calculated to facilitate the transfer of his power.

Although by all the insiders' accounts Stalin had suffered a stroke earlier than the officially announced date of March 2, probably the day before,[93] days passed before the news was made public on March 4 in an anouncement whose wording betrayed the hand of politicians rather than of the physicians whose signatures it bore. One of those physicians, Professor Miasnikov, later reminisced how they were still trying to confirm Stalin's illness as cerebral hemorrhage when a young woman doctor appeared from nowhere, took a cardiogram, and wrongly diagnosed a heart attack.[94] They were terrified, remembering their colleagues so recently accused by another female expert of murdering their Kremlin patients by prescribing them the wrong treatment for that very condition. Yet the mysterious diagnostician disappeared as suddenly as she had appeared, causing no harm.

While the dying man was still breathing, Beriia masterminded the disposal of his power at the secret combined session of the party central committee, the council of ministers, and the presidium of the Supreme Soviet on March 5.[95] Beriia moved that the post of premier be given to Malenkov, who in turn obliged him by proposing the merger under his command of the ministries of the interior and state security, the latter of which under Ignatev had supervised the unfolding of the deadly doctors' plot. Khrushchev received satisfaction by having his protégé Ignatev safely moved to the secretariat of the central committee. Molotov resumed the direction of foreign policy and, signaling its most pressing priority, had his deputy Vasilii V. Kuznetsov promptly designated to replace Paniushkin as the new ambassador to China.[96]

Less than an hour before Stalin expired at 9:50 P.M., the assembly effectively deposed him by leaving him no other function than an ordinary membership in the party presidium. This was simultaneously reorganized to undo the changes introduced by him four months earlier in preparation for the ouster of most of its members. The large presidium he had created was reduced to its previous size and refilled with the lucky survivors. Its inner bureau, reserved by the dictator for his new favorites, was abolished along with the special agencies he had designed to maximize his control—the presidium's standing commissions on foreign affairs and on defense.

The Soviet Union's new rulers delayed releasing the notice of Stalin's death to the public until the next morning, March 6, and that of their succession for yet another day. Meanwhile, they took special care to preventively dispel any suspicions of impropriety. The medical bulletins they had issued abounded in professional jargon, and once the patient was dead they had an alternative team of physicians summoned to certify that he had been treated properly. They included the doctor-hunter Timashuk for added reassurance.

As Stalin's funeral was being hastily arranged, the authorities grew nervous, and for good reason. Despite the precautions against unrest, there were disturbances in Moscow, which the police could not control. Masses of people milled aimlessly through the streets, causing injuries and even deaths, estimates of which range from a few to hundreds, if not thousands—surely exaggerated numbers.[97] In the end, however, what mattered more than the numbers was what struck the U.S. naval attaché, Captain Lang, as the remarkable absence of any "dangerously anti-police or [anti-]government" attitude.[98] Ordinary Russians were as much awestruck at the dead Stalin as they had been at the living one.[99]

So were, despite the fate he had in store for them, his former associates, with the notable exception of Beriia. According to Vsevolod N. Merkulov, the minister of state control and his former crony, Beriia had wanted Stalin to die and was elated at composing his funeral oration.[100] During the burial rites on March 9, he joined others in singing paeans to the deceased leader, but not Khrushchev and Molotov in shedding tears. In another month Beriia spoke his mind to his protégé Mikhail Chiaureli, the film director known for his Stalin spectaculars: "Forget this son of a bitch! Stalin was a filthy rascal, a scoundrel, and a tyrant. He kept us all scared, the blood-sucker. The whole nation trembled before him. That was his only strength. Luckily we got rid of him."[101]

Its facade of totalitarian perfection notwithstanding, Stalin's regime had reached its dead end even before he did his. His trying, and succeeding, to make everyone around him absolutely insecure may have even cost him life. Yet this critical moment of Soviet vulnerability passed without the West's trying to open the window of opportunity, which may have been the only one for ending the Cold War earlier than it eventually did. Once Stalin was gone, the situation was no longer the same, although it was not immediately clear how different it was. Would it provide his successors with incentives to seek because of their weakness the end of the confrontation that the West had failed to seek despite its strength? This was but one of the many questions that they had to face.

10

Coping with the Stalin Legacy March–July 1953

In having to manage a predecessor's difficult legacy, in 1953 Stalin's heirs faced a predicament similar to the one Brezhnev's successors would face thirty years later. They all sought to adapt an obsolescent system to new domestic requirements amid rising international tension without losing control, and perhaps power, in the process. On both occasions they were handicapped by being both the products and the victims of the system they had inherited. Nor was their task made easier by the presence in Washington of a hostile Republican administration seemingly bent on exploiting their distress to its advantage. Whereas the second time the system collapsed, in 1953 it had survived. In ensuring this outcome—and the resulting nearly four decades of continued East-West rivalry—the five months after Stalin's death were crucial.

The Prevailing Consensus

The post-Stalin leadership was less fractious than its enemies hoped. Rather than please them by falling out with each other, the autocrat's successors divided up his mantle among themselves so that no one would be able to wear it whole. As agreed when he lay dying, on March 15 the Supreme Soviet rubberstamped Beriia's nomination of Malenkov for premiership after he had the day before relinquished the party secretaryship to Khrushchev.[1] Molotov again became foreign minister, while the crucial, if passive, role of the armed forces at the moment of the regime's vulnerability was acknowledged by the inclusion of the unremarkable Marshal Bulganin in the ruling quintet.

Some of the Soviet signals that have long been tantalizing can now be better explained from inside sources. The conspicuous publication in *Pravda* the day after Stalin's funeral of the retouched photograph from

1949 on which Malenkov was inserted next to the generalissimo and Mao as something of the Holy Ghost in a divine trinity[2] was not indicative of his grab for power, as it was widely thought to be. He himself was reportedly furious at the obsequious journalists who had done the pasting job on their own.[3] What the collage was indicative of was the high priority the new men in Kremlin assigned to the settlement of the Korean War together with the Chinese.

On the question of who took the first step, the Russian and Chinese sources differ. The Russian sources say the Chinese did, and the Chinese, the Russians. At least it is clear that on March 19 the Soviet government made the fundamental decision to reverse Stalin's policy by secretly adopting a plan to end the war.[4] This gave substance to Malenkov's well-known speech three days before which attracted worldwide attention because of his having stated that his country had no dispute with any other that "could not be settled by peaceful means, on the basis of mutual agreement."[5]

Drafted by Molotov after consultations with Zhou Enlai, who had been in town since Stalin's funeral, the decision, approved without changes, justified the turnabout by asserting that "it would be inappropriate to continue the line that had been pursued in the question until now, not introducing changes corresponding to the present political moment."[6] It proceeded outlining a detailed plan of action, to be closely coordinated with the Chinese and North Koreans, leading to an armistice. After a long meeting of the whole Soviet leadership with Zhou Enlai on March 21, the plan was put into effect.[7]

Once the Stalin stumbling block was gone, Molotov, with the backing of all of his colleagues, prepared the plan together with the one Chinese leader whom he could expect to be most supportive. Zhou Enlai was the person who in October 1950 had communicated to Stalin in Molotov's presence Beijing's reluctance to intervene in Korea in the first place.[8] And in 1952 he had tried but not succeeded to impress upon Stalin the desirability of postponing the decision about the exchange of prisoners of war after armistice has been concluded, thus removing the main obstacle to its conclusion.[9]

This was the position Moscow now adopted in trying to expedite the end of the war. Attesting to undiminished Soviet supremacy in managing the Korean conflict, within days Beijing and Pyongyang started implementing the agreed script. They signaled their readiness to exchange sick and wounded prisoners of war without forcing those unwilling to return home. On March 30 Zhou proposed publicly—as he had already done in 1952 to Stalin secretly—that such persons might temporarily be transferred to a neutral country, thus paving the road toward the acceptance by the communists of the principle of voluntary repatriation on which U.S. consent to the armistice hinged.[10] Following the same script, Molotov in his April 2 statement dwelling on the need for peace in Korea associated the Soviet Union with the actions of its allies.[11]

Much like Molotov was entrusted by his peers to take the lead in foreign policy, so was Beriia in domestic policy. The later insinuation that he acted on his own in defiance of others is not supported by contemporary evidence. Feeling, in Khrushchev's pregnant phrase, "terribly vulnerable,"[12] they all sought to keep not only the appearance but also the substance of solidarity. The separation of the top government and party jobs, supervised by Beriia, was followed by a reorganization of the party apparatus to bolster the institutions of the government.[13] The shift served to make all of Stalin's successors more secure by reversing the process that had previously assisted him in establishing his tyranny. Nor did the transfer at Beriia's initiative of foreign policy matters from the party presidium to the council of ministers alter the balance within the ruling team, all of whose members continued to sit on both.[14]

As the person responsible for security, Beriia presided over the critical first steps toward reform, which was both unavoidable and potentially dangerous. Initially, Khrushchev remembered, "we went on as before, out of inertia. Our boat just continued to float down the stream, along the same course that had been set by Stalin, even though we all sensed that things were not right."[15] By mid-March, however, the abrupt termination of the vigilance campaign and the anti-semitic slurs in the media, after which also public references to Stalin all but ceased,[16] signified the intention to right some of his worst wrongs.

The proclamation over Beriia's signature of an amnesty on March 27, which ended the arbitrary confinement of millions of people in prisons and labor camps, was a landmark by past standards but otherwise a timid harbinger of true reform. Providing for the release of prisoners sentenced to terms shorter than five years, it excluded most political offenders.[17] Additionally, lest they cause trouble to him or anyone else in power, Beriia tried to ensure that released political prisoners would be strictly supervised and preferably made to collaborate with the police.[18]

Stalin's former minions were mainly concerned with those of his abuses that affected them personally. On April 4 Beriia reversed the outrageous doctors' plot, though not entirely.[19] The announcement by his interior ministry rehabilitated most, but not all, of the alleged culprits —besides additional individuals who had not been publicly incriminated before.[20] Riumin, the chief executor of the fraud, was arrested and subsequently shot, whereas Ignatev, his former supervisor as minister of state security—the man whom Khrushchev had the month before preventively removed to the party secretariat—merely lost his new job because of nothing worse than "blindness and gullibility."[21]

The handling of this sordid affair showed the disposition of the oligarchs, all accessories to Stalin's crimes, to accommodate each other the best they could for the sake of their collective safety. If Beriia was the most visible among them, this was because of his responsibility for security matters that were the top priority. In April his peers also entrusted to him the overhauling of the administration in his native Georgia, allowing him to reinstate there his local retainers, sacked by Stalin in his "Mingrelian" purge.[22]

Whatever the inevitable disagreements among the oligarchs about such tricky policies, their positions and alignments have never been reliably identified. Malenkov's son Andrei describes his father as the main force for reform and an enemy of Beriia.[23] Yet assuming that he and Molotov were already at this time, respectively, the main proponent and opponent of reform at home as well as of accommodation abroad, means projecting their later positions backward.[24] According to Mikoian's son Sergo, it was in turn his father and Khrushchev who championed reform, while Beriia and others tried to drag their feet as much as they could.[25] Malenkov's former aides, however, have attributed the desire to resist change to Khrushchev, describing him as close to Beriia.[26] In this maze of self-serving attributions, the extent of the prevailing consensus gets lost. There was more of accord than of discord. The reform, or what substituted for it, rested on the oligarchs' common conviction that no more than minimal change was either desirable or prudent. This set the limits to their foreign policy as well.

Presuming Soviet readiness to settle not only the Korean War but also other important issues, Churchill wanted to test it at the highest level.[27] To Dulles, however, the Soviet signals did not give "any great comfort."[28] He was right, for the desire for change they conveyed was, indeed, strictly limited. Behind the closed doors of the central committee, Molotov himself disparaged his "so-called 'peace initiative'" as merely a ruse, explaining it to insiders as being designed to sow "confusion in the ranks of our aggressive adversaries."[29] Nor was Churchill's proposal for a summit regarded by the Kremlin leaders as anything but a Western ruse, Khrushchev reports, calculated to "wring some concessions" out of them before they had their "feet firmly on the ground."[30]

The immediate goal of Stalin's successors was retrenchment—recoiling for better jumping. Having decided to heal the Asian sore, they concentrated on their real priority, Europe, where Yugoslavia, Greece, and Turkey had recently joined in the Balkan Pact, indirectly linked to NATO. In April Molotov invited Yugoslav ambassador Dragoje Djurić for a surprise interview, after which the two governments patched up their dispute enough to allow for the resumption of diplomatic relations.[31] To the central committee Molotov explained what he was doing: "Since we did not succeed in settling the particular problem [of Yugoslavia] by a frontal assault, it became necessary to resort to other methods. It was decided to establish with Yugoslavia the same relations as with the other bourgeois states tied with the aggressive North Atlantic bloc."[32]

Wooing the other two members of the none too solid Balkan grouping, Moscow renounced the territorial claims that Stalin had gratuitously raised against Turkey in 1945 and restored diplomatic relations with Greece as well.[33] After difficult secret negotiations started at Soviet initiative in Bulgaria, it did the same with Israel, thus undoing the break that the dictator had so needlessly forced at the height of his anti-Jewish drive half a year before.[34] Thus his successors sought to rectify some of his most egregious mistakes, for which the country had been paying a price.

Given the Soviet premises, any hopes for a breakthrough in the East-West rivalry were bound to be disappointed even if the West had been accommodating, which it was not. This was more by default than by design. Although officials in Washington and London had been impatiently awaiting Stalin's exit from the scene, they were utterly unprepared when it came.[35] Having set their minds at its long-term consequences—the inevitable terminal crisis of the system no one else could effectively run—they were at a loss about what to do in the meantime. Without contingency plans for the not so unpredictable event of Stalin's incapacitation and death, they scrambled, drafting plans of action only after his illness had became public knowledge, though not implementing any one of them.

This was just as well, for each of the plans was less suitable than the other. They ranged from Churchill's pet project for a friendly chat with Stalin's successors to "turn the leaf"[36]—as if the previous experiences with this sort of personal diplomacy had not been sufficiently discouraging—to the wild ideas of Harold Stassen, the director of the Mutual Security Agency, about how to bring the communists to their knees. In his view the Kremlin would be seized with panic if only Washington could spread the rumor "that Mao Tse-tung will be the first one liquidated by Malenkov," or "that other leaders, naming them, are in fact marked for liquidation."[37]

On the dubious assumption that the new men in the Kremlin were ready for substantive negotiations, Churchill vainly tried to enlist Eisenhower for a tête-à-tête with them. As an alternative, presidential adviser C. D. Jackson had been promoting the plan for a conference of foreign ministers intended—as Khrushchev rightly guessed—to extract from the Russians substantial concessions because of their momentary weakness. Ironically, the State Department vetoed the plan—on the grounds that it might succeed, and cause the Western Europeans' lukewarm support for the projected European Defense Community to vanish altogether.[38] Thus the U.S. perception of persisting Soviet military threat, which the project was designed to counter, ruled out political bargaining with Moscow.

Eisenhower's long-delayed "Chance for Peace" speech of April 16, with its mixture of vague homilies and ill-chosen demands, was anything but an invitation for bargaining. On the one hand, it tried to lure the cynical Soviet chieftains to join hands with the West in stopping the arms race and waging instead a "new kind of war" on the world's "brute forces of poverty and need"—an idea sounding like a bad joke to them.[39] On the other hand, the particular conditions Eisenhower insisted they must fulfill if they wanted to prove their good faith— ending the Korean War, signing the Austrian state treaty, releasing the remaining German prisoners of war—were ill-suited to give the desired proof. Aside from the Korean War, which Moscow had already decided to end for its own reasons, the fulfillment of the other two conditions was contingent on its believing that it served its interests better than the Western ones. This could not possibly be in 1953. The two were only to be fulfilled in 1955 by Khrushchev after he would come to believe that by doing so he could make greater gains elsewhere.

When Bohlen, four days after Eisenhower's speech, presented his credentials as the new U.S. ambassador to Voroshilov flanked by Molotov's aide Pushkin, he referred to Korea as the "litmus test."[40] Yet the speech did not make the Moscow leaders more accommodating and may have even made them less so because of their concern about looking weak. Its unprecedented verbatim publication in the Soviet press on April 25 did not mark any concerted Soviet campaign for genuine rapprochement with the West, as has sometimes been argued.[41] Internally, Soviet officials judged the speech "irritating and provocative" rather than worthy of a serious response.[42] Compounded two days later by Dulles's public demand that the Soviet Union give up dominating Eastern Europe[43]—and thus in effect capitulate—the impression Moscow received was that of a U.S. administration unwilling to negotiate. But this correct impression merely reinforced the Soviet unwillingness to do so either. There was simply not enough trust on either side to give the other the necessary benefit of the doubt.

Following its own schedule, rather than responding to the presidential speech, the Soviet government had already before its delivery informed U.S. chargé d'affaires Beam of the forthcoming release of American citizens held by North Korea—another step on the road to armistice.[44] By then, regardless of the hitches that would occur before it would actually be signed on July 27, for the Soviet Union its signing was a foregone conclusion. While Washington's eyes were still fixed on Korea, Moscow was beginning to look beyond, preparing a new initiative in Europe. When Pushkin was listening to Bohlen talk about the litmus test, he and another of Molotov's helpers, Malik, had already been commissioned by the foreign minister to prepare a proposal that, far from intended to accommodate the Americans, envisaged exploiting the apparent disarray in their alliance.

On the premise that the recent ratification by Bonn of the treaty on the European Defense Community had provoked "serious disagreements" between Washington and its allies, the plan was to present the West with a new proposal on Germany that would presumably be difficult to resist.[45] Its novelty consisted in proposing the formation of a provisional government for the whole country, to be followed by elections that would lead to the creation of a permanent one, until which time the governments currently in power in the two parts of Germany would continue functioning.

In preparation, East German representatives were to be invited to Moscow for a ceremony where they would receive the trappings of sovereignty for their state, whose pretension of independence would be simultaneously boosted by the elevation of the Soviet diplomatic mission to an embassy. The officials working on the plan expected as its final outcome a four-power conference that they thought could be convened as early as June to conclude with Germany a peace treaty satisfactory to their bosses.

The plan indicated that Molotov, like Churchill, entertained illusions about his ability to induce the other side to negotiate, although the expected results of the negotiations he hoped to set into motion were quite different from, and certainly incompatible with, the Western expectations. In again trying to coax the Western governments to collaborate against their will in a German settlement the Soviet Union wanted, the plan was little more than a rehash of Stalin's futile attempt to achieve the same with his March 1952 note. It attested to the persistence in Moscow of the old illusions as well as their purveyors, notably Semenov, Stalin's and now Molotov's chief German expert, who revealingly commented on the merits of draft.

Semenov admitted that even though this time the Western powers appeared to be weaker and more divided, and consequently less resistant, they were unlikely to yield easily. If he nevertheless recommended the plan to Molotov, he did so in trying to justify the continuation in East Germany of the "accelerated construction of socialism," started by Ulbricht in July 1952 and since then promoted under Semenov's principal supervision. As a result of the policy with which Semenov had been so prominently associated, in his view the communist state had reached the level of maturity and strength that it finally deserved to be treated as a "socialist" state.[46] He was thereby advocating East Germany's full integration into the Soviet bloc, an ardent wish of Ulbricht's and his companions' that under Stalin Moscow had been notably reluctant to grant.[47] Semenov's summons to Moscow on April 22,[48] to participate there in the ceremonies marking the planned transfer of power to the East Berlin regime, implied official endorsement of his views.

Rather than seek a compromise settlement about Germany as a whole, the Kremlin leadership thus reaffirmed the course Semenov had been supervising in its eastern part. This was the old Stalinist course that Ulbricht had been so eagerly ramming through. He was being praised for doing so by the local representative of the Soviet party central committee, who singled out for particular commendation his regime's intention to implement more rigorously the "lessons" of the Slánský trial, which it had so diligently spelled out in a special document already the previous December, while Stalin was still alive.[49] The more rigorous implementation, as outlined by the central committee in its follow-up declaration on May 14, meant an expanded hunt for "saboteurs" and other miscreants who had supposedly wormed their way into the party. Mentioned by name was disgraced politburo member Franz Dahlem, thus earmarked as the sacrificial lamb in a prospective show trial.[50]

What caused the Soviet Union to subsequently reverse the Stalinist course it had been abetting in East Berlin is as intriguing as it is crucial for explaining the larger question of why the leadership consensus in Moscow did not last. As late as May 20, the newspaper of the Soviet military administration, *Tägliche Rundschau*, still endorsed the old course, praising the recent party declaration and exhorting communists to be vigilant and smite all enemies.[51] Yet the replacement on May 22 of Semenov as political adviser to the military administration inaugurated not only a precipitous dismantling of the most resented policies and institutions established by the Ulbricht regime during the past year, but also a possible incipient repudiation of the substance of communist rule, as suggested by the officially instigated revival of other political parties.[52]

By then East Germany's mounting economic crisis had made a change of course imperative. The massive flight of its citizens to the West—100,000 people in March alone—thoroughly refuted Semenov's rosy estimate of the Ulbricht regime's stability.[53] Since Semenov was closer to Molotov than was his successor, Iudin, who was reputedly close to Beriia, the reshuffle might be cited in support of the appealing theory that Beriia, acting as "rational and purposeful ... policeman turned liberal,"[54] was behind the change. Yet not only was Iudin a Stalinist to his core but so was Beriia, as shown in his conduct in the erupting crisis, thus making the explanation of what happened not that simple.

Beriia and the German Crisis

Ever since the beginning of May, if not before, trouble had been brewing for the security chief. For the first time ever, the Soviet prison empire was shaken by a rebellion of its inmates, thus putting his competence as its manager in doubt. Not suppressed swiftly enough at a time when the Soviet empire's new rulers were feeling, as Khrushchev put it, "terribly vulnerable," and to that extent unsure of themselves, the rebellion spread; accordingly, as soon as they felt better, they had no qualms crushing it with the customary brutality.[55] Beriia was no exception in their trying to keep Stalinist practices as intact as they could get away with. For example, in trying to reform the outrageous practice of summary proceedings that gave the police the authority to interrogate, prosecute, sentence, as well as imprison anyone at will, he proposed no more than the reduction by ten years of the twenty-year maximum imprisonment allowable under the system.[56] As if nothing had changed since Stalin's times, he tried to keep the power of his apparatus, and thereby his own, unimpaired and as free as possible from both government and party interference.

Unrest was also spreading in the Soviet Union's ethnic republics, straining the quasi-colonial relationship between the Russian-dominated central government and the nominal republics of other nationalities, among which the Baltic ones were the most advanced and the Ukraine the largest. More sensitive than his colleagues to the explosive potential of pent-up nationalism, Beriia acted to suppress the nationalist underground still operating in the restive provinces while aiming to defuse the pressure by putting more non-Russians into responsible positions there, as he had done in his Georgian fiefdom.[57] He was instrumental in having the central committee pass on May 26 resolutions providing for more employment of the native cadres in Lithuania and the western Ukraine.[58]

The policy was reasonable enough, but its implementation was a disaster. Beriia tried to replace not only the local security personnel, which he was entitled to do, but also the party bureaucracy, which he was not. He attempted to remove the mainly Russian officials long ensconced in their places while reinstating those, many of them Jewish, purged in Stalin's recent anti-Semitic drive. In the Ukraine the reshuffle threatened people appointed by Khrushchev while he was in charge promoting the drive there. One of them in Lvov, named Timofei Strokach, complained to him about having been shouted down by Beriia and threatened to be "thrown out, locked up, made to rot in the camps, and ground into the camp dust."[59] In Beriia's treatment of subordinates, such was the obverse of his servility to Stalin.

According to one of his cronies, the deputy chief administrator of those camps Amaiak Z. Kobulov, Beriia wanted to "decisively break the old order not only in our country but also in the [people's] democracies."[60] True enough, as long he was in charge, the purge trials in Eastern Europe, which threatened his agents there, ceased. Yet there is no evidence of his having a concept, much less a policy, of reform beyond trying with his collegues to limit it to the necessary minimum. He was anything but a closet liberal. A spoiled product of a rotten system, Beriia was manifestly unfit to be a champion of its radical transformation. Having spent a lifetime as a despot's administrator of terror, he lacked the prerequisites of a politician, much less of a statesman. He was one of history's more unsuitable persons ever given an opportunity to shape it.

As the manager of the Soviet intelligence and counterintelligence networks, Beriia was strategically placed to utilize them for better or worse. He chose to do the latter—with singular incompetence. Having confided to Malenkov his view that the Soviet Union's recent normalization of governmental relations with Yugoslavia should be extended to collaboration between their security services, he received no encouraging response. He nevertheless proceeded to prepare on his own a message to his Belgrade counterpart Ranković, proposing a secret rendezvous with Tito, while trying to keep the rest of the leadership in the dark about his intended escapade.[61] His other meddling with foreign affairs betrayed the same spymaster's notion of politics as little more than backstage intrigue.

Beriia showed a particular interest in the German question.[62] He requested from his intelligence analysts an assessment of the foreign policy views of West Germany's opposition social democrats and was dismayed to receive the safe, if inaccurate, reply that the views did not substantially differ from those of the government.[63] If, as Sudoplatov maintains, Beriia was interested in exploring the feasibility of a united neutral Germany,[64] he was doing so most clumsily. Summoning most of his foreign operatives to Moscow for consultations, he reduced drastically the number left in the country precisely when they were needed there the most to monitor the deteriorating East German situation.[65]

The Soviet estimate of that situation was accurate enough to warrant the convocation of an emergency session of the ruling group on May 27. During this important meeting, Beriia shocked his colleagues by suggesting not only that "the course of building socialism in the GDR be abandoned at the present time" but also that a neutralized Germany would be preferable to a socialist one. Although this was nothing other than Stalin's original idea, it had by then been superseded by Soviet commitment to the build-up of the communist German state, most recently reiterated in the proposal Molotov had prepared in his ministry. Both he and Khrushchev fiercely opposed Beriia's suggestion, insisting upon a mere slowdown rather than the abandonment in East Germany of the "construction of socialism"—a change to be epitomized by dropping the word "accelerated" from its official description.[66]

The disagreement was not as serious as it looked. Having found himself in the minority of one, Beriia withdrew his proposal and submitted to the majority.[67] He still fretted to Bulganin the next day about the danger Germany might pose if it were not neutralized, grumbling that some of the presidium members should be removed from the government to make this possible—a broad allusion to Molotov.[68] Yet Beriia took no action beyond his dangerously loose talk, thus substantiating Molotov's retrospective assessment of him as an arrogant man but a poor politician, "lacking deeper interest in fundamental policy questions."[69]

Having regained consensus, the Soviet leadership proceeded to unilaterally implement in East Germany the measures aimed to lead to the solution of the German question as outlined in the foreign ministry proposal, which had originally been intended to be dispatched by then to the Western capitals but was shelved instead.[70] Stabilization of the shaky Ulbricht regime had meanwhile become more topical than the destabilization of the overall German situation that was the gist of the proposal. On May 28 Semenov received his promotion by being appointed to the newly created post of high commissioner, while Iudin kept his new job, too, thus adding to the evidence that any dispute with Beriia had been laid to rest. Further following the blueprint, three days later a top-level East German delegation arrived in Moscow.

Rather than to prepare for any subsequent negotiations with the West about Germany as a whole, however, the purpose of the visit now was merely to prevent the impending collapse of the regime in its eastern part. Summoned rather than invited, its representatives arrived under a cloud, further darkened by the riots erupting in both neighboring Czechoslovakia and distant Bulgaria, Moscow's staunchest allies. Making matters even worse, it was workers who had gone rioting— both in the Bulgarian tobacco factories and at the huge Škoda armament plant in Plzeň.[71]

Provoked by the draconian currency reform of May 30, the unrest in Czechoslovakia spread throughout the usually quiescent country until as many as 129 strikes were reported. On the day the East German delegation reached the Soviet capital, twenty thousand demonstrators marched through Plzeň, raising American flags in memory of its liberation by the U.S. Army in 1945. Having lost control of the situation, the authorities had to bring in special police in order to restore it.[72] All this looked as if the collapse from within that the Western secret services had for so long been hoping for and trying to precipitate might finally be coming.

Compensating in his way for the recent blows to his prestige as the chief guarantor of public order, Beriia treated the visiting East Germans so despicably "that it was embarrassing to listen."[73] He rubbed it in that theirs was "not even a real state. It's only kept in being by Soviet troops, even if we do call it the 'German Democratic Republic.'"[74] Dressed down, as used to be Stalin's habit, for having only done what Moscow had wanted them to do, they were asked to put down in writing how they proposed to deal with the shambles. No sooner did they finish than the impossible Beriia exploded again, dismissing the product of their labors as nothing but a "bad replica of our own document."[75] More politely, Kaganovich explained that "our document favors a turn, yours a reform."[76]

The difference was not that between keeping and not keeping communist East Germany, which had been at issue in Beriia's clash with his colleagues at their meeting the week before; after he had realigned himself with them, what mattered rather was the distinction between real and cosmetic change. In preparing the final document[77] as the basis for a resolution to be adopted in Berlin by the party central committee and subsequently published, the Soviet leaders wanted the text to set unequivocal guidelines for the recovery of the sick state, without being mistaken for wanting something else. Hence they resisted their guests' desperate attempts to camouflage the turnabout as a tactical concession to impress the West,[78] lest it backfire by being misread as an invitation for negotiations on German reunification. This was not a topical subject until the expected recovery would put Moscow into a position to negotiate from strength rather than from weakness.

The foreign ministry proposal had been shelved after Churchill, responding to what he misinterpreted as Moscow's encouraging reaction to the Eisenhower speech, had on May 11 preempted the pending Soviet initiative by himself calling for a conference of the great powers.[79] Insisting that their security needs were compatible, he outlined a possible settlement modeled after the 1925 treaty of Locarno, meaning international guarantees of Germany's frontiers, to prevent any future aggression on its part. The lukewarm Soviet response disappointed the British.[80] Never keen on Locarno—which had failed to discourage German aggression before—Moscow welcomed the idea of compatible security but otherwise merely repeated its routine call for a speedy German settlement without adding anything new about how to achieve it. In view of East Germany's dire condition, nothing more could be expected.

The "New Course," inaugurated less dramatically also in other countries of the Soviet realm in anticipation of a coming crisis, sought, as did the limited reform course in the Soviet Union proper, to overcome the crisis by revitalizing the regimes in power. A week after the East Germans, the Hungarians were called to Moscow and told to start reforming, giving Beriia another occasion to show off his antics. Mocking the Budapest party boss Rákosi as a "Jewish king of Hungary,"[81] he wisecracked that the country may have had "Turkish sultans, Hapsburg emperors, Tatar kings, and Polish princes but . . . never . . . a Jewish king," and vowed that "we won't allow it." In the end, he and his colleagues did allow the disreputable Rákosi to keep his job of party secretary but made him turn over the responsibility for reform to the less ambitious and more attractive Imre Nagy as the head of the government.

Deferring to Soviet wishes, the draft of the resolution submitted for approval to the East German politburo meeting on June 5 envisaged slowing down forced collectivization of agriculture but not dismantling the farms already collectivized. Taking its cue from Lenin's New Economic Policy of the 1920s, it was calculated to encourage private enterprise but reserve the "commanding heights" to the state. It promised to end the Stalinist terror and rehabilitate its victims but not punish any of its principal carriers, including particularly Ulbricht.[82] It still amounted to "the most far-reaching change of policy that ever took place in a communist country" until that date, as well as the riskiest.[83]

The politburo, convened under the watchful eye of the ubiquitous Semenov, was acutely aware of the risks, and so was he. Its members, notably Rudolf Herrnstadt, the editor-in-chief of the party daily *Neues Deutschland*, lambasted Ulbricht for his famous arrogance but stopped short of stripping him of power. Semenov soft-pedalled the criticism of Ulbricht, which applied just as well to him. Herrnstadt, having drafted the final text of the resolution finally adopted on June 9, worried lest its immediate publication shake the state to its foundation. He therefore urged that it be kept confidential for at least another fourteen days, to give party propagandists enough time to condition the population for the coming shock. But Semenov, warning that "in fourteen days you may not have a state any more," would have no delay.[84]

Semenov was right. The publication of the document on the next day set off shock waves culminating a week later in the revolt of the Berlin workers that indeed threatened the rickety state with extinction. At the same time, the uprising prepared the acid test of both the viability of the policy of controlled reform on which Stalin's successors had staked the survival of the Soviet system and the credibility of the proclaimed American readiness to challenge that system by exploiting its vulnerabilities. The outcome was determined by Soviet action and Western inaction.

Once masses of demonstrators swept through East Berlin on June 17, CIA officials in the western part of the city pondered supplying them with rifles and sten guns, but were quickly vetoed by subordinate officials in Washington while the head of the agency, Allen Dulles, was out of town.[85] Leaving aside its merit, and there was certainly enough merit in not trying to precipitate unnecessary violence, the decision—or the lack of it—proved how little both he and his brother John Foster, the Secretary of State, believed in the possible crumbling of Soviet power that their declaratory policy of rollback and liberation was supposedly designed to encourage. The British, not to speak of the French, at least made no pretensions. More than that, Churchill stated bluntly, if privately, that if "the Soviet Government, as the occupying Power, were faced ... with widespread movements of violent disorders they surely have the right to declare Martial Law in order to prevent anarchy."[86]

Belatedly, the U.S. National Security Council, convened in a hurry, directed the Psychological Strategy Board to prepare for exploiting the situation, and the president was prompted to approve the formation of the half-forgotten Volunteer Freedom Corps of Eastern European émigrés, authorized by the Kersten amendment nearly two years earlier.[87] Yet before anything worth doing could be done, the uprising was crushed, after which Secretary Dulles publicly went on record to deny not only any U.S. responsibility but, awkwardly, also any intent to instigate unrest that might lead to violence (as if one could tell in advance), thus exposing his policy as a sham.[88]

In retrospect, it is easy to see that the outcome was predetermined; at the time, however, neither the American inaction could be reliably predicted, nor could the forceful Soviet action after Beriia had floated the idea that the East German regime was not worth propping up. Now the danger of its being toppled made him a good soldier again, ready to support its rescue without reservations and—if Semenov's self-cleaning memoirs, full of tall stories, could be used as a source—even scold this down-to-bone opportunist for waiting too long in "saving the bullets."[89]

As a result, Stalin's heirs weathered their most severe crisis since his death united rather than divided. This did not make them any more, but if anything less, accommodating. Hence also the indefinite postponement of Churchill's plan for a summit because of his incapacitation by the stroke he suffered on June 26 was no opportunity lost. If previously the Soviet leaders had not been sufficiently secure to dare to negotiate, now they were not sufficiently insecure to feel compelled to. This time the U.S. intelligence estimate got it right in grasping that Moscow would "offer no real concessions to effect a settlement" with the West.[90]

The basic continuity of Soviet policy on East Germany before and after the suppressed uprising conveyed enduring consensus at the highest level. Instead of discarding there the new course that had failed to prevent the disturbances, Moscow abetted preparations for trying harder to implement it. The special commission formed in East Berlin to carry out the reform program outlined in the June 9 resolution entrusted to Ulbricht's critic Herrnstadt the elaboration of a plan of action.

The resulting document rightly attributed the party's recent debacle to its policies under Ulbricht, setting as the goal East Germany's transformation into "a model democratic state"[91]—more precisely, a state governed with the consent of the people that Ulbricht had so disastrously forfeited. The critical question was whether the cynical men in the Kremlin would share Herrnstadt's naive optimism about the party's aptitude to accomplish such a feat and be willing to reject Ulbricht's alternative interpretation of the June 17 uprising as nothing but a Western plot. That even some Soviet officials on the spot who could see with their eyes what was happening were prepared to believe this fairy tale was not encouraging.

Without mentioning Ulbricht by name, the Herrnstadt commission recommended his demotion by abolishing his party job. Independently, the head of the military administration Sokolovskii, Semenov, and Iudin proposed to their Moscow superiors merely his dismissal from his government post of deputy prime minister while allowing him to run the party. The special subcommittee in the Soviet foreign ministry that was to decide took no action on either proposal.[92] As a result, Ulbricht kept his power despite his failure to prevent the debacle of his policy; the one who lost power despite Moscow's success in overcoming it was Beriia.[93] His ouster on June 26 came as a bolt out of the blue.

The Saving of the Soviet System

The ejection of so senior a leadership figure as Beriia by his peers was unprecedented in Soviet history since Stalin had engineered the downfall of Trotskii, which, however, had come gradually and could be predicted. It would be easier to explain why Beriia could not possibly be ousted than it is why he actually was. The risky removal of the man who held the levers of the formidable Soviet security apparatus by a group at pains to demonstrate its solidarity could only have been undertaken for the weightiest of reasons. Yet these have never been satisfactorily explained and can only be reconstructed from fragmentary and distorted evidence unlikely to be ever supplemented with anything more substantive.

The arrest of Beriia at a secret session of the party presidium on June 26 was organized by Khrushchev on a very short notice—perhaps no more than a week or two—with the help of trusted military officers after he had obtained more or less willing consent by key members of that body.[94] Rather than being immediately disposed of, however, the dangerous captive was kept alive in jail for six months while time was wasted on interrogating him for the dubious purpose of preparing against him the same kind of phony charges that he and his henchmen had been so proficient in concocting against others in Stalin's times. It looked as if those who deposed him had subsequently found the reasons for their having done so not sufficiently substantiated.

The few people who have been allowed to peek into the forty-odd volumes of the interrogations and other documents collected to incriminate Beriia have found there little of interest.[95] Eventually, the material was not needed anyway. At the end of December 1953, he was tried in camera by a kangaroo court under the Stalinist procedure that did not even require the presence of the defendant, much less any serious examination of evidence. What it did provide for was the prompt execution of the death sentence prepared in advance by the court.[96] The case against Beriia, if any, thus seems to have been built on the flimsiest of foundations, thus making the reasons for his misfortune even more puzzling.

A week after Beriia's arrest, its organizers went to great lengths trying to explain themselves to the secret meeting of the central committee they had convened for this purpose. A much shorter version—much too short—appeared on July 10 in the official notice of his demise, intended for the public. Yet another explanation was subsequently offered in special briefings for the understandably distraught representatives of foreign communist parties, summoned to Moscow to be reassured.[97]

The accusations leveled against Beriia included nothing less than "plans to seize the leadership of the party and the country" in trying to put into effect a "capitulationist policy which would in the last analysis lead to the restoration of capitalism."[98] Yet rather than focusing upon these grave charges and supporting them by convincing evidence, his detractors diverted attention from them by adding a hodgepodge of disjointed and improbable wrongdoings, to which his suitably manipulated interrogation was presumably expected to later give a semblance of coherence and plausibility.

The real issues of the case have to be read between the lines of the record of what was said and left unsaid behind the closed doors of the meeting of the central committee. As used to be Stalin's habit, Beriia was affixed with responsibility for assorted past misdeeds which those currently in power needed to retrospectively account for. They concerned mostly what had happened before rather than after Stalin's death. Many of the charges, such as Beriia's complicity in the 1949 Leningrad purge, were fair enough; others, notably his alleged collaboration with Western secret services ever since the 1920s, were ludicrous. Studiously ignored were those of his wrongdoings that implicated also his accusers, particularly Khrushchev, whose silences were more revealing than his explications.

Conspicuously missing from the long list of Beriia's abominations was any hint of foul play during the murky last weeks of Stalin's life, although little imagination would have been needed to invent some. It was never officially suggested that he might have had a hand in shortening the tyrant's life; the farthest Khrushchev ever went in that direction was to report his indignation at Beriia's openly rejoicing at Stalin's death.[99] If the scoundrel did have a hand in precipitating it, and his former companions knew that they were indebted to him for the deed, the less said about it the better.

The evidence that was offered to prove Beriia's alleged later conspiracy was of the most spurious kind. It consisted of half-truths turned against him rather than outright lies. His alleged manipulation of collective decisions for his evil purposes concerned intentions rather than their implementation. He was portrayed as the caricature of a conspirator rather than a real one. Depicted as having consulted about his schemes in advance with the very persons against whom these were presumably directed, he was shown as especially inept at trying to win any confederates.

Within the ruling group, Khrushchev alone had a personal grudge against Beriia. Leaving aside what the best-informed man in the country may have known about Khrushchev's possible complicity in Stalin's murderous last machinations, Khrushchev's protégés in the Ukraine had been Beriia's targets. Of all the collective decisions that he was said to have abused, the only one that was subsequently rescinded was the decision providing for more employment of local cadres in the ethnic republics.[100] Still, the written complaints from the provinces—all dated after Beriia's arrest—that were collected by Khrushchev to document the alleged abuses have the appearance of plants commissioned retroactively.[101]

The available evidence is that of a conspiracy *against* Beriia rather than *by* him. Its time frame was the crisis that climaxed in the East German uprising. Yet Beriia was never blamed for it, nor was he held responsible for any of the policies the leadership had pursued while he was still in good standing. He himself alluded to the common responsibility for anything that had gone wrong by telling the visiting East German officials, "We have all made the mistakes together, so no reproaches."[102] And Khrushchev seconded by telling their Hungarian counterparts on another occasion: "We were there, too, when these errors were ascertained, every one of us!"[103]

While he was still ascendant, Beriia made himself vulnerable by his actual or suspected actions, which, if allowed to run their course, were bound to be alarming for the rest of the leadership. They were hinted at in Khrushchev's repeatedly referring to him by using the word "adventurer" (*avantiurist*) as the one most descriptive of what was wrong with Beriia. Apart from his controversial activities in the ethnic republics, he was said to have been gathering material about the country's defenses, particularly its fleet, and summoned his foreign operatives to Moscow —all this for no obvious reasons.[104]

According to Sudoplatov, Beriia acted on what he reports was Stalin's intent to bolster the Soviet capacity for sabotage and subversion abroad while also trying to increase the armed forces' readiness for preemptive strikes against Western targets. Sometime in May, he supposedly went so far as to probe NATO's defenses by dispatching a reconnaissance plane to survey its installations in northern Europe.[105] If he was indeed prone to indulging in such risky experiments, his colleagues were right to be concerned. So were they if they suspected, as they said they did, his wanting to take it upon himself to secretly proceed with the test of the first Soviet hydrogen device.[106] In any case, on the very day of Beriia's arrest, the responsibility for the nation's nuclear program was transferred from the committee he had been chairing to a newly created ministry of "medium machine-building."

Khrushchev told the central committee that Beriia, "devil knows, may have even been receiving assignments from foreign intelligence representatives," vaguely relating this suspicion to the dispute about Germany.[107] Preposterous as this may seem, Beriia would not have been the first person of his ilk to entertain exaggerated notions about his suitability as a credible interlocutor with the West; so did Hitler's no less disreputable security chief Heinrich Himmler when he approached the Allies with an offer to mediate in the final days of World War II. From the Western side, however, no indication of any approaches by Beriia has so far come to light, except, perhaps, that document purportedly from Stalin's office, showing Soviet complicity in launching the Korean War, received by the U.S. intelligence just at the right time—in April 1953—albeit without any known attempt at a follow-up.[108]

If the intention was to reassure the West about Soviet good will—which the post-Stalin leadership was so notably reluctant to do—a more suitable messenger would have been Ivan M. Maiskii, Moscow's well-regarded wartime ambassador to London, arrested shortly before Stalin's death. Again, according to Sudoplatov, Beriia wanted to haul him out of prison to replace Molotov as foreign minister in one of the most absurd intrigues ever heard of—by soliciting assistance from Molotov's own wife![109] Whatever were the limits of stupidity ascribed to this amateur conspirator, even after Beriia's fall from power his enemies still had reasons to keep Maiskii incarcerated and indeed sentence him for treason. But his release soon afterward, followed by his elevation to the Academy of Sciences, amounted to the retrospective admission that his putative involvement in any scheme by Beriia had been nothing but a figment of their imagination.[110]

What Beriia intended to do, however, was less important than what he had become capable of doing. He had gathered an impressive power, and under the system inherited from Stalin no one could be certain about what he might try to do with it. Even if he did not want to use it against his peers, and no conclusive evidence has been offered that he did, there was enough disconcerting evidence of his ineptitude in wielding it. In the last analysis, it was not so much his lack of scruples —which was taken for granted—as this dangerous mixture of power and incompetence that the embattled leadership could not afford to tolerate.

By keeping after Beriia's downfall the essentials of the course he had been instrumental in inaugurating, his enemies confirmed that any threat he may have posed to them was potential rather than actual. In particular, the crucial separation of the party and the government responsibilities, implemented at his initiative to make excessive accumulation of power and its abuse more difficult, remained in effect and was extended to Eastern Europe.[111] As a secondary, though by no means negligible, benefit of Beriia's removal, shifting onto Stalin's most detestable henchman most of the responsibility for his criminal rule that his remaining other henchmen shared helped them to divest themselves from those of the despot's practices that most urgently needed to be repudiated—for their own safety, if not for loftier reasons.[112]

The changes attributable to Beriia's demise concerned merely persons, not policies or institutions. No cleansing of the Soviet security apparatus, except for his most intolerable cronies, took place.[113] Even in East Germany, where most of the changes occurred, they only entailed careers of individuals. Adroitly exploiting Beriia's downfall to ensure his own political survival, Ulbricht dropped dark hints about liaisons with Beriia of his nemesis Herrnstadt and, more plausibly, Wilhelm Zaisser, the East German security chief, on whose neglect, if not something worse, could be blamed the responsibility for the June 17 disaster that the document prepared by Herrnstadt had properly placed on him. Having received Moscow's approval for the expulsion of both Zaisser and Herrnstadt from the party, Ulbricht managed to persuade the central committee to scuttle the document and substitute it with his own version. He succeeded in slowing down and eventually side-tracking the implementation of the June 9 resolution, although he did not reinstate the old course.[114] In Hungary, too, Rákosi tried, though unsuccessfully, to sabotage the new course and get rid of Nagy by insinuating his connections with Beriia.[115]

The successful plot against Beriia allowed his former companions to continue more safely without him on the same course they had previously embarked upon with him. Their successful handling of the German crisis helped—as did the willingness of their Western adversaries to leave them alone. Exposing the emptiness of the confrontational Republican rhetoric, the Eisenhower administration proved at a critical moment more reluctant to challenge the Soviet leaders than they previously had reasons to expect. The defeat of the East German uprising reinforced in Washington the belief that "the detachment of any major European satellite from the Soviet bloc does not now appear feasible except by Soviet acquiescence or by war."[116]

This was true but misleading. After intensive study, U.S. intelligence experts concluded that, regardless of persisting popular dissatisfaction, communist control in East Germany remained firm—a conclusion which served them poorly to be better prepared for the next revolutionary outbreaks that were on their way in other parts of the Soviet bloc.[117] The president, regarding aggressive political warfare incompatible with America's traditional concepts of war and peace, rejected as unacceptable the "Solarium" plan that he had secretly commissioned to examine the possibility of using nonmilitary means in trying to weaken the Soviet grip on Eastern Europe.[118] If the U.S. budget for covert operations nevertheless increased, this was only "to keep the pot simmering but not to bring it to a boil."[119]

Nor was Washington's rigid new strategy of massive retaliation, which went to the other extreme by proclaiming a readiness to respond to any Soviet aggression by all means at America's disposal, including nuclear weapons, much for the Soviet leaders to worry about if they had no intention to attack in the first place. Instead, the American preoccupation with military contingencies allowed them more flexibility in themselves challenging the West by more aggressive political warfare while taking another look at the utility of their own growing nuclear assets. In August 1953, they conspicuously advertised the successful testing of the hydrogen device that Beriia had supposedly wanted to hide.[120]

What was not advertised was the simultaneous review behind the scenes of the Soviet security doctrine with a growing emphasis on the importance of preemption—another reported Beriia initiative that was implemented rather than discarded should war become imminent for any reason.[121] Yet in the Soviet perception its likelihood diminished as the review rejected Stalin's fallacious "law" of its inevitability as well as his dated theory of "permanently operating factors," with its reliance upon conventional military forces backed by a regimented domestic front.

The theory had made Soviet security hostage to the country's old weakness—its repressive political system—rather than allowing its new leaders to offset that weakness by building on what was becoming its new strength: its growing nuclear might. Their incipient departure from Stalin's intransigent, but static, security doctrine toward the more dynamic strategy of a nuclear competition precluding neither limited accommodation abroad nor limited reform at home heralded the beginning of a new era, but not the end of the Cold War.

In the critical months that followed Stalin's death, his heirs proved both unable and unwilling to fundamentally reform the system whose products they were. Nor did they try for any breakthrough in the East-West confrontation that, too, was one of its products. They may have been compelled to act differently if their internal crisis culminating in the East German revolt had coincided with irresistible Western pressure. Barring this, they managed to overcome the crisis and, having transferred to the dead Beriia the onus of their dirty past, gave the Soviet system a new lease on life. In the fullness of time, the system would mellow as Stalin's criminal legacy would turn into the more benignly corrupt legacy of Brezhnev, bringing into power Gorbachev, a reformer no longer burdened by Stalinism, who would in turn bring the Cold War to an end. But before this could happen, the Soviet system first had to be disposed of, not merely reformed.

Conclusion:
The Soviet Threat in Retrospect

The Soviet sense of insecurity that bred the Cold War also provided the constraints that kept it within bounds. The inside evidence of Moscow's capabilities and intentions no longer leaves a doubt that its leaders never wanted to overstep the limits. This is not to say that the threat the West perceived was an empty one. In their quest for security, Stalin and his successors were inclined to take greater risks whenever they saw the correlation of forces turning in their favor. In estimating their own strengths and the weaknesses of their adversaries, they were prone to miscalculations. These were enhanced by their ideological preconceptions, which postulated the ultimate victory of their system despite temporary setbacks.

In the course of the forty-year contest, domestic considerations determined Soviet international behavior far more than most contemporaries, misled by the Kremlin's not having to account for its action to anyone, were prepared to believe. As long as Stalin was in charge, those considerations were more general than specific—his need to maintain his autocratic power and an economy that would sustain it. Only later did they entail a clash of specific interests resulting in alternative policies articulated by different individuals or groups. During Stalin's lifetime, Soviet policy was for all intents and purposes his policy, on the whole conforming to Khrushchev's description of the despot as someone behaving "like Almighty God with a host of angels and archangels. He might listen to us, but the main thing was that he spoke and we listened. He did not explain his reasoning, but passed down the word to lesser mortals. They did what they were told when he wanted it done."[1] This was the time when the Soviet system came the closest to the ideal model of totalitarian autocracy; in the real world, however, the autocrat's control was less than total, his policy often inconsistent, and its results not always commensurate with expectations.

As the Soviet empire spread, it became more difficult to manage simply because of its sheer size and complexity. There were more problems than someone of more systematic working habits than Stalin could effectively handle in one day—and his preferred nighttime schedule was anything but normal nor was his reliance on superior memory rather than on paperwork a guarantee of efficiency. The situation sometimes allowed subordinates at the intermediate level—rather than those in his main entourage—to exert a greater impact on policy than might be expected from the nature of the system, though never against his will.

In trying to divine Stalin's wishes from the not always clear signals those "lesser mortals" were receiving from him, subordinates applied the Marxist categories that they held in common with their superiors. In doing so while preventively protecting themselves against accusations of slackness, they were prone to err on the militant side. Their resulting initiatives were sometimes rejected as too risky, sometimes adopted; when the often indecisive despot was unable or unwilling to decide, they might be implemented by default.

In waging the Cold War, Stalin relied mainly on the *political* cohesion of the terroristic system he had created and its presumed comparative advantage over what he regarded in Marxist terms as inherent disarray among his capitalist enemies—only to see these unite as never before and himself bequeath to his heirs a system both untenable in its original form and unreformable in the long run. Afterward Khrushchev counted more on what he believed was the *economic* superiority of Soviet-style socialism—its supposed ability to outperform Western capitalism—only to strain the inherited system by lopsided expenditure of scarce resources for short-term effects and risky improvisations that undercut its performance and eventually cost him power. Under Brezhnev and his weak successors, the Soviet Union finally came to bank for all intents and purposes mainly on the strength inherent in its *military* power—to attain parity and preferably superiority in advanced weaponry in relation to the United States, thus offsetting its own diminishing capability to compete in nearly all other fields—only to strain the backward system to the extent of effectively paralyzing it. In the end its last manager, Gorbachev, presided over its collapse after realizing that the prohibitively expensive military might could not be translated into political strength.

Ironically, the Soviet Union was the least dangerous when it appeared to be the most—at the peak of the Cold War, when it acted the most hostile. The paradox was due not so much to U.S. nuclear superiority, whose utility for policy Stalin rightly discounted, as to his sense of internal and external vulnerability that existed for other reasons. He could neither bring his despotism to the level of perfection he considered necessary in order to feel safe nor could he believe that his capitalist enemies would ever be appeased. Somehow and some-where, the insurmountable tension would therefore lead to a show-down, for which his country was not yet prepared.

Stalin's foreign policy has been aptly described as not so much "inexplicable in its parts as incoherent in its whole."[2] This was the result of its being so thoroughly dominated by a single man who managed to project the image of prescience and omnipotence but was in fact a mere caricature of the Almighty God whom he tried to imitate. The more that is known about Stalin, the less he looks like the shrewd calculator and hard-boiled realist, for which supposed qualities he was respected even by his adversaries. He was given to illusions and wishful thinking to an extraordinary degree, repeated-ly overestimating the extent of discord among his adversaries that might allow him to dictate his terms from a position of relative strength—something that his country never attained and could not have attained.

This chronic error of judgment accounted particularly for the per-sistence of Stalin's belief that Germany could be united on the foun-dations established in the eastern part of it that he controlled. Admittedly, the German question would have taxed the dexterity of any Russian statesman. More important than the magnitude of Stalin's problems, however, was his propensity to pile them up and make them worse by bungling. Such blunders as the Berlin blockade or the row with Tito would have been sufficient for Stalin to be sacked if he had been the head of a responsible government. Since he was not, he could afford to merely cut his losses and proceed with more blunders—none greater than his lending support to Kim Il Sung's adventure in Korea. His opening of the "second front" in Asia in collusion with the Chinese was a development of much greater importance than outsiders had grown accustomed to believe before the vast new evidence of its inner workings became available.

The Korean War was important not so much because it strained Soviet relations with the Chinese—which it did only in the long run, whereas at the time Stalin's control over them and their willing submission to him were much greater that they later cared to admit—as because Stalin was conducting by proxy a real war, not only the cold one, against the United States—and could not win it. The benefits he derived from its stalemate paled in comparison with his growing frus-tration, so evident during his last years of life.

By then, Soviet foreign policy all but came to a standstill as he concentrated on solving all his real and imaginary security problems at home. His manner of doing so was singularly counterproductive. Hav-ing decimated his faithful disciples in Eastern Europe, he generated even among his closest associates in the Kremlin a fear for their lives that may well have precipitated his death as a result of a conspiracy, though not soon enough to prevent considerable damage to the cause of communism and to the empire he had created.

In his insatiable quest for security at the price of everybody else's insecurity, Stalin made the people he controlled accessories to his crimes. In Eastern Europe the complicity destroyed any moral claim for leadership by the communist parties—the pillars of Soviet power in the area. The attempt by Stalin's successors to shift the responsibility for his and their own crimes onto their chief executor, Beriia, helped them to proceed with a partial repudiation of Stalinism's debilitating legacy but did not enable them to overcome it.

The unbridgeable chasm between Stalinist and Western values was also the ultimate reason why the post-Stalin leadership could not be brought to entertaining the idea of genuine accommodation with the West even if the West had tried. Dismal though this may sound, Acheson and other U.S. policymakers were unfortunately right not allowing into their "minds the idea that unilateral concessions ... would change, by ameliorating, Soviet policy."[3] What he and his contemporaries were not right about was estimating the scope and nature of the Soviet threat.

If that threat has been widely misunderstood, there cannot be any misunderstanding about the sheer evil of Stalinism. The depth of that evil was, if anything, underestimated even by its foreign critics, who otherwise correctly saw it as inherent in the system and its sustaining ideology, and could only be adequately grasped once the particulars have been attested to by witnesses and documents from its homeland. These have fully substantiated the notion that the Soviet Union under Stalin was not a normal state but one run by a criminal syndicate at the service of a bloody tyrant hungry for power and ready to abuse it whenever he could do so without paying too high a price.

If the empire Stalin created was in fact every bit as evil as suspected, and much more, then those who waged the Cold War against it need not apologize for the effort. The pertinent question, rather, is whether they did the best they could. Certainly, the U.S. policy based on Kennan's premise that Soviet expansionism could be contained until the system that propelled it would collapse under the weight of its own sins has been vindicated—but only after forty years. The cost of delay might seem to make more commendable the ostensibly realistic alternative once advocated by the liberal publicist Walter Lippmann and shared as well by Winston Churchill, who believed that because of its weakness Moscow could be compelled much earlier to accept a deal favorable to the West.[4] Yet not only did the way the Soviet power eventually collapsed in 1989–91 defy the canons of realism, but Stalin was also not realistic enough to entertain such a deal in the first place.

There was, in theory, the third alternative of the United States using its superior military and moral power to topple Stalin's tyranny by war or a threat of war—just as he feared it might. And the whole rotten edifice could well have come tumbling down—just as he feared it might. Such an outcome would have been most probable during the time of paralyzing great fear that preceded his death. But it could never be certain, for the house Stalin built had withstood, after all, the assault by Hitler, who, too, had wielded great military power though not the moral power the West had. In any case, preventive war was never part of U.S. policy, nor would the American system of government have allowed it.

The United States did challenge the enemy by conducting unconventional war through its propaganda and covert operations. While their precise extent remains uncertain—because of the American rather than Russian archival restrictions—they were more extensive than has so far been officially admitted. Although they were not effective enough to achieve the intended destruction of Soviet power, they sufficiently advertised this intention as to generate the impression of being more dangerous than they actually were. Magnified by the communists' presumption of implacable capitalist hostility and the subversion they invented to justify preventive repression, the fear in their hearts was genuine, not merely pretended.

That fear was not the original cause but an important contributing one to the terror Stalin unleashed in Eastern Europe against mostly imaginary subversives at the height of the Cold War. Its conspicuous victims were not so much his real enemies, who had already been made harmless before, as the communist elites. Although their periodic purging was part and parcel of his system, the Western psychological warfare added to its impact by nourishing—both deliberately and unwittingly—the despot's growing insecurity. Thus the West, though not destroying the enemy power, succeeded in impairing it. Yet the United States did not try to exploit this success during the critical time before Stalin's death, and later on the conditions were no longer so favorable.

If the Soviet threat had been disposed of already in Stalin's time rather than in 1991, with it would have gone not only the additional costs of the Cold War but also some of its benefits. The Eastern Europeans, to be sure, could not claim any, having had to endure communism against their will for another thirty-five years. At least the unsuccessful attempts during this period to reform the system to reconcile it with their preferences as well as with modern times proved that this was impossible. In contrast, for the West, the years of the Cold War were those of unprecedented accomplishments and rewards—though ultimately less so for the United States than for Western Europe.

Meeting the Soviet threat as a moral challenge was congenial to the American mentality. What was not congenial was doing so over a prolonged period of time, with self-restraint, and by avoiding military force to achieve a clear decision. In view of these limitations, the Marshall Plan and NATO were monuments to creative and imaginative statesmanship, which in effect decided the outcome of the Cold War already by 1949, although the accomplishment would not become clear until all was over. In the meantime, the futile arms race exacted a price from the United States as the key member of the alliance that disproportionately bore the brunt of the military competition with the Soviet adversary. Its unexpected disappearance found Americans unprepared, disoriented about the purpose of their redundant military power, and groping for a better foundation of their role in the world.

In Western Europe, the perceived Soviet threat prompted an historic reckoning by a generation that had in its lifetime experienced firsthand an extraordinary series of catastrophes: the suicide of nationalist Europe during World War I, the debacle of capitalism in the Great Depression, the subsequent defeat of democracy by fascism amid the destruction of World War II, and finally Stalin's travesty of socialism. The pragmatic response to these calamities amalgamated what was the best in nationalism and internationalism, capitalism and socialism, liberalism and conservatism, democracy and meritocracy. It promoted national reconciliation within the movement for European unity, helped launch the first successful democracy on German soil, and gave Western Europe an unprecedented prosperity, which after the Cold War would serve as the enviable model for Eastern Europe as well.

A larger benefit of the Cold War was the demonstration that a conflict so intense could nevertheless be managed without getting out of hand, thus contributing to the growing belief that war as such may have become obsolete. At first, its avoidance was a matter of choice more for the Americans than for Stalin, who could less afford it. He was therefore never tempted to use his military force, particularly not the atomic bomb, against the West, while the United States was prepared to use even the bomb if provoked, although the Soviet self-perception of inferiority never provided the provocation. So, despite Stalin's Korean miscalculation, not so much the Soviet threat as that of a major war was contained.

In ensuring the critical margin of stability, the penetration of Western governments by Soviet spies, which the counterintelligence services tried so desperately but unsuccessfully to prevent, was a benefit in disguise. The information thus received offered Stalin the necessary minimum of reassurance about enemy intentions to convince him that war was not imminent but only impending. The longer it failed to materialize, the more difficult it was sustaining the belief that it ever would.

After Stalin the situation was reversed. The advent of the "nuclear balance of terror" effectively ruled out America's resorting to the doomsday weapon in defending Europe. In contrast, the perceived strategic parity enabled the Soviet Union, or its increasingly influential military men, to entertain dangerous notions about the utility of conventional, if not nuclear, weapons in advancing Soviet power and influence also into Western Europe.[5] And the gratuitous attempts to advance them by force in other parts of the world, motivated by ideological preconceptions to a much greater degree than suspected, made the Cold War stability a good deal more precarious than the superficial détente of those years made Europeans to believe.

With the Soviet threat gone, both its perils and its benefits have come under a cloud. The vaunted new world order has yet to come— although its absence is certainly preferable to the one resting on the obscene Cold War notion of "mutual assured destruction." Predictions of the obsolescence of war as an tool of policy proved premature— although the tool may well have become unacceptable for the growing number of nations that have chosen to permanently reduce their demographic reserves by other means.[6] And Western European unity, born and nourished in adversity, still needs to show its vitality in a more benign international environment—although it is plainly here to stay. So does the Western welfare state, which triumphed as an alternative to communism but went into a deep crisis soon after its collapse, much like the slowdown of European integration followed the disintegration of the Soviet realm.

If the lessons to be drawn from the Cold War are therefore inconclusive, those that should *not* be learned from it can be stated with more confidence. Its survivors are as susceptible as generals to learn the wrong lessons from the last war they fought. One such lesson would be to assume that, just because the Soviet Union proved in the end to have been such a paper tiger, all regimes driven by universalist ideologies—secular or religious—must be the same. This would make coping with their growing number that much more difficult and avoiding those "wars of civilizations" that Samuel Huntington has prophesized similarly so.

Moreover, if Stalin and all his successors proved so much more restrained and responsible than they could have been in wielding their military power, the nuclear weapons especially, it does not follow that all unaccountable leaders can be expected to behave that way. Nor, just because the Soviet ones in the end abdicated without a fight despite having enough power to give one and create havoc before going down, should it be taken for granted that all international troublemakers would always be so obliging. Such happy endings have been exceedingly rare in history. Hence the lessons of the Cold War are not those of complacency and the redundancy of force in keeping the world safe.

The bipolar order that made peace impossible and war improbable until the Soviet threat disposed of itself is most likely to be remembered as a historical curiosity, highlighting the aberration which began with the avoidable destruction of the old order in World War I, continued with the resistible rise, and led to the eventual fall of both fascist and communist totalitarianism. Something similar is most unlikely to ever happen again, thus making it that much harder to imagine that there could ever again be a threat so acutely felt and yet so warped as that which the Soviet Union projected in its time. For this reason, the study of the Cold War as a topic in the pathology, rather than theory, of international relations will remain not any less essential to keep them sane.

Abbreviations Used in the Notes

APRF
: Arkhiv Prezidenta Rossiiskoi Federatsii [Archives of the President of the Russian Federation]

AVPRF
: Arkhiv vneshnei politiki Rossiiskoi Federatsii [Foreign Policy Archives of the Russian Federation]

AÚV KSČ
: Archív Ústředního výboru Komunistické strany Československa [Archives of the Central Committee of the Communist Party of Czechoslovakia], Prague

CPN
: Communistische Partij van Nederland [Communist Party of the Netherlands]

FRUS
: Foreign Relations of the United States: Diplomatic Papers, U.S.Government Printing Office, Washington, D.C.

HICOG
: Office of the U.S. High Commissioner in Germany

IG
: Istituto Gramsci, Rome

MZV
: Ministerstvo zahraničních věcí [Ministry of Foreign Affairs of the Czech Republic], Prague

NA
: National Archives, Washington, D.C.

RG
: Record Group

RTsKhIDNI
: Rossiiskii Tsentr Khraneniia i Izucheniia Dokumentov Noveishei Istorii [Russian Center for the Preservation and Study of Documents of Recent History], Moscow

SAPMO-BA
: Stiftung Archiv der Parteien und Massenorganisationen der DDR im Bundesarchiv, Berlin

TASS   Telegrafnoe agentstvo Sovetskogo Soiuza, Soviet press agency

TsKhSD   Tsentr Khraneniia Sovremennoi Dokumentatsii [Storage Center for Contemporary Documentation], Moscow

ZK SED   Zentralkomitee der Sozialistischen Einheitspartei Deutschlands

ZPA   Zentrales Parteiarchiv, Berlin

Notes

Introduction: The Cold War as History

1. Francis Fukuyama, "The End of History?" *The National Interest*, 1989, no. 16: 3–18, looks into the future rather than trying to come to terms with the past.

2. The concept was anticipated by Louis J. Halle, *The Cold War as History* (New York: Harper & Row, 1967).

3. The term of John L. Gaddis, "The Long Peace: Elements of Stability in the Postwar International System," in his *The Long Peace: Inquiries into the History of the Cold War* (New York: Oxford University Press, 1987), pp. 215–45.

4. The U.S. perception is admirably analyzed in Robert H. Johnson, *Improbable Dangers: U.S. Conceptions of Threat in the Cold War and After* (New York: St. Martin's, 1994).

5. Vojtech Mastny, *Russia's Road to the Cold War: Diplomacy, Warfare, and the Politics of Communism, 1941–1945* (New York: Columbia University Press, 1979).

6. John L. Gaddis, "Theory and the End of the Cold War," *International Security* 17, no. 3 (1992–93): 5–58.

7. Anton W. DePorte, *Europe between the Superpowers: The Enduring Balance* (New Haven, Conn.: Yale University Press, 1979).

8. Vojtech Mastny, review of *After Brezhnev: Sources of Soviet Conduct in the 1980s*, ed. Robert F. Byrnes, *Russian Review* 43 (1984): 186–88, at p. 188.

9. The development of the concept is explored in Abbott Gleason, *Totalitarianism: The Inner History of the Cold War* (New York: Oxford University Press, 1995), and its continued utility in Alexander J. Motyl, "The End of Sovietology: From Soviet Studies to Post-Soviet Studies," in *The Post-Soviet Nations: Perspectives on the Demise of the USSR*, ed. Alexander J. Motyl (New York: Columbia University Press, 1992), pp. 302–16, at pp. 306–14.

10. The CSCE and Expansion of European Security," and "The New Framework of European Security" in Vojtech Mastny, *Helsinki, Human Rights and European Security, 1975–1985: Analysis and Documentation* (Durham, N.C.: Duke University Press, 1986), pp. 1–36, and *The Helsinki Process and the*

Reintegration of Europe, 1986–1991: Analysis and Documentation (New York: New York University Press, 1992), pp. 1–49, respectively.

11. A.J.P. Taylor, *The Origins of the Second World War* (New York: Atheneum, 1961), p. 72.

12. John Kautsky, quoted in *Political Development in Eastern Europe*, ed. Jan F. Triska and Paul M. Cocks (New York: Praeger, 1977), p. 4.

13. Mark Kramer, "Archival Research in Moscow: Progress and Pitfalls," *Cold War International History Project Bulletin*, 1993, no. 3: 1, 18–39.

14. Amy Knight, "The Fate of the KGB Archives," *Slavic Review* 52 (1993): 582–86.

15. Such is the case of the memoirs by the eighty-five-year-old disreputable ex-operative of Soviet secret services Pavel Sudoplatov, based on no more than twenty-odd hours of recordings of what he remembered, subsequently inflated and embellished by his ambitious son, before being further doctored by a pair of U.S. journalists, eager to make the patchwork sell. Pavel Sudoplatov and Anatoly Sudoplatov, *Special Tasks: The Memoirs of an Unwanted Witness—A Soviet Spymaster* (Boston: Little, Brown, 1994).

1   Stalin's Quest for Soviet Security

1. Daniel Rancour-Laferriere, *The Mind of Stalin: A Psychoanalytical Study* (Ann Arbor, Mich.: Ardis, 1988).

2. Kevin McDermott, "Stalin and the Comintern during the 'Third Period,' 1928–33," *European History Quarterly* 25 (1995): 409–29.

3. Conrad Brandt, *Stalin's Failure in China, 1924–1927* (Cambridge, Mass.: Harvard University Press, 1958); C. Martin Wilbur and Julie Lien-ying How, *Missionaries of Revolution: Soviet Advisers and Nationalist China, 1920–1927* (Cambridge, Mass.: Harvard University Press, 1989), pp. 385–412.

4. In the sense of Karl A. Wittfogel, *Oriental Despotism: A Comparative Study of Total Power* (New Haven, Conn.: Yale University Press, 1957).

5. Robert C. Tucker, "The Emergence of Stalin's Foreign Policy," *Slavic Review* 36 (1977): 563–89.

6. John P. Sontag, "The Soviet War Scare of 1926–1927," *Russian Review* 34, 1 (1975): 66–77, at p. 71.

7. William C. Wohlworth, *The Elusive Balance: Power and Perceptions during the Cold War* (Ithaca, N.Y.: Cornell University Press, 1993), pp. 46–51.

8. Frederic S. Burin, "The Communist Doctrine of the Inevitability of War," *American Political Science Review* 57 (1963): 334–54.

9. Nikita V. Zagladin, *Istoriia uspekhov i neudach sovetskoi diplomatii: Politologicheskii aspekt* [A History of the Successes and Failures of Soviet Diplomacy from the Political Science Point of View] (Moscow: Mezhdunarodnye otnosheniia, 1990), p. 102.

10. Wolfgang Pfeiler, "Das Deutschlandbild und die Deutschlandpolitik Josef Stalins," *Deutschland-Archiv* 12 (1979): 1258–82.

11. S.A. Gorlov and S.V. Ermachenkov, "Voenno-uchebnye tsentry Reikhsvera v Sovetskom Soiuze: Sotrudnichestvo SSSR i Germanii v 20-e gody" [Military Training Centers of the Reichswehr in the Soviet Union: The Soviet-German Collaboration in the Twenties], *Voennoistoricheskii zhurnal*, 1993, nos. 6: 39–44, 7: 41–44, 8: 36–42.

12. Nicholas N. Kozlov and Eric D. Weitz, "Reflections on the Origins of the 'Third Period': Bukharin, the Comintern, and the Political Economy of Weimar Germany," *Journal of Contemporary History* 24 (1989): 387–410.

13. Robert C. Tucker, *Stalin in Power: The Revolution from Above, 1928–1941* (New York: Norton, 1990), pp. 228–32, 275, 409–15, 512.

14. Zbigniew Brzezinski, *The Permanent Purge* (Cambridge, Mass.: Harvard University Press, 1956).

15. J. Arch Getty, *Origins of the Great Purge: The Soviet Communist Party Reconsidered, 1933–1938* (New York: Cambridge University Press, 1985).

16. Jiri Hochman, *The Soviet Union and the Failure of Collective Security* (Ithaca, N.Y.: Cornell University Press, 1984).

17. Verne Newton, *The Cambridge Spies* (Washington, D.C.: Madison Books, 1991).

18. R. Dan Richardson, *Comintern Army: The International Brigades and the Spanish Civil War* (Lexington: University Press of Kentucky, 1982), pp. 3–15; M.T. Meshcheriakov, "Sudba Interbrigad v Ispanii po novym dokumentam" [New Documents on the Fate of the International Brigades in Spain], *Novaia i noveishaia istoriia*, 1993, no. 5: 18–41.

19. David T. Cattell, *Soviet Diplomacy and the Spanish Civil War* (Berkeley: University of California Press, 1957), p. 70.

20. Donald C. Watt, "The Initiation of Negotiations Leading to the Nazi-Soviet Pact: A Historical Problem," in *Essays in Honour of E.H. Carr*, ed. C. Abramsky (Hamden, Conn.: Archon, 1974), pp. 152–70.

21. S.A. Gorlov, "Sovetsko-germanskii dialog nakanune pakta Molotova-Ribbentropa 1939 g." [Soviet-German Dialogue on the Eve of the 1939 Molotov-Ribbentrop Pact], *Novaia i noveishaia istoriia*, 1993, no. 4: 13–34.

22. "Sovetsko-germanskie dokumenty 1939–1941 gg. iz arkhiva TsK KPSS" [Soviet-German Documents on 1939–1941 from the Archives of the CPSU Central Committee], *Novaia i noveishaia istoriia*, 1993, no. 1: 83–95.

23. Cf. new documents from Soviet archives in "Poezdka Molotova v Berlin v noiabre 1940 g." [Molotov's Trip to Berlin in November 1940], *Novaia i noveishaia istoriia*, 1993, no. 5: 64–99.

24. Soviet minutes in O.A. Rzheshevskii, "Vizit A. Idena v Moskvu v dekabre 1941 g.: Peregovory s I.V. Stalinym i V.M. Molotovym" [Eden's Visit to Moscow in December 1941: Conversations with Stalin and Molotov], *Novaia i noveishaia istoriia*, 1994, no. 2: 85–102.

25. Joan Barth Urban, *Moscow and the Italian Communist Party: From Togliatti to Berlinguer* (Ithaca, N.Y.: Cornell University Press, 1986), pp. 154–61.

26. Record of Stalin-Thorez conversation, November 19, 1944, document in Moscow presidential archives, cited in Vladimir Volkov, "The Soviet Leadership and Certain Problems of Southeastern Europe in the Closing Year of the War (1944–1945)" (paper presented at the conference "The Establishment of Communist Regimes in Eastern Europe," Moscow, March 29–31, 1994), pp. 16–18.

27. In the Moscow headquarters there were no more than 150–200 officials. Vadim A. Kirpichenko, "Razvedka vykhodit iz zony molchaniia" [The Intelligence Services End Their Silence], *Voennoistoricheskii zhurnal*, 1995, no. 2: 80–87, at p. 84.

28. Josef Klečka, "O zradě" [On Treason], *Odboj a revoluce* [Prague] 5, no. 4 (1966): 47–52, and Vilém Kahan, "O některých nedostatcích v konspirativní činnosti" [Some Shortcomings in the Conspiratorial Activities], *ibid.* 5, no. 4 (1967): 97–98.

29. Władysław Gomułka, *Pamiętniki* [Memoirs], vol. 2 (Warsaw: BGW, 1994), pp. 160–257, 349–59.

30. Milorad M. Drachkovitch, "The Comintern and the Insurrectional Activity of the Communist Party of Yugoslavia in 1941–1942," in the book edited by him and Branko Lazitch, *The Comintern: Historical Highlights* (New York: Praeger, 1966), pp. 184–213.

31. "Statement of the Praesidium of the E.C.C.I. on the Dissolution of the Communist International," May 21, 1943, in Jane Degras, ed., *The Communist International, 1919–1943*, vol. 3 (London: Oxford University Press, 1965), pp. 477–79.

32. Grant M. Adibekov, *Kominform i poslevoennaia Evropa, 1947–1956 gg.* [The Cominform and Postwar Europe], (Moscow: Rossiia Molodaia, 1994), pp. 7–10.

33. Vojtech Mastny, "Stalin and the Prospects of a Separate Peace in World War II," *American Historical Review* 77 (1972): 1365–88.

34. Aleksei Filitov, "Problemy poslevoennogo ustroistva v sovetskikh vneshepoliticheskikh kontseptsiakh perioda vtoroi mirovoi voiny" [Problems of the Postwar Order in Soviet Foreign Policy Conceptions during World War II] (paper presented at the conference "The Soviet Union and Europe in the Cold War (1943–1953)," Cortona, September 23–24, 1994), p. 3.

35. Vojtech Mastny, "Soviet War Aims at the Moscow and Teheran Conferences of 1943," *Journal of Modern History* 47 (1975): 481–504.

36. Jochen Laufer, "Die UdSSR und die Zonenteilung Deutschlands (1943/ 44)," *Zeitschrift für Geschichtswissenschaft* 45 (1995): 309–31, at pp. 328–29.

37. Sven G. Holtsmark, *A Soviet Grab for the High North? USSR, Svalbard, and Northern Norway 1920–1953* (Oslo: Institutt for Forvarsstudier, 1993), pp. 57–64.

38. Idem, "The Soviet Pursuit of Strategic Interests in Norway and Denmark 1944–47: The Limits to Soviet Influence" (paper presented at the conference "The Soviet Union and Europe in the Cold War (1943–1953)," Cortona, September 23–24), 1994, pp. 6–19.

39. G.P. Kynin, "Germanskii vopros vo vzaimootnosheniiakh SSSR, SShA i Velikobritanii 1941–1943 gg.: Obzor dokumentov" [The German Question in the Relations between the U.S.S.R., U.S., and Great Britain in 1941–43: A Survey of Documents], *Novaia i noveishaia istoriia*, 1995, no. 1: 91–113, at p. 93.

40. Maiskii to Molotov, March 3, 1944, and November 10, 1943, Filitov, "Problemy poslevoennogo ustroistva," pp. 6–7.

41. *Ibid.*, 29–32.

42. Horst Laschitza, *Kämpferische Demokratie gegen Faschismus* (Berlin: Deutscher Militärverlag, 1969), pp. 136–38. Cf. Peter Erler, *et al.*, ed., *"Nach Hitler kommen wir": Dokumente zur Programmatik der Moskauer KPD-Führung 1944/45 für Nachkriegsdeutschland* (Berlin: Akademie, 1994).

43. Dimitrov to Finder, April 2, 1943, "Perepiska Generalnogo Sekretaria IK KI G.M. Dimitrova s rukovodstvom Polskoi Rabochei Partii (1942–43)" [Correspondence between Secretary General of the Executive Committee of the

Communist International G.M. Dimitrov and the leadership of the Polish Workers' Party], *Novaia i noveishaia istoriia*, 1964, no. 5: 122.

44. As related by Czechoslovak president Edvard Beneš in Vojtech Mastny, "The Beneš-Stalin-Molotov Conversations in 1943: New Documents," *Jahrbücher für Geschichte Osteuropas* 20 (1972): 367–402, at p. 379.

45. Ludomir Smosarski, ed., "Dyskusje w PPR w sprawie zjednoczenia sił demokratycznych (notatki protokolarne z posedzień KC PPR maj-czerwiec 1944 r." [The Debates in the Polish Workers' Party on the Unification of Democratic Forces: Notes from the Central Committee Sessions in May–June 1944], *Archiwum ruchu robotniczego*, vol. 2 (Warsaw: Książka i Wiedza, 1975), pp. 154–63.

46. Minutes of Stalin's meeting with the Polish Committee of National Liberation, October 9, 1944, Antony Polonsky and Bolesław Drukier, ed., *The Beginnings of Communist Rule in Poland: December 1943–June 1945* (London: Routledge & Kegan Paul, 1980), p. 298.

47. Albert Resis, "The Churchill-Stalin Secret 'Percentages' Agreement on the Balkans, Moscow, October 1944," *American Historical Review* 83 (1977–78): 368–87.

48. Branko Petranović, *Balkanska Federacija, 1943–1948* [The Balkan Federation] (Belgrade: Edicija Svedočanstva, 1991), pp. 128–35.

49. Vojtech Mastny, *Russia's Road to the Cold War: Diplomacy, Warfare, and the Politics of Communism, 1941–1945* (New York: Columbia University Press, 1979), p. 211.

50. Jukka Nevakivi, "A Decisive Armistice 1944–1947: Why Was Finland Not Sovietized?" *Scandinavian Journal of History* 19 (1994): 91–115, at p. 114.

51. Entry for August 17, 1975, in Feliks Chuev, *Sto sorok besed s Molotovym* [Hundred and Forty Conversations with Molotov] (Moscow: Terra, 1991), p. 76.

52. "On the Question of Soviet-American Relations," July 14, 1944, quoted in Vladimir O. Pechatnov, "The Big Three after World War II: New Documents on Soviet Thinking about Post War Relations with the United States and Great Britain," Working Paper no. 13, *Cold War International History Project* (Washington, D.C.: Woodrow Wilson International Center for Scholars, 1995), p. 6.

53. Charles Gati, *Hungary and the Soviet Bloc* (Durham, N.C.: Duke University Press, 1986), pp. 33–37.

54. Vladimir Dedijer, *Novi prilozi za biografiju Josipa Broza Tita* [New Supplements to the Biography of Josip Broz Tito], vol. 3 (Belgrade: Rad, 1984), p. 325.

55. Edvard Kardelj, *Boj za priznanje in neodvisnost nove Jugoslavije: Spomini* [The Struggle for the Recognition and Independence of New Yugoslavia: Memoirs] (Ljubljana: Državna založba Slovenije, 1980), pp. 103–104.

56. Maiskii to Molotov, January 11, 1944, quoted in Pechatnov, "The Big Three after World War II," p. 3.

57. Mastny, "The Beneš-Stalin-Molotov Conversations," pp. 367–402.

58. Speech by Molotov, February 1, 1944, *Izvestiia*, February 2, 1944; Mario Toscano, "Resumption of Diplomatic Relations between Italy and the Soviet Union in World War II," in his *Designs in Diplomacy* (Baltimore, Md.: Johns Hopkins Press, 1970), pp. 294–95.

59. Andrei A. Gromyko, *Pamiatnoe* [Memories], vol. 1 (Moscow: Izdatelstvo politicheskoi literatury, 1988), p. 189.

60. Entry for August 15, 1975, in Chuev, *Sto sorok besed*, p. 76.

61. Molotov as reported by Yugoslav ambassador Vladimir Popović, cited in Leonid Ia. Gibianskii, "Doneseniia iugoslavskogo posla v Moskve ob otsenkakh rukovodstvom SSSR potsdamskoi konferentsii i polozheniia v vostochnoi Evrope (avgust–noiabr 1945 g.)," [Reports by the Yugoslav Ambassador about the Assessments by the Soviet Leadership of the Potsdam Conference and the Situation in Eastern Europe], *Slavianovedenie*, 1994, no. 1: 3–13, at p. 6.

62. George F. Kennan, *Memoirs, 1925–1950* (Boston: Little, Brown, 1967), p. 284.

63. Molotov's evaluation of the conference, cited in Vladimir V. Shustov, "A View on the Origins of the Cold War and Some Lessons Thereof," in *Beyond the Cold War: New Dimensions in International Relations*, ed. Geir Lundestad and Odd Arne Westad (Oslo: Scandinavian University Press, 1993), pp. 23–37, at pp. 30–31.

64. Arnold Wolfers, *Discord and Collaboration* (Baltimore: Johns Hopkins Press, 1965), pp. 92, 150–51.

65. Vojtech Mastny, "The Cassandra in the Foreign Commissariat: Maxim Litvinov and the Cold War," *Foreign Affairs* 54 (1975–76): 366–76.

66. Entry for November 28, 1974, in Chuev, *Sto sorok besed*, p. 86.

67. *Izvestiia*, February 11, 1992, p. 6.

68. John O. Iatrides, "Civil War, 1945–1949: National and International Aspects," in *Greece in the 1940s: A Nation in Crisis*, ed. John O. Iatrides (Hanover, N.H.: University Press of New England, 1981), p. 203.

69. *Khrushchev Remembers: The Last Testament* (Boston: Little, Brown, 1974), pp. 295–96.

70. Louise L'Estrange Fawcett, *Iran and the Cold War: The Azerbaijan Crisis of 1946* (Cambridge: Cambridge University Press, 1992), pp. 83–107; N.I. Egorova, "'Iranskii krizis' 1945–1946 gg. po rassekrechennym arkhivnym dokumentam" [The "Iranian Crisis" of 1945–1946 from Declassified Archival Documents], *Novaia i noveishaia istoriia*, 1994, no. 3: 24–42.

71. Pechatnov, "The Big Three after World War II," p. 20.

72. Quoted in Yuri Aksyutin, "Why Stalin Staked on Confrontation Rather than Cooperation with the Wartime Allies after the Victory (Some Socio-Psychological Aspects of the Cold War Origins)" (paper presented at the conference "New Evidence on Cold War History," Moscow, January 12–15, 1993), p. 17.

73. They are thoroughly examined and persuasively explained in Norman Naimark, *The Russians in Germany: The History of the Soviet Zone of Occupation, 1945–1949* (Cambridge, Mass.: Harvard University Press, 1995).

74. *Khrushchev Remembers: The Glasnost Tapes* (Boston: Little, Brown, 1990), p. 99.

75. Note by Pieck on conversation with Stalin, June 4, 1945, in *Wilhelm Pieck: Aufzeichnungen zur Deutschlandpolitik, 1945–1953*, ed. Rolf Badstübner and Wilfried Loth (Berlin: Akademie, 1994), pp. 50–51, at p. 50.

76. Dietrich Staritz, "Parteien für ganz Deutschland? Zu den Kontroversen über ein Parteiengesetz im Alliierten Kontrollrat 1946/47," *Vierteljahrshefte für Zeitgeschichte* 32 (1984): 240–68; Elisabeth Kraus, *Ministerien für ganz Deutschland? Der Alliierte Kontrollrat und die Frage gesamtdeutscher Zentralverwaltungen* (Munich: Oldenbourg, 1990).

77. Report by Ulbricht on consultation with Stalin, February 6, 1946, Badstübner and Loth, ed., *Wilhelm Pieck: Aufzeichnungen zur Deutschlandpolitik*, pp. 68–69.

78. Alexei M. Filitov, "The Soviet Union, the German Question and the Origins of the Cold War, 1945–1949," (paper presented at the conference "New Evidence on Cold War History," Moscow, January 12–15, 1993), pp. 6–9, 21, in contrast to Fraser Harbutt, "American Challenge, Soviet Response: The Beginning of the Cold War, February–May 1946," *Political Science Quarterly* 96 (1981–82): 623–39.

79. Notes by Pieck on conversation with Stalin, January 31, 1947, Pieck NL 36/694, SAPMO-BA; see also Bernd Bonwetsch and Gennadij Bordjugov, "Stalin und die SBZ: Ein Besuch der SED Führung in Moskau vom 30. Januar–7. Februar 1947, *Vierteljahrshefte für Zeitgeschichte* 42 (1994): 279–303.

80. Staritz, "Ein 'besonderer deutscher Weg' zum Sozialismus?" *Aus Politik und Zeitgeschichte,* 1982, no. 51–52: 15–31.

81. Ruth A. Rosa, "The Soviet Theory of 'People's Democracy,'" *World Politics* 1 (1948–49): 489–510.

82. Milovan Djilas, *Conversations with Stalin* (New York: Harcourt, Brace & World, 1962), p. 113.

83. Karel Kaplan, "Il piano di Stalin," *Panorama* [Milan] 15, no. 575 (April 26, 1977): 169–89.

84. Peter Coleman, *The Liberal Conspiracy: The Congress for Cultural Freedom and the Struggle for the Mind of Postwar Europe* (New York: Free Press, 1989), p. 8.

85. Leonid Ya. Gibianskii, "Kak voznik Kominform: Po novym arkhivnym materialam" [The Origins of the Cominform: New Archival Materials], *Novaia i noveishaia istoriia,* 1993, no. 4: 131–52, at pp. 134–35.

86. Yugoslav minutes of Stalin-Tito conversation, May 27, 1946, in Leonid Ya. Gibianskii, "Poslednii vizit I. Broza Tito k I.V. Stalinu: Sovetskaia i iugoslavskaia zapisy besedy 27–28 maia 1946 g." [Tito's Last Visit to Stalin: Soviet and Yugoslav Minutes of the Meetings on May 27–28, 1946], *Istoricheskii Arkhiv,* 1993, no. 2: 16–35, at p. 28.

87. Artiom A. Ulunian, "SSSR i 'grecheskii vopros' (1946–1953): Problemy i otsenki" [The USSR and the "Greek Question": Problems and Interpretations] (paper presented at the conference "The Soviet Union and Europe in the Cold War (1943–1953)," Cortona, September 23–24, 1994), pp. 8–10.

88. Scott D. Parrish, "The USSR and the Security Dilemma: Explaining Soviet-Self-Encirclement, 1945–1985" (Ph.D. diss. Columbia University, 1993), pp. 199–201.

89. Othmar Nikola Haberl, "Die sowjetische Außenpolitik im Umbruchsjahr 1947," in *Der Marshall-Plan und die europäische Linke,* ed. Othmar Nikola Haberl and Lutz Niethammer (Frankfurt: Europäischer Verlagsanstalt, 1986), pp. 75–96, at pp. 79–83; Dean Acheson, *Present at the Creation: My Years in the State Department* (New York: Norton, 1969), p. 225.

90. Cf. John L. Gaddis, "Was the Truman Doctrine a Real Turning Point?" *Foreign Affairs* 52 (1973–74): 386–402.

91. Martina Kessel, *Westeuropa und die deutsche Teilung: Englische und französische Deutschlandpolitik auf den Außenministerkonferenzen von 1945 bis 1947* (Munich: Oldenbourg, 1989), pp. 235–46.

92. Wilfried Loth, "Frankreichs Kommunisten und der Beginn des Kalten Krieges: Die Entlassung der kommunistischen Minister im Mai 1947," *Vierteljahrshefte für Zeitgeschichte* 26 (1978): 9–65.

93. Abraham Boxhorn, *The Cold War and the Rift in the Governments of National Unity: Belgium, France and Italy in the Spring of 1947, A Com-*

parison (Amsterdam: Historisch Seminarium van de Universiteit van Amsterdam, 1993), pp. 248–49.

94. Zhdanov to Thorez, June 2, 1947, in Nataliia I. Egorova, "Stalinskaia vneshniaia politika i Kominform, 1947–1953" [Stalin's Foreign Policy and the Cominform] (paper presented at the conference "The Soviet Union and the Cold War (1943–1953)," Cortona, September 23–24, 1994), p. 4.

95. Alan S. Milward, "Was the Marshall Plan Necessary?" *Diplomatic History* 12 (1989): 231–53.

96. Michael Hogan, *The Marshall Plan: America, Britain, and the Reconstruction of Western Europe, 1947–1952* (Cambridge: Cambridge University Press, 1987), pp. 40–45.

97. John L. Gaddis, "Dividing Adversaries: The United States and International Communism, 1945–1958," in his *The Long Peace: Inquiries into the History of the Cold War* (New York: Oxford University Press, 1987), p. 156.

98. Geoffrey Roberts, "Moscow and the Marshall Plan: Politics, Ideology and the Onset of the Cold War, 1947," *Europe-Asia Studies* 46 (1994): 1371–86.

99. Vladimir I. Erofeev, "Desat let v sekretariate Narkomindela" [Ten Years in the Secretariat of the Foreign Commissariat], *Mezdunarodnaia zhizn*, 1991, no. 9: 108–16, at p. 108; Vladislav M. Zubok, "Soviet Intelligence and the Cold War: The 'Small' Committee of Information, 1952–53," Working Paper no. 4, *Cold War International History Project* (Washington, D.C.: Woodrow Wilson International Center for Scholars, 1992), p.6.

100. Molotov to Bodrov, June 22, 1947, in Galina A. Takhnenko, "Anatomiia odnogo politicheskogo resheniia: K 45-letiiu plana Marshalla" [The Anatomy of a Political Decision: The 45th Anniversary of the Marshall Plan], *Mezhdunarodnaia zhizn*, 1992, no. 5: 113–27, at pp. 119–20.

101. "Direktivy Sovetskoi delegatsii na soveshchanii ministrov inostrannykh del v Parizhe" [Directives for the Soviet Delegation at the Paris Conference of Ministers of Foreign Affairs], June 25, 1947, 06/1947/9/214/18, pp. 4–6, AVPRF.

102. Varga to Molotov, June 1947, 06/1947/9/213, p. 18, AVPRF.

103. Soviet minutes of conference of foreign ministers, Paris, June 27, 1947, 06/9/215/18, pp. 13–14, AVPRF; cf. Caffery (Paris) to Secretary of State, June 28, 1947, *FRUS*, 1947, vol. 3, pp. 297–98.

104. Soviet minutes of conference of foreign ministers, Paris, June 28 and 30, 1947, 06/9/215/18, pp. 37–39, 58, and 216/18, pp. 97–98, AVPRF; cf. Caffery (Paris) to Secretary of State, June 28 and 29, 1947, *FRUS*, 1947, vol. 3, pp. 299–300.

105. Caffery (Paris) to Secretary of State, July 1, 1947, *FRUS*, 1947, vol. 3, pp. 303–4.

106. Ibid., pp. 301–2.

107. Scott D. Parrish and Mikhail M. Narinsky, "New Evidence on the Soviet Rejection of the Marshall Plan, 1947: Two Reports," Working Paper no. 9, *Cold War International History Project* (Washington, D.C.: Woodrow Wilson International Center for Scholars, 1994), p. 45.

108. Mikhail M. Narinskii, "SSSR i plan Marshala: Po materialam Arkhiva Prezidenta RF" [The USSR and the Marshall Plan: From Materials from the Archives of the President of the Russian Federation], *Novaia i noveishaia istoriia*, 1993, no. 2: 11–19.

109. Molotov to Bodrov, July 5, 1947, in Takhnenko, "Anatomiia," p. 125.

110. Molotov to Bodrov, July 7 and 8, 1947, in Takhnenko, "Anatomiia,", p. 126.

111. Jan Masaryk (Prague) to Czechoslovak Embassy in Moscow, June 19, 1947, Minister's Secretariat, 1945–55, Box 94, MZV; cf. Karel Krátký, "Czechoslovakia, the Soviet Union and the Marshall Plan," in *The Soviet Union in Eastern Europe, 1945–89*, ed. Odd Arne Westad et al. (New York: St. Martin's Press, 1994), pp. 9–25.

112. Vojtech Mastny, "Stalin, Czechoslovakia, and the Marshall Plan: New Documentation from Czechoslovak Archives," *Bohemia* [Munich] 32 (1991): 139–44.

113. Lebedev (Warsaw) to Molotov, July 8, 1947, 06/1947/9/37/20, p. 7, AVPRF.

114. "Návrh na řeč československého delegáta J. Noska v Paříži přijatý na zasedání předsednictva vlády 9.VII.1947" [Draft of the Paris Speech of the Czechoslovak Representative J. Nosek Approved at the Session of the Council of Ministers on July 9, 1947], Minister's Secretariat, 1945–55, Box 94, MZV.

115. Lebedev (Warsaw) to Molotov, July 10, 1947, 06/1947/9/237/20, pp. 8–12, AVPRF.

116. Smith (Moscow) to Secretary of State, July 10, 1947, *FRUS*, 1947, vol. 3, p. 327.

2   The Specter of Communism

1. Wilfried Loth, "Frankreichs Kommunisten und der Beginn des Kalten Krieges: Die Entlassung der kommunistischen Minister in Mai 1947," *Vierteljahrshefte für Zeitgeschichte* 26 (1978): 9–65, at p. 60.

2. L. Baranov, "Zapiska Zhdanovu o plane organizatsionnykh meropriiatii dlia Soveshchaniia" [Memorandum for Zhdanov on the Planned Organizational Measures for the Conference], August 15, 1947, 575/2/38 (1), RTsKhIDNI.

3. "Mezhdunarodnoe polozhenie Sovetskogo Soiuza" [The International Situation of the Soviet Union], 575/2/38 (1), pp. 40–41, RTsKhIDNI.

4. Natalya Yegorova, "From the Comintern to the Cominform: Ideological Dimension of the Cold War Origins (1945–1948)" (paper presented at the conference "New Evidence on Cold War History," Moscow, January 12–15, 1993), pp. 5, 28.

5. Jukka Nevakivi, "The Control Commission in Helsinki: A Finnish View," in *Finnish-Soviet Relations 1944–1948*, ed. Jukka Nevakivi (Helsinki: Department of Political History, University of Helsinki, 1994), pp. 67–79.

6. The complete text is printed in *The Cominform: Minutes of the Three Conferences, 1947/1948/1949*, ed. Giuliano Procacci (Milan: Feltrinelli, 1994), pp. 216–51.

7. Leonid Ia. Gibianskii, "Kak voznik Kominform: Po novym arkhivnym materialam" [The Origins of the Cominform: New Archival Materials], *Novaia i noveishaia istoriia*, 1993, no. 4: 131–52, at p. 146.

8. Speech by Zhdanov, September 25, 1947, in *The International Situation and Soviet Foreign Policy: Key Reports by Soviet Leaders from the Revolution to the Present*, ed. Myron Rush (Columbus, Ohio: Merrill, 1970), pp. 125–39, at pp. 129–30, 138.

9. Minutes of meeting, September 24, 1947, p. 152, 575/1/1, RTsKhIDNI.

10. Krystyna Kersten, *The Establishment of Communist Rule in Poland, 1943–1948* (Berkeley: University of California Press, 1991), pp. 407–8; Leonid Ia. Gibianskii, "Problems of East European International-Political Structuring in the Period of Formation of the Soviet Bloc in the 1940," (paper presented at the conference "New Evidence on Cold War History," Moscow, January 12–15, 1993), p. 53.

11. Minutes of meetings, September 25–26, 1947, pp. 195–208, 209–17, 575/1/1, RTsKhIDNI.

12. Minutes of meeting, September 26, 1947, pp. 225–31, 575/1/1, RTsKhIDNI.

13. "Intervento del compagno Luigi Longo sul rapporto del comp. A. Jdanov," undated, pp. 139–48, 0192/0136–0151, central committee archives, IG.

14. "Discorso di conclusione del comp. Jdanov," notes by Longo, undated, 0192/0152–0155, and minutes of the meeting of the central committee, November 11–13, 1947, pp. 570sq., 039/528–842, central committee archives, IG.

15. *New York Times*, October 23, 1947.

16. Minutes of meetings, October 7 and 10, 1947, Verbali della Direzione P.C.I., 1947, IG.

17. Duclos at the October session of the French party politburo, cited in Grant M. Adibekov, *Kominform i poslevoennaia Evropa, 1947–1956 gg.* [The Cominform and Postwar Europe] (Moscow: Rossiia Molodaia, 1994), p. 61.

18. Ronald Tiersky, *French Communism, 1920–1972* (New York: Columbia University Press, 1974), pp. 166–71; John W. Young, *France, the Cold War and the Western Alliance, 1944–49: French Foreign Policy and Post-War Europe* (Leicester, England: Leicester University Press, 1990), p. 174.

19. "Zakonomernost rosta vliianiia kommunisticheskikh partii v evropeiskikh stranakh" [The Laws Governing the Growth of the Communist Parties' Influence in European Countries], 575/2/38 (1), RTsKhIDNI.

20. H. Gordon Skilling, "'People's Democracy' in Soviet Theory," *American Slavic and East European Review* 10 (1951): 100–116, and *Soviet Studies* 3 (1951–52): 131–49.

21. "Discorso di conclusione del comp. Jdanov," 0192/0153, central committee archives, IG.

22. Minutes of meeting, September 26, 1947, 575/1/1, p. 265, RTsKhIDNI.

23. Radomír V. Luža, "February 1948 and the Czechoslovak Road to Socialism," *East Central Europe* 4, no. 1 (1977): 44–55, at pp. 51–52.

24. Josif Broz Tito, *Speech at the Second Congress of the People's Front of Yugoslavia, September 27, 1947* (Washington, D.C.: Yugoslav Embassy, 1947); A. Ross Johnson, *The Transformation of Communist Ideology: The Yugoslav Case, 1945–1953* (Cambridge, Mass.: MIT Press, 1972), pp. 39–40.

25. Vesselin Dimitrov, "The Cominform and the Bulgarian Communist Party: Embarking on a New Course?" (paper presented at the conference "The Soviet Union and the Cold War in Europe (1943–1953)," Cortona, September 23–24, 1994), pp. 10–15.

26. Speech by Zhdanov, September 22, 1947, in Rush, *The International Situation and Soviet Foreign Policy*, p. 130.

27. Gibianskii, "Problems of East European International-Political Structuring," p. 63.

28. Leonid Ia. Gibianskii, "Sovetsko-iugoslavskii konflikt i sovetskii blok" [The Soviet-Yugoslav Conflict and the Soviet Bloc] (paper presented at the

conference "The Soviet Union and Europe in the Cold War (1943–1953)," Cortona, September 23–24, 1994), pp. 5–6.

29. Ibid., p. 13.

30. Slobodan Nešović, *Bledski sporazumi: Tito-Dimitrov (1947)* [The Bled Agreements] (Zagreb: Globus, 1979).

31. Ivo Banac, *With Stalin against Tito: Cominformist Splits in Yugoslav Communism* (Ithaca, N.Y.: Cornell University Press, 1988), p. 37.

32. Stalin to Tito, August 12, 1947, quoted in Iurii S. Girenko, *Stalin-Tito* (Moscow: Izdatelstvo politicheskoi literatury, 1991), pp. 325–26.

33. Leonid Ia. Gibianskii, "U nachala konflikta: Balkanskii uzel" [The Origins of the Conflict: The Balkan Knot], *Rabochii klass i sovremennyi mir*, 1990, no. 2: 171–85, at pp. 173–74.

34. R.H. Markham, *Tito's Imperial Communism* (Chapel Hill: University of North Carolina Press, 1947), pp. 275–92.

35. Stephen D. Kertesz, "American and Soviet Negotiating Behavior," in *Diplomacy in a Changing World*, ed. Stephen D. Kertesz (Notre Dame, Ind.: University of Notre Dame Press, 1959), pp. 133–71, at p. 151.

36. Teresa Torańska, *"Them": Stalin's Polish Puppets* (New York: Harper & Row, 1987), p. 285.

37. Leonid Gibianski, "The 1948 Soviet-Yugoslav Conflict and the Formation of the 'Socialist Camp' Model," *The Soviet Union in Eastern Europe, 1945–89*, ed. Odd Arne Westad et al. (New York: St. Martin's Press, 1994), pp. 26–46, at p. 31.

38. Alexander Werth, *Russia: The Post-War Years* (London: Hale, 1971), p. 306.

39. Artiom A. Ulunian, "SSSR i 'grecheskii vopros' (1946–1953): Problemy i otsenki" [The USSR and the "Greek Question": Problems and Interpretations] (paper presented at the conference "The Soviet Union and Europe in the Cold War (1942–1953)," Cortona, September 23–24, 1994), p. 10.

40. Lawrence S. Wittner, *American Intervention in Greece, 1943–1949* (New York: Columbia University Press, 1982), pp. 260–61; Kenneth W. Condit, *The History of the Joint Chiefs of Staff: The Joint Chiefs of Staff and National Policy*, vol. 2, *1947–1949* (Wilmington, Del.: Glazier, 1979), pp. 42–50.

41. Jon Halliday, *The Artful Albanian: The Memoirs of Enver Hoxha* (London: Chatto & Windus, 1986), pp. 73, 105. Heath (Sofia) to Secretary of State, December 31, 1947, RG-59, 875.00/12-1847, NA.

42. Stalin to Tito, December 23, 1947, quoted in Girenko, *Stalin-Tito*, p. 330.

43. Milovan Djilas, *Rise and Fall* (London: Macmillan, 1985), pp. 148–49.

44. Lavrentev to Molotov, January 8, 1948, quoted in Girenko, *Stalin-Tito*, pp. 351–52; I.V. Bukharin and Leonid Ya. Gibianskii, "Pervye shagi konflikta" [The First Steps into Conflict], *Rabochii Klass i Sovremennyi Mir*, 1990, no. 5: 152–63, at pp. 160–63.

45. Djilas, *Rise and Fall*, p. 153; Gibianskii, "U nachala," pp. 180–81.

46. Gibianskii, "Sovetsko-iugoslavskii konflikt," p. 19.

47. Extracts from statement by Dimitrov, Bucharest, January 17, 1948, in *Documents on International Affairs, 1947–1948*, ed. Margaret Carlyle (London: Oxford University Press, 1952), p. 297; Cyrus L. Sulzberger, "U.S. Held Target of Pacts That Bind Soviet Satellites," *New York Times*, January 17, 1948, p. 7.

48. Ranković to Ðilas, January 19, 1948, cited in Gibianskii, "U nachala," p. 182.

49. Speech by Bevin, January 22, 1948, in Carlyle, *Documents on International Affairs, 1947–1948*, pp. 201–21.

50. John Kent and John W. Young, "The 'Western Union' Concept and British Defence Policy, 1947–8," in *British Intelligence Strategy and the Cold War, 1945–51*, ed. Richard J. Aldrich (London: Routledge, 1992), pp. 166–92.

51. Bukharin and Gibianskii, "Pervye shagi konflikta," p. 153.

52. Halliday, *The Artful Albanian*, p. 112.

53. Quoted in Dmitrii S. Chuvakhin, "S diplomaticheskoi missiei v Albanii, 1946–1952 gg." [With the Diplomatic Mission in Albania in 1946–52], *Novaia i noveishaia istoriia*, 1995, no. 1: 114–31, at p. 128.

54. Gibianskii, "U nachala," pp. 182–83.

55. TASS statement, January 28, 1948, in Carlyle, *Documents on International Affairs, 1947–1948*, pp. 297–98.

56. *Prime Minister Georgi Dimitrov's Historic Report at the Second Conference of the Fatherland Front* (Sofia: Ministry of Foreign Affairs, 1948), p. 21.

57. Molotov to Tito, February 1, 1948, in Girenko, *Stalin-Tito*, p. 337.

58. Leonid Ia. Gibianskii, "Vyzov v Moskvu" [The Summons to Moscow], *Politicheskie Issledovaniia*, 1991, no. 1: 195–207.

59. Leonid Ia. Gibianskii, "K istorii sovetsko-iugoslavskogo konflikta 1948–1953 gg.: Sekretnaia sovetsko-iugoslavo-bolgarskaia vstrecha v Moskve 10 fevralia 1948 goda" [On the History of the Soviet-Yugoslav Conflict in 1948–1953: The Secret Soviet-Yugoslav-Bulgarian Meeting in Moscow on February 10, 1948], *Sovetskoe Slavianovedenie*, 1991, no. 4: 27–36, at p. 28.

60. Schoenfeld (Bucharest) to Secretary of State, February 2, 1948, RG-59, 871.00/2-248, NA; Carlyle, *Documents on International Affairs, 1947–1948*, pp. 298–99.

61. Vladimir Dedijer, *Tito* (New York: Simon & Schuster, 1952), p. 318; Milovan Djilas, *Conversations with Stalin* (New York: Harcourt, Brace & World, 1962), p. 178; Edvard Kardelj, *Boj za priznanje in neodvisnost nove Jugoslavije: Spomini* [The Struggle for the Recognition and Independence of New Yugoslavia] (Ljubljana: Državna založba Slovenije, 1980), pp. 105–6.

62. Dedijer, *Tito*, pp. 321–22; Kardelj, *Boj za priznanje*, pp. 112–15; Vladimir Dedijer, *Novi prilozi za biografiju Josipa Broza Tita* [New Supplements to the Biography of Josip Broz Tito], vol. 3 (Belgrade: Rad, 1984), p. 284.

63. Gibianskii, "Problems of East European International-Political Structuring," pp. 34–35.

64. Charles G. Stefan, "The Emergence of the Soviet–Yugoslav Break: A Personal View from the Belgrade Embassy," *Diplomatic History* 6 (1982): 387–404, at p. 393; Leonid Ia. Gibianskii, "Sekretnaia sovetsko-iugoslavskaia perepiska 1948 goda" [The Secret Soviet-Yugoslav Correspondence in 1948], *Voprosy istorii*, 1992, no. 4–5: 119–36, at p. 121.

65. Lavrentev to Molotov, quoted in Girenko, *Stalin-Tito*, pp. 348–50.

66. Minutes of central committee meeting, March 1, 1948, in Dedijer, *Novi prilozi*, 3, pp. 303–8; Gibianskii, "Sekretnaia sovetsko-iugoslavskaia perepiska," p. 135.

67. Molotov to Tito, March 13, 1948, in Gibianskii, "Sekretnaia sovetsko-iugoslavskaia perepiska," p. 124.

68. Molotov to Tito, March 18, 1948, ibid., pp. 124–25.

69. Dietrich Staritz, "Einheits- und Machtkalküle der SED (1946–1948)," in *"Provisorium für längstens ein Jahr": Protokoll des Kolloquiums Die Gründung der DDR*, ed. Elke Scherstjanoi (Berlin: Akademie, 1993), pp. 15–31, at pp. 27–28.

70. Note by Horák on conversation with Molotov, October 22, 1947, Zprávy ZÚ Moskva [Reports of Foreign Posts, Moscow], 1947, 2947/B/47, MZV.

71. A. Paniushkin, "Politicheskii otchet Posolstva SSSR v SShA za 1947 g." [Political Summary by the USSR Embassy in the U.S. for 1947], p. 13, 0129/1947/31A/1, AVPRF.

72. Statement by Marshall, December 10, 1947, in Carlyle, *Documents on International Affairs, 1947–1948*, pp. 514–15; U.S. delegation at the CFM to Truman, Acting Secretary of State and others, December 10, 1947, *FRUS*, 1947, vol. 2, pp. 762–64.

73. Secretary of State to Acting Secretary of State, December 11, 1947, *FRUS*, 1947, vol. 2, pp. 764–65.

74. U.S. delegation at the CFM to Truman, Acting Secretary of State, and others, December 15, 1947, *FRUS*, 1947, vol. 2, pp. 770–72.

75. Statement by Molotov, December 12, 1947, in Carlyle, *Documents on International Affairs, 1947–1948*, pp. 515–21.

76. Harold Nicolson, *Diaries and Letters*, vol. 3 (New York: Atheneum, 1968), p. 116.

77. Piers Dixon, *Double Diploma: The Life of Sir Pierson Dixon, Don and Diplomat* (London: Hutchinson, 1968), p. 192.

78. Molotov's statement to the press, December 31, 1947, in Carlyle, *Documents on International Affairs, 1947–1948*, pp. 545–54.

79. "Archivio Pietro Secchia, 1945–1973," *Annali Feltrinelli*, vol. 19 (Milan: Feltrinelli, 1979), pp. 611–27, 425–27.

80. Norman Naimark, *The Russians in Germany: The History of the Soviet Zone of Occupation, 1945–1949* (Cambridge, Mass.: Harvard University Press, 1995), pp. 341–49.

81. Murphy (Berlin) to Secretary of State, March 3, 1947, *FRUS*, 1948, vol. 2, pp. 878–79.

82. Paniushkin to Acting Secretary of State (Lovett), February 13, 1948, *FRUS*, 1948, vol. 2, pp. 338–39.

83. Minutes of Czechoslovak-Polish-Yugoslav conference of foreign ministers, Prague, February 17–18, 1948, Secretariat of the Minister, 1945–55, box 89, MZV.

84. Communiqué, March 6, 1948, in Carlyle, *Documents on International Affairs, 1947–1948*, pp. 556–58.

85. George F. Kennan, "Resumé of World Situation," November 6, 1947, *FRUS*, 1947, vol. 1, pp. 770–77, at p. 771.

86. Karel Kaplan, *Poslední rok prezidenta: Edvard Beneš v roce 1948* [The President's Last Year: Edvard Beneš in 1948] (Brno: Doplněk, 1994), pp. 36–37.

87. Steinhardt (Prague) to Secretary of State, April 30, 1948, *FRUS*, 1948, vol. 4, p. 750.

88. Miroslav Bouček and Miloslav Klimeš, *Dramatické dny února 1948* [The Dramatic Days of February 1948] (Prague: Naše vojsko, 1973), pp. 142, 304–5. Jindřich Veselý, *Kronika únorových dnů* [A Chronicle of the February Days] (Prague: Svoboda, 1958), p.164.

89. Cited in Karel Kaplan, *The Short March: The Communist Takeover of Power in Czechoslovakia, 1945–1948* (New York: St. Martin's Press, 1987), p. 179.

90. Roy Allison, *Finland's Relations with the Soviet Union, 1944–1984* (New York: St. Martin's Press, 1985), p. 21.

91. Richard Pakenham's statement in the House of Lords, quoted in Jusi Hanhimäki, "'Containment' in a Borderland: The United States and Finland, 1948–49," *Diplomatic History* 18 (1994): 353–74, at p. 357.

92. Alan Bullock, *Ernest Bevin: Foreign Secretary, 1945–1951* (London: Heinemann, 1983), p. 528.

93. Djilas, *Rise and Fall*, p. 155.

94. A. F. Upton, *Communism in Scandinavia and Finland: Politics of Opportunity* (Garden City, N.Y.: Doubleday, 1973), p. 290.

95. Quoted in David Dilks, "The British View of Security: Europe and a Wider World, 1945–1948," in *Western Security, the Formative Years: European and Atlantic Defence, 1947–1953*, ed. Olav Riste (Oslo: Universitetsforlaget, 1985), pp. 25–59, at p. 52.

96. Clay to Chamberlin, March 5, 1948, in *The Papers of General Lucius D. Clay: Germany, 1945–1949*, vol. 2, ed. Jean Edward Smith (Bloomington: Indiana University Press, 1974), pp. 568–69; Bidault, quoted in Pierre Melandri, "France and the Atlantic Alliance 1950–1953: Between Great Power Policy and European Integration," in Riste, *Western Security*, pp. 266–82, at p. 266.

97. Sheila Kerr, "The Secret Hotline to Moscow: Donald Maclean and the Berlin Crisis of 1948," in *Britain and the Cold War*, ed. Ann Deighton (London: Macmillan, 1990), pp. 71–87, at p. 75.

98. Cees Wiebes and Bert Zeeman, "The Pentagon Negotiations March 1948: The Launching of the North Atlantic Treaty," *International Affairs* [London] 59 (1982–83): 351–63, at p. 363.

99. Murphy (Berlin) to Secretary of State, March 20 and April 1, 1948, *FRUS*, 1948, vol. 2, pp. 883–86.

100. Aleksei M. Filitov, *Germanskii vopros: Ot raskola k obedineniiu* [The German Question: From Partition to Unification] (Moscow: Mezhdunarodnye otnosheniia, 1993), p. 104.

101. See note 68 above.

102. Cannon (Belgrade) to Secretary of State, December 24, 1948, RG-59, 875.00/12-2448, NA; Chuvakhin, "S diplomaticheskoi missiei v Albanii," p. 125.

103. Wittner, *American Intervention in Greece*, p. 263.

104. Tito to Molotov, March 18 and 20, 1948, in Gibianskii, "Sekretnaia sovetsko-iugoslavskaia perepiska," pp. 124–26.

105. Stalin to Tito, March 27, 1948, in Gibianskii, "Sekretnaia sovetsko-iugoslavskaia perepiska," pp. 127–29.

106. Adam B. Ulam, *The Communists: The Story of Power and Lost Illusions, 1948–1991* (New York: Scribner's, 1992), p. 23.

107. Jukka Nevakivi, "American Reactions to the Finnish-Soviet Friendship Treaty of 1948," *Scandinavian Journal of History* 13 (1988): 279–91, at p. 283.

108. Maxim L. Korobochkin, "USSR and the Treaty of Friendship, Cooperation and Mutual Assistance with Finland," in *Finnish-Soviet Relations, 1944–1948*, ed. Jukka Nevakivi (Helsinki: Department of Political History, University of Helsinki, 1994), pp. 169–89, at p. 183.

109. Treaty of friendship, co-operation and mutual assistance between the USSR and Finland, April 6, 1948, in Carlyle, *Documents on International Affairs, 1947–1948*, pp. 315–17.

110. James E. Miller, "Taking Off the Gloves: The United States and the Italian Elections of 1948," *Diplomatic History* 7 (1983): 35–55.

111. Tito to Stalin and Molotov, April 13, 1948, in *Yugoslavia and the Soviet Union, 1939–1973: A Documentary Survey*, ed. Stephen Clissold (London: Oxford University Press, 1975), pp. 174, 177.

112. Gibianskii, "Sovetsko-iugoslavskii konflikt," pp. 25–27.

113. Central Committee of the Communist Party of the Soviet Union to Central Committee of the Communist Party of Yugoslavia, May 4, 1948, in *The Soviet-Yugoslav Controversy*, pp. 23–35, at p. 34.

114. Statement by Smith to Molotov, May 4, and Molotov's reply, May 9, 1948, in Carlyle, *Documents on International Affairs, 1947–1948*, pp. 153–59; TASS statement, May 11, 1948, in *Vneshniaia politika Sovetskogo Soiuza, 1948 god: Dokumenty i materialy* [Foreign Policy of the Soviet Union in 1948: Documents and Materials], pt. 1 (Moscow: Gospolitizdat, 1950), pp. 195–202.

3 A Harvest of Blunders

1. Murphy (Berlin) to Secretary of State, June 17, 1948, *FRUS*, 1948, vol. 2, pp. 908–9.

2. Mikhail M. Narinsky, "Soviet Policy and the Berlin Blockade, 1948–1949" (paper presented at the conference "The Soviet Union, Germany, and the Cold War, 1945–1962," Essen, June 28–30, 1994).

3. Clay (Berlin) to Sokolovskii, June 18, 1948, *FRUS*, 1948, vol. 2, pp. 909–10.

4. "Proclamation of the Soviet Military Administration to the German Population," June 19, 1948, in *The Soviet Union and the Berlin Question* (Moscow: Ministry of Foreign Affairs, 1948), pp. 21–24.

5. Sokolovskii (Berlin) to Clay, June 20, 1948, in *The Soviet Union and the Berlin Question*, pp. 25–27.

6. Letter by Marshall to his colleagues on the Control Council, June 20, 1948, in *Documents on International Affairs, 1947–1948*, ed. Margaret Carlyle (London: Oxford University Press, 1952), pp. 579–80.

7. Sokolovskii (Berlin) to Clay, June 22, 1948, in *The Soviet Union and the Berlin Question*, pp. 28–29.

8. Murphy (Berlin) to Secretary of State, June 23, 1948, *FRUS*, 1948, vol. 2, pp. 212–14.

9. "Order No. 111," June 23, 1948, in *Documents on Germany under Occupation, 1945–1954*, ed. Beate Ruhm von Oppen (London: Oxford University Press, 1955), pp. 295–300.

10. Murphy (Berlin) to Secretary of State, June 23, 1948, *FRUS*, 1948, vol. 2, p. 915.

11. Statement by Prague conference of foreign ministers, June 24, 1948, in *The Soviet Union and the Berlin Question*, pp. 32–41.

12. Quoted in minutes of the meeting of the Czechoslovak party presidium, June 28, 1948, p. 11, 02/1/2/125, AÚV KSČ.

13. Cyrill Buffet, *Mourir pour Berlin: La France et l'Allemagne, 1945–1949* (Paris: Colin, 1991), p. 190.

14. Mikhail Narinskii, "Sovetskii Soiuz i berlinskii krizis 1948–1949 gg." [The Soviet Union and the Berlin Crisis of 1948–49] (paper presented at the conference "The Soviet Union and Europe in the Cold War (1943–1953)," Cortona, September 23–24, 1994), pp. 9–13.

15. Cf. Hannes Adomeit, *Soviet Risk-Taking and Crisis Behavior: A Theoretical and Empirical Analysis* (London: Allen & Unwin, 1982), pp. 164–66.

16. Clay (Berlin) to Draper, June 27, 1948, in *The Papers of General Lucius Clay: Germany, 1945–1949*, vol. 2, ed. Jean E. Smith (Bloomington: Indiana University Press, 1974), pp. 707–8; Robert Murphy, *Diplomat among Warriors* (Garden City, N.Y.: Doubleday, 1956), p. 174.

17. Viktor M. Gobarev, "Soviet Military Plans and Activities during the Berlin Crisis, 1948–1949" (paper presented at the conference "The Soviet Union, Germany, and the Cold War, 1945–1962," Essen, June 28–30, 1994).

18. Kenneth W. Condit, *The History of the Joint Chiefs of Staff: The Joint Chiefs of Staff and National Policy*, vol. 2, *1947–1949* (Wilmington, Del.: Glazier, 1979), p. 125; Walter Millis, ed., *The Forrestall Diaries* (New York: Viking, 1951), pp. 454–55.

19. Richard K. Betts, *Nuclear Blackmail and Nuclear Balance* (Washington, D.C.: Brookings, 1987), pp. 23–31.

20. John L. Gaddis, "The Origins of Self-Deterrence: The United States and the Non-Use of Nuclear Weapons, 1945–1958," in his *The Long Peace: Inquiries into the History of the Cold War* (New York: Oxford University Press, 1987), p. 110.

21. Murphy (Berlin) to Secretary of State, June 30, 1948, *FRUS*, 1948, vol. 2, p. 232; Statement by Marshall, June 30, 1948, *Department of State Bulletin*, July 11, 1948, p. 54.

22. Murphy (Berlin) to Secretary of State, July 1, and to U.S. Embassy in France, July 3, 1948, *FRUS*, 1948, vol. 2, pp. 941, 948–50.

23. Soviet note to U.S. government, July 14, 1948, in *The Soviet Union and the Berlin Question*, pp. 42–46.

24. Smith (Moscow) to Secretary of State, July 24, 1948, *FRUS*, 1948, vol. 2, pp. 984–85.

25. "Richtlinien für die Verfassung der deutschen demokratischen Republik," August 4, 1948, Pieck NL 36/763, ZK SED, SAPMO–BA.

26. Narinskii, "Sovetskii Soiuz i berlinskii krizis," p. 17.

27. Rolf Steininger, "Wie die Teilung Deutschlands verhindert werden sollte: Der Robertson-Plan aus dem Jahre 1948," *Militärgeschichtliche Mitteilungen* 33 (1983): 49–89.

28. Avi Shlaim, "Britain, the Berlin Blockade and the Cold War," *International Affairs* [London] 60 (1984): 1–14.

29. Anne Deighton, *The Impossible Peace: Britain, the Division of Germany and the Origins of the Cold War* (Oxford: Clarendon, 1990), pp. 122–25.

30. Smith (Moscow) to Secretary of State, August 3, 1948, *FRUS*, 1948, vol. 2, pp. 999–1006.

31. Frank Roberts, "Stalin, Khrushchev i Berlinskie krizisy" [Stalin, Khrushchev, and the Berlin Crises], *Mezhdunarodnaia zhizn*, 1991, no. 10: 130–43, at p. 135.

32. Smith (Moscow) to Secretary of State, August 3, 1948, *FRUS*, 1948, vol. 2, p. 1006.

33. Ibid., August 6, 1948, *FRUS*, 1948, vol. 2, p. 1018.

34. Ibid., August 12, 17, and 24, 1948, *FRUS*, 1948, vol. 2, pp. 1035–38, 1042–47, 1065–68.

35. Draft proposal by Stalin, August 24, 1948, *FRUS*, 1948, vol. 2, pp. 1069–70.

36. Smith (Moscow) to Secretary of State, August 27 and 30, 1948, *FRUS*, 1948, vol. 2, pp. 1088–90 (quote on p. 1090) and 1092–97.

37. Murphy (Berlin) to Secretary of State, August 13, 1948, *FRUS*, 1948, vol. 2, pp. 1038–40.

38. Smith (Moscow) to Secretary of State, September 16, 1948, *FRUS*, 1948, vol. 2, pp. 1160–62.

39. Vyshinskii (Moscow) to Smith, September 12 and 13, 1948, 0129/1948/32/1/011/204, pp. 22–23, AVPRF; Dietrich Staritz, "Die SED, Stalin und die Gründung der DDR," *Aus Politik und Zeitgeschichte*, January 25, 1991, pp. 3–16, at pp. 6–7.

40. Narinskii, "Sovetskii Soiuz i berlinskii krizis," p. 24.

41. Wilfried Loth, *Stalins ungeliebtes Kind: Warum Moskau die DDR nicht wollte* (Berlin: Rowohlt, 1994), pp. 135–42.

42. U.S. aide-mémoire, September 12, Soviet aide-mémoire, and Smith (Moscow) to Secretary of State, September 18, 1948, *FRUS*, 1948, vol. 2, pp. 1152–55, 1162–73.

43. Minutes of Marshall's meeting with Bevin and Schuman, September 21, 1948, *FRUS*, 1948, vol. 2, p. 1178; Acting Secretary of State to Paniushkin, September 22, 1948, *FRUS*, 1948, vol. 2, pp. 1180–81.

44. Paniushkin to Acting Secretary of State, September 25, 1948, *FRUS*, 1948, vol. 2, pp. 1181–84, at p. 1183.

45. TASS communiqué, September 26, 1948, referred to in *FRUS*, 1948, vol. 2, p. 1184.

46. Acting Secretary of State to Paniushkin, September 26, and resolution by neutral powers at the Security Council, October 25, 1948, mentioned in Editorial Note, *FRUS*, 1948, vol. 2, pp. 1187–93 and 1233–34.

47. Molotov to Kohler, October 3, 1948, *FRUS*, 1948, vol. 2, pp. 1201–10, at p. 1208.

48. "On the Situation in the Communist Party of Yugoslavia," June 28, 1949, in *The Cominform: Minutes of the Three Conferences, 1947/1948/1949*, ed. Giuliano Procacci (Milan: Feltrinelli, 1994), pp. 610–21.

49. Cannon (Belgrade) to Secretary of State, November 5, 1948, RG-59, 860H.00 (W)/11-548, NA.

50. Cannon (Belgrade) to Secretary of State, July 31, 1948, *FRUS*, 1948, vol. 4, pp. 634–37.

51. See chap. 2, n. 15.

52. Reams (Belgrade) to Secretary of State, September 27, 1948, RG-59, 860H.00/9-2748, NA.

53. Ivo Banac, *With Stalin against Tito: Cominformist Splits in Yugoslav Communism* (Ithaca, N.Y.: Cornell University Press, 1988), pp. 129–30.

54. *Khrushchev Remembers* (Boston: Little, Brown, 1970), p. 600.

55. Nathan Leites and Elsa Bernant, *The Ritual of Liquidation: The Case of the Moscow Trials* (Glencoe, Ill.: Free Press, 1954), p. 388.

56. G[hita] I[onescu], "The Evolution of the Cominform, 1947–1950," *World Today* 6 (May 1950): 213–28, at pp. 217–25.

57. See chap. 2, n. 24.

58. Petur Avramov, "Razvitie na vuzgleda za Devetoseptemvriiskoto vustanie i narodno-demokratichnata vlast (1944–1948 g.)" [The Evolution of the Views on the September 9 Uprising and the People's Democracy], Istoricheski pregled, 1982: 78–94, at p. 92; Georgi Dimitrov, Political Report Delivered to the V Congress of the Bulgarian Communist Party (Sofia: Ministry of Foreign Affairs, 1949).

59. Iu. S. Aksenov, "Apogei stalinizma: Poslevoennaia piramida vlasti" [The Apogee of Stalinism: The Postwar Pyramid of Power], Voprosy istorii KPSS, 1990, no. 11: 90–104, at p. 101.

60. Marshall D. Shulman, Stalin's Foreign Policy Reappraised (Cambridge:, Mass.: Harvard University Press, 1963), pp. 41–42; Yaacov Ro'i, Soviet Decision Making in Practice: The U.S.S.R. and Israel, 1947–1954 (New Brunswick, N.J.: Transaction Books, 1982), p. 317.

61. Iurii N. Zhukov, "Borba za vlast v rukovodstve SSSR v 1945–1952 godakh" [The Struggle for Power in the USSR Leadership in 1945–1952], Voprosy istorii, 1995, no. 1: 23–39, at p. 27.

62. See chap. 9, n. 23, and chap. 10, n. 15.

63. Paul Marantz, "Soviet Foreign Policy Factionalism under Stalin," Soviet Union 2, pt. 1 (1976): 91–107; Gavriel D. Ra'anan, International Policy Formation in the U.S.S.R.: Factional "Debates" during the Zhdanovshchina (Hamden, Conn.: Archon, 1983).

64. Werner G. Hahn, Postwar Soviet Politics: The Fall of Zhdanov and the Defeat of Moderation, 1946–53 (Ithaca, N.Y.: Cornell University Press, 1982), pp. 98–103; Shulman, Stalin's Foreign Policy Reappraised, p. 47.

65. Aksenov, "Apogei stalinizma," p. 101.

66. Tanigawa Yoshihiko, "The Cominform and Southeast Asia," in The Origins of the Cold War in Asia, ed. Yonosuke Nagai and Akira Irye (New York: Columbia University Press, 1977), pp. 372–74.

67. Interview with Shanti Bose, an organizer of the conference, Calcutta, April 4, 1990.

68. Geoffrey Jukes, The Soviet Union in Asia (Sydney: Angus & Robertson, 1973), p. 103.

69. Odd Arne Westad, Cold War and Revolution: Soviet-American Rivalry and the Origins of the Chinese Civil War, 1944–1946 (New York: Columbia University Press, 1993), pp. 173–74.

70. John H. Kautsky, Moscow and the Communist Party of India: A Study in the Postwar Evolution of International Communist Strategy (New York: Wiley, 1956), pp. 33–34.

71. Ruth T. McVey, The Soviet View of the Indonesian Revolution: A Study in the Russian Attitude towards Asian Nationalism (Ithaca, N.Y.: Department of Far Eastern Studies, Cornell University, 1957), pp. 58–73.

72. Uri Bialer, "The Czech-Israeli Arms Deal Revisited," Journal of Strategic Studies 8 (1985): 307–15.

73. Jiří Dufek, Karel Kaplan, and Vladimír Šlosar, Československo a Izrael v letech 1947–1953: Studie [Czechoslovakia and Israel, 1947–1953: Studies] (Brno: Doplněk, 1993), pp. 40–80, 11–89.

74. Arnold Krammer, The Forgotten Friendship: Israel and the Soviet Bloc, 1947–53 (Urbana: University of Illinois Press, 1974), p. 77.

75. Pavel Sudoplatov and Anatoly Sudoplatov, Special Tasks: The Memoirs of an Unwanted Witness—A Soviet Spymaster (Boston: Little, Brown, 1994), pp. 285–98.

76. Gregor Aronson, "The Jewish Question during the Stalin Era," in *Russian Jewry, 1917–1967,* ed Aronson *et al.* (New York: Yoseloff, 1969), pp. 171–208, at pp. 192–95. Léon Leneman, *La tragédie des Juifs en U.R.S.S.* (Paris: Desclée de Brouwer, 1959), pp. 112–38.

77. Amy Knight, *Beria: Stalin's First Lieutenant* (Princeton: Princeton University Press, 1993), pp. 146–47.

78. Note by Suslov to Stalin, November 26, 1946, cited in the central committee memorandum "Ob antikonstitutsionnoi praktike 30–40-kh i nachala 50-kh godov" [The Unconstitutional Practices of the Nineteen-Thirties, Forties, and Early Fifties], December 25, 1988, *Istochnik* 1955, no. 1: 117–32, at p. 128.

79. *Khrushchev Remembers,* p. 260.

80. Golda Meir, *My Life* (London: Weidenfeld & Nicolson, 1975), pp. 205–207.

81. Benjamin Pinkus, *The Soviet Government and the Jews, 1948–1967: A Documented Study* (Cambridge: Cambridge University Press, 1984), pp. 39–42, 195–96.

82. "O tak nazyvaemom 'dele Evreiskogo Antifashistskogo Komiteta'" [The So-called Affair of the Jewish Anti-Fascist Committee], *Izvestiia TsK KPSS,* 1989, no. 12: 35–40, at p. 37.

83. Yaacov Ro'i, "Soviet-Israeli Relations, 1947–1954," in *The U.S.S.R. and the Middle East,* ed. Michael Confino and Shimon Shamir (Jerusalem: Israel Universities Press, 1973), pp. 123–46, at p. 132.

84. R.J.A. Schlesinger, "Some Materials on the Recent Attacks against Cosmopolitanism," *Soviet Studies* 1 (1949–50): 178–88.

85. "Report on Internal Political and Social Development in the Soviet Union for February 1949," Kohler (Moscow) to Secretary of State, March 4, 1949, RG-59, 861.00/3-449, NA; Pinkus, *The Soviet Government and the Jews,* pp. 156–57, 172–78, 185–89.

86. Kommunistische Partei Österreichs, Landesleitung Wien, "Aktionsplan," November 15, 1948, box 1 U 16, dossier 2 (mai 1945–janvier 1955), Archives of the Ministry of Defense, Service Historique de l'Armée de Terre, Vincennes. I am indebted to Professor Günter Bischof of the University of New Orleans for a copy of this document. See also his "'Prag liegt westlich von Wien: 'Internationale Krisen im Jahre 1948 und Ihr Einfluß auf Österreich," in Günter Bischof und Josef Leidenfrost, ed., *Die bevormundete Nation: Österreich und die Alliierten, 1945–1949* (Innsbruck: Haymon, 1988), pp. 315–45, at pp. 336–38.

87. Vladimir Dedijer, *Novi prilozi za biografiju Josipa Broza Tita* [New Supplements to the Biography of Josip Broz Tito], vol. 3 (Belgrade: Rad, 1984), p. 268.

88. Artiom A. Ulunian, "SSSR i 'grecheskii vopros' (1946–1953): Problemy i otsenki" [The USSR and the "Greek Question": Problems and Interpretations] (paper presented at the conference "The Soviet Union and Europe in the Cold War (1943–1953)," Cortona, September 23–24, 1994), p. 13.

89. Yugoslavia began to curtail its deliveries in October 1948. Nicholas Pappas, "The Soviet-Yugoslav Conflict and the Greek Civil War," in *At the Brink of War and Peace: The Tito-Stalin Split in a Historical Perspective,* ed. Wayne S. Vucinich (New York: Columbia University Press, 1982), pp. 219–37, at pp. 224–25.

90. Elisabeth Barker, "Yugoslav Policy towards Greece, 1947–1949," in *Studies in the History of the Greek Civil War, 1945–1949,* ed. Les Baerentzen *et al.* (Copenhagen: Museum Tusculanum Press, 1987), pp. 263–95, at pp. 282–83.

220 NOTES TO PAGES 57–60

91. Henryk Różański, "Początki RWPG" [The Beginnings of the Council of Mutual Economic Assistance], *Polityka* [Warsaw], May 9, 1987.

92. Quoted in Karel Kaplan, "RVHP a Československo 1949–1955" [The Council of Mutual Economic Assistance and Czechoslovakia], *Svědectví* [Paris] 15, no. 58 (1979): 284–96, at pp. 287–88.

93. Paul Marer, "Soviet Economic Policy in Eastern Europe," in *Reorientation and Commercial Relations of the Economies of Eastern Europe*, ed. John P. Hardt. (Washington: U.S. Government Printing Office, 1974), p. 136.

94. Geir Lundestad, "Empire by Invitation? The United States and Western Europe, 1945–1952," *Journal of Peace Research* 23 (1986): 263–77.

95. Thomas G. Paterson, *Meeting the Communist Threat: Truman to Reagan* (New York: Oxford University Press, 1988), pp. 45–53.

96. "Estimate of Soviet Intentions and Capabilities 1948–1954," January 2, 1948, CCS 092 USSR (3–27–45), sec. 27, RG-218, NA.

97. NSC 10/2, June 18, 1948, in *CIA Cold War Records: The CIA under Harry Truman*, ed. Michael Warner (Washington, D.C.: Central Intelligence Agency, 1994), pp. 213–16.

98. Warner Schilling, Paul Hammond, and Glenn Snyder, *Strategy, Politics, and Defense Budgets* (New York: Columbia University Press, 1962), pp. 29–30.

99. Harry S. Truman, *Memoirs*, vol. 2, *Years of Trial and Hope* (Garden City, N.Y.: Doubleday, 1956), p. 171.

100. David A. Rosenberg, "U.S. Nuclear Stockpile, 1945 to 1950," *Bulletin of the Atomic Scientists* 38 (May 1982): 25–30, at p. 26; Melvyn P. Leffler, *A Preponderance of Power: National Security, the Truman Administration and the Cold War* (Stanford, Calif.: Stanford University Press, 1992), pp. 225–26.

101. Matthew Evangelista, *Innovation and the Arms Race: How the United States and the Soviet Union Develop New Military Technologies* (Ithaca, N.Y.: Cornell University Press, 1988), pp. 92–94.

102. "United States Policy on Atomic Warfare," NSC-30, September 10, *FRUS*, 1948, vol. 1, pp. 624–28, at p. 626.

103. According to KGB General Sergei A. Kondrashev, *SALT II and the Growth of Mistrust*, transcript of the second conference of The Carter-Brezhnev Project (Providence, R.I.: Center for Foreign Policy Development of Brown University, 1994), p. 43.

104. Vladislav M. Zubok, "Soviet Intelligence and the Cold War: The 'Small' Committee of Information, 1952–53," *Diplomatic History* 18 (1994): 353–74, at p. 455.

105. Quoted in Michael MccGwire, *Military Objectives in Soviet Foreign Policy* (Washington, D.C.: Brookings Institution, 1987), pp. 19–20.

106. Referred to in "Politicheskii otchet Posolstva SSSR v SShA" [The Political Report of the USSR Embassy in the USA], July 21, 1949, pp. 28–30, 0129/1949/33/12/218, AVPRF.

107. *Khrushchev Remembers*, pp. 260–61; Teresa Torańska, *"Them": Stalin's Polish Puppets* (New York: Harper & Row, 1987), pp. 170–71.

108. See chap. 2, n. 70.

109. Kaplan, "RVHP a Československo," p. 285. On contemporary Soviet perceptions of the likelihood of war, cf. Michael MccGwire, "Rethinking War: The Soviets and European Security," *The Brookings Review*, spring 1988, pp. 3–12, at p. 4.

110. Matthew A. Evangelista, "Stalin's Postwar Army Re-appraised,"

International Security 7, no. 3 (1982–83): 110–38, at pp. 132–33; Condit, *The History of the Joint Chiefs of Staff*, 2: 315–16.

111. Thomas W. Wolfe, *Soviet Power and Europe, 1945–1970* (Baltimore, Md.: Johns Hopkins Press, 1970), pp. 38, 45–46.

112. See chap. 1, n. 102.

113. William C. Wohlworth, *The Elusive Balance: Power and Perceptions during the Cold War* (Ithaca, N.Y.: Cornell University Press, 1993), pp. 115–18.

114. Stalin to Wallace, May 18, 1948, in Carlyle, *Documents on International Affairs, 1947–1948*, pp. 163–64.

115. Truman, *Memoirs*, 2: 215; Smith (Moscow) to Secretary of State, November 9, 1948, *FRUS*, 1948, vol. 4, pp. 931–33.

116. "Punten voor een rede ter herdenking van de Oktober-revolutie in leden- en kadervergaderingen der C.P.N." [Points for an Address Commemorating the October Revolution at Meetings of Party Members and Activists], November 6, 1948, p. 4, CPN.

117. "Position of the United States with Respect to Germany Following the Breakdown of Moscow Discussions," by Policy Planning Staff, November 2, 1948, *FRUS*, 1948, vol. 2, pp. 1240–47, at pp. 1241–42.

118. Stalin to Kingsbury Smith, January 30, 1949, *FRUS*, 1949, vol. 5, pp. 562–63.

119. Nataliia Egorova, "Stalinskaia vneshniaia politika i Kominform, 1947–1953" [Stalin's Foreign Policy and the Cominform] (paper presented at the conference "The Soviet Union and Europe in the Cold War (1943–1953)," Cortona, September 23–24, 1994), pp. 13–14.

120. Irwin M. Wall, *French Communism in the Era of Stalin: The Quest for Unity and Integration, 1945–1962* (Westport, Conn.: Greenwood Press, 1983), p. 97.

121. Verbali Direzione P.C.I., February 2, 1949, p. 1, IG.

122. Wall, *French Communism*, p. 77; Shulman, *Stalin's Foreign Policy Reappraised*, p. 17; Ger Harmsen, *Nederlands kommunisme: Gebundelde opstellen* [Netherlands Communism: Collected Essays] (Nijmegen: Sociaalistiese Uitgeverij, 1982), p. 55.

123. Wall, *French Communism*, p. 97.

124. Shulman, *Stalin's Foreign Policy Reappraised*, p. 63.

125. Olav Riste, "Was 1949 a Turning Point? Norway and the Western Powers, 1947–1950," in *Western Security, the Formative Years: Europe and Atlantic Defence 1947–1953*, ed. Olav Riste (Oslo: Universitetsforlaget, 1985), pp. 128–49, at p. 139.

126. Zhukov, "Borba za vlast," 34.

127. Kennan to Bohlen, March 15, 1949, *FRUS*, 1949, vol. 5, pp. 592–94, at p. 594.

128. Zhukov, "Borba za vlast," p. 34.

4 Retreat and Consolidation

1. Notes by Pieck for a report on conversations with Stalin on December 18, 1948, dated December 27, 1948, in *Wilhelm Pieck: Aufzeichnungen zur Deutschlandpolitik, 1945–1953*, ed. Rolf Badstübner and Wilfried Loth (Berlin: Akademie, 1994), pp. 265–74.

2. Stalin to Kingsbury Smith, January 30, 1949, *FRUS*, 1949, vol. 5, pp. 562–63.

3. On its effects, see "Ogranicheniia vvedennye anglo-amerikanskimi i frantsuzskimi vlastiami" [The Restrictions Introduced by the Anglo-American and French Authorities], pp. 9–15, Vyshinskii's Secretariat, 07/1949/126/4/68, AVPRF.

4. Ann Tusa and John Tusa, *The Berlin Blockade* (London: Hodder & Stoughton, 1989), pp. 387–411.

5. Memorandum by Jessup on conversation with Malik, March 15, 1949, *FRUS*, 1949, vol. 3, pp. 695–98.

6. Memorandum by Jessup on conversation with Malik, March 21, 1949, *FRUS*, 1949, vol. 3, pp. 701–4.

7. Memorandum by Jessup on conversation with Malik on April 10, dated April 11, 1949, *FRUS*, 1949, vol. 3, pp. 717–20.

8. Statement by TASS, April 26, 1949, in *Berlin: Quellen und Dokumente, 1945–1951*, vol. 2 (Berlin: Spitzing, 1964), pp. 1545–46.

9. Communiqué of the Council of Foreign Ministers, May 5, 1949, referred to in *FRUS*, 1949, vol. 3, pp. 750–51.

10. George F. Kennan, *Memoirs, 1925–1950* (Boston: Little, Brown, 1967), pp. 423–24, 444–46.

11. The United States Delegation at the Council of Foreign Ministers to President Truman and the Acting Secretary of State, Paris, May 24, 1949, *FRUS*, 1949, vol. 3, pp. 917–19.

12. Memorandum by the Deputies for Austria of the United Kingdom, the United States and France to their Foreign Ministers, May 30, 1949, *FRUS*, 1949, vol. 3, pp. 931–34.

13. Communiqué of the Sixth Session of the Council of Foreign Ministers, June 20, 1949, *FRUS*, 1949, vol. 3, pp. 1062–65.

14. Audrey Kurth Cronin, *Great Power Politics and the Struggle over Austria, 1945–1955* (Ithaca, N.Y.: Cornell University Press, 1986), pp. 75–76.

15. Note by Pieck on conversation with Semenov, July 19, 1949, in Badstübner and Loth, *Wilhelm Pieck*, pp. 287–91.

16. Michael Lemke, "'Doppelte Alleinvertretung': Die nationalen Wiedervereinigungskonzepte der beiden deutschen Regierungen und die Grundzüge ihrer politischen Realisierung in der DDR (1949–1952/53)," *Zeitschrift für Geschichtswissenschaft* 40 (1992): 531–43, at p. 537.

17. Günter Albrecht, ed., *Dokumente zur Staatsordnung der Deutschen Demokratischen Republik*, vol. 1 (Berlin: Deutscher Zentralverlag, 1959), pp. 420–52.

18. "Disposition für die Regierungserklärung," September 8, 1949, Pieck NL 36/768, SAPMO-BA.

19. Alfred Grosser, *Germany in Our Time: A Political History of the Postwar Years* (New York: Praeger, 1970), pp. 72–74.

20. The United States Representative at the United Nations (Austin) to Acting Secretary of State, September 30, 1949, *FRUS*, 1949, vol. 3, pp. 1167–68.

21. Cronin, *Great Power Politics and the Struggle over Austria*, pp. 89–90.

22. Caffery (Paris) to Secretary of State, August 12, 1948, RG–59, 864.00/8–1248, NA.

23. Robert Conquest, *Stalin and the Kirov Murder* (New York: Oxford University Press, 1989).

24. Jože Pirjevec, *Il gran rifiuto: Guerra fredda e calda tra Tito, Stalin e l'Occidente* (Trieste: Stampa Triestina, 1990), pp. 138–39.

25. Leonhard (Belgrade) to Secretary of State, December 13, 1948, RG–59, 860H.00/12–1348, NA.

26. Vladimir Tismaneanu, "Ceauşescu's Socialism," *Problems of Communism* 34 (January–February 1985): 50–66, at pp. 56–57.

27. Hamilton Fish Armstrong, *Tito and Goliath* (New York: Macmillan, 1955), p. 231; William E. Griffith, *Albania and the Sino-Soviet Rift* (Cambridge, Mass.: MIT Press, 1963), p. 21.

28. Krystyna Kersten, *The Establishment of Communist Rule in Poland, 1943–1948* (Berkeley: University of California Press, 1991), pp. 438–50.

29. Andrzej Werblan, *Władysław Gomułka, Sekretarz Generalny PPR* [Władysław Gomułka, General Secretary of the Polish Workers' Party] (Warsaw: Książka i Wiedza, 1988), pp. 207–8, 571–81; see also chap. 1, n. 45.

30. Jan Ptasiński, *Pierwszy z trzech zwrotów czyli rzecz o Władysławie Gomułce* [The First of the Three Turnabouts, or the Case of Władysław Gomułka] (Warsaw: Krajowa Agencja Wydawnicza, 1984), pp. 117–20.

31. Minutes of the meeting of the presidium of the central committee of the Communist Party of Czechoslovakia, September 9, 1948, 02/1/3/132, AÚV KSČ.

32. Minutes of the meeting of the presidium of the central committee of the Communist Party of Czechoslovakia, June 28, 1948, p. 6, 02/1/2/125, AÚV KSČ.

33. Milan Churaň et al., *Kdo byl kdo v našich dějinách ve 20. století* [Who's Who in Our History in the 20th Century] (Prague: Libri, 1994), pp. 271–72.

34. Karel Kaplan, *Sovětští poradci v Československu, 1949–1956* [Soviet Advisers in Czechoslovakia] (Prague: Ústav pro soudobé dějiny AV ČR, 1993), pp. 10–11, 19.

35. Cannon (Belgrade) to Secretary of State, April 8, 1949, NG-59, 860H.002/4-849, NA.

36. Nicholas Pappas, "The Soviet-Yugoslav Conflict and the Greek Civil War," in *At the Brink of War and Peace: The Tito-Stalin Split in a Historical Perspective*, ed. Wayne S. Vucinich (New York: Columbia University Press, 1982), pp. 219–37, at p. 227.

37. Hristo H. Devedijev, *Stalinization of the Bulgarian Society, 1949–1953* (Philadelphia: Dorrance, 1975), pp. 15–16; Robert Conquest, *Power and Policy in the U.S.S.R.: The Study of Soviet Dynastics* (New York: St. Martin's Press, 1961), p. 165.

38. Adam B. Ulam, *Titoism and the Cominform* (Cambridge, Mass.: Harvard University Press, 1952), pp. 204–7.

39. "Protokol soveshchanii Informatsionnogo biuro kommunisticheskikh partii, 19–23 iuniia 1948" [Minutes of the Conference of the Information Bureau of Communist Parties, June 19–23, 1948], pp. 58–59, 575/1/2, RTsKhIDNI.

40. Petur Semerdzhiev, *Sudebnyi protsess Traicho Kostova v Bolgarii* [The Trial of Traicho Kostov in Bulgaria] (Jerusalem: Soviet and East European Research Centre, Hebrew University, 1980), pp. 30–38.

41. Vladimir Kolarov, "Struggle against Nationalism in Bulgarian Communist Party," *For a Lasting Peace, For a People's Democracy!* [Bucharest], May 15, 1949 (emphasis added); John D. Bell, *The Bulgarian Communist Party from Blagoev to Zhivkov* (Stanford, Calif.: Hoover Institution Press, 1986), pp. 105–107.

42. Cannon (Belgrade) to Secretary of State, April 8, 1949, RG-59, 860H.00 (W)/4-849, NA; "Political Summary Report on Hungary for June 1949," U.S. Legation Budapest to Secretary of State, July 5, 1949, RG-59, 864.00/7.549, NA.

43. Leonhart (Belgrade) to Secretary of State, March 29, 1949, RG-59, 860H.00/3-2949, NA; Cannon (Belgrade) to Secretary of State, April 2, 1949, RG-59, 860H.00(W)/4-149, NA.

44. Heath (Sofia) to Secretary of State, May 16, 1949, RG-59, 875.00/5, 1649, NA.

45. Dmitrii S. Chuvakhin, "S diplomaticheskoi missiei v Albanii, 1946–1952 gg." [With the Diplomatic Mission in Albania in 1946–1952], *Novaia i noveishaia istoriia*, 1995, no. 1: 114–31, at pp. 128–29.

46. Karel Kaplan, *Die politischen Prozesse in der Tschechoslowakei, 1948–1954* (Munich: Oldenbourg, 1986), pp. 189–90.

47. State Department to U.S. Legation in Budapest, October 7, 1949, *FRUS*, 1949, vol. 5, p. 58; evidence by Hungarian historian Mária Schmidt from Budapest archives reported in Sam Tanenhaus, "A 'Smoking Gun' in Hungary Points to Alger Hiss," *International Herald Tribune*, October 16–17, 1993.

48. Noel Field, "Hitching Our Wagon to a Star," *Mainstream* 14, no. 1 (January 1961): 3–17.

49. Flora Lewis, *The Man Who Disappeared: The Strange History of Noel Field* (London: Barker, 1966), p. 191.

50. Quoted in Armstrong, *Tito and Goliath*, p. 253.

51. Paul Ignotus, *Political Prisoner* (New York: Collier, 1964), p. 60.

52. Former Hungarian Prime Minister András Hegedüs, quoted in William Shawcross, *Crime and Compromise: János Kádár and the Politics of Hungary since Revolution* (New York: Dutton, 1974), p. 58.

53. Leonid Gibianskii, "The Last Conference of the Cominform," in *The Cominform: Minutes of the Three Conferences, 1947/1948/1949*, ed. Giuliano Procacci (Milan: Feltrinelli, 1994), pp. 645–67, at p. 657.

54. V. L. Musatov, "SSSR i vengerskie sobytiia 1956 g.: Novye arkhivnye materialy" [The USSR and the Hungarian Events of 1956: New Archival Materials], *Novaia i noveishaia istoriia*, 1993, no. 1: 3–22, at p. 7.

55. Charles Gati, "A Note on Communists and the Jewish Question in Hungary," in his *Hungary and the Soviet Bloc* (Durham, N.C.: Duke University Press, 1986), pp. 100–107; Raymond Taras, "Gomułka's 'Rightist-Nationalist Deviation,' the Postwar Jewish Communists, and the Stalinist Reaction in Poland, 1945–1950," *Nationalities Papers* 22, supplement 1 (1994): 11–27.

56. Michael Checinski, *Poland: Communism, Nationalism, Anti-Semitism* (New York: Karz-Cohl, 1982), pp. 11, 76–82.

57. András Hegedüs, *Im Schatten einer Idee: Eine Befragung von Zoltán Zsille zur Vergangenheitsbewältigung eines Stalinisten* (Zurich: Ammann, 1986) p. 95.

58. Milovan Djilas, *Rise and Fall* (London: Macmillan, 1985), p. 237.

59. Cannon (Belgrade) to Secretary of State, July 1, 1949, RG-59, 860H.00(W)/7-149, NA; Gilbert (Zagreb) to Secretary of State, July 14, 1949, RG-59, 860H.00/7-1449, NA.

60. Vladimir Dedijer, *Novi prilozi za biografiju Josipa Broza Tita* [New Supplements to the Biography of Josip Broz Tito], vol. 3 (Belgrade: Rad, 1984), pp. 464–86.

61. Pirjevec, *Il gran rifiuto*, pp. 356–59; extracts from Soviet note, August 18, 1949, in Carlyle, *Documents on International Affairs 1949–50*, pp. 463–65.

62. *Pravda*, February 13, 1957, referred to in Philip Windsor, "Yugoslavia, 1951, and Czechoslovakia, 1968," in *Force without War: U.S. Armed Forces as a Political Instrument*, ed. Barry M. Blechman and Stephen S. Kaplan (Washington, D.C.: Brookings, 1978), pp. 440–512, at p. 451.

63. Ivo Banac, *With Stalin against Tito: Cominformist Splits in Yugoslav Communism* (Ithaca, N.Y.: Cornell University Press, 1988), pp. 242–43.

64. Russian denunciation of its treaty of April 11, 1945, with Yugoslavia, September 28, 1949, in Carlyle, *Documents on International Affairs 1949–50*, pp. 473–74; Peter Calvocoressi, *Survey of International Affairs 1949–50* (London: Oxford University Press, 1953), pp. 265–66.

65. Memorandum by Robert P. Joyce on meeting of Policy Planning Staff, April 1, 1949, *FRUS*, 1949, vol. 5, p. 12.

66. "U.S. Policy toward the Soviet Satellite States in Eastern Europe," P.P.S. 59, August 25, 1949, *FRUS*, 1949, vol. 5, pp. 21–26, at p. 25.

67. "Conclusions and Recommendations of the London Conference of October 24–26 of U.S. Chiefs of Mission to the Satellite States," undated, *FRUS*, 1949, vol. 5, pp. 28–35.

68. Stefan Troebst, "Vernichtungsterror und 'Säuberungen' in der Bulgarischen Kommunistischen Partei 1936 bis 1953," in *Kommunisten verfolgen Kommunisten: Stalinistischer Terror und "Säuberungen" in den kommunistischen Parteien Europas seit den dreißiger Jahren*, ed. Hermann Weber and Dietrich Staritz (Berlin: Akademie., 1993), pp. 470–86, at p. 480.

69. See chap. 1, n. 47.

70. Testimony of Lazar Brankov at the Rajk trial, in Marton Horváth, "Rajk Gang Exposed and Smashed," *For a Lasting Peace, For a People's Democracy!* September 23 and 30, 1949; indictment of Kostov, *Le procès de Traïtcho Kostov et de son groupe* (Sofia: Ministère des Affaires Etrangères, 1949).

71. *Le procès de Traïcho Kostov*, pp. 91–125, 492–98.

72. Ibid., p. 109.

73. Ibid., pp. 113–15; *László Rajk and His Accomplices before the People's Court: A Transcript of the Rajk Trial* (Budapest: Budapest Printing Press, 1949), p. 255.

74. Banac, *With Stalin against Tito*, pp. 332–33.

75. *Le procès de Traïtcho Kostov*, p. 109.

76. Beatrice Heuser, *Western "Containment" Policies in the Cold War: The Yugoslav Case, 1948–53* (London: Routledge, 1989), pp. 75–88, 117–24.

77. Ernest R. May, John D. Steinbrunner, and Thomas Wolfe, "History of the Strategic Arms Competition 1945–1972," part 1 (Washington, D.C.: Office of the Secretary of Defense, 1981), pp. 246–50, NSA.

78. Thomas W. Wolfe, *Soviet Power and Europe, 1945–1970* (Baltimore, Md.: Johns Hopkins Press, 1970), pp. 42–49; Ithiel de Sola Pool et al., *Satellite Generals: A Study of Military Elites in the Soviet Sphere* (Stanford, Calif.: Stanford University Press, 1955), passim.

79. Christian Greiner, "The Defence of Western Europe and the Rearmament of West Germany, 1947–1950," in *Western Security: The Formative Years: European and Atlantic Defence, 1947–1953*, ed. Olav Riste (Oslo: Universitetsforlaget, 1985), pp. 150–77, at pp. 151–52.

80. David Holloway, *Stalin and the Bomb: The Soviet Union and Atomic Energy, 1939-1956* (New Haven, Conn.: Yale University Press, 1994), p. 240.

81. Kenneth W. Condit, *The History of the Joint Chiefs of Staff: The Joint Chiefs of Staff and National Policy*, vol. 2, *1947-1949* (Wilmington, Del.: Glazier, 1979), pp. 288-314.

82. Holloway, *Stalin and the Bomb*, pp. 229-30.

83. Maj. Gen. V. Khlopov, "O kharaktere voennoi doktriny amerikanskogo imperializma" [The Nature of the Military Doctrine of American Imperialism], *Voennaia mysl*, 1950, no. 6: 67-78, referred to in Holloway, *Stalin and the Bomb*, pp. 238-39.

84. Kirk (Moscow) to Secretary of State, July 21, 1949, ibid., pp. 632-34.

85. Quoted in Ivan V. Kovalev, "Dialog Stalina s Mao Tszedunom" [Stalin's Dialog with Mao Zedong], *Problemy Dalnego Vostoka*, 1991, no. 6: 83-93, at p. 87.

86. William C. Wohlworth, *The Elusive Balance: Power and Perceptions during the Cold War* (Ithaca, N.Y.: Cornell University Press, 1993), pp. 115-18.

87. See chap. 3, n. 113.

88. "O tak nazyvaemom 'Leningradskom dele,'" [The So-called Leningrad Affair], *Izvestiia TsK KPSS*, 1989, no. 2: 126-34, at pp. 128-29, 133-34.

89. Conquest, *Power and Policy*, pp. 95-111; William O. McCagg Jr., *Stalin Embattled, 1943-48* (Detroit, Mich.: Wayne State University Press, 1978), pp. 118-46.

90. See chap. 9, n. 20.

91. Cf. Sidney I. Ploss, *Conflict and Decision-Making in Soviet Russia: A Case Study of Agricultural Policy, 1953-1963* (Princeton, N.J.: Princeton University Press, 1965), pp. 57-58.

92. "O tak nazyvaemom 'Leningradskom dele,'" pp. 129-31; V. A. Kutuzov, "Tak nazyvaemoe 'leningradskoe delo,' [The So-called Leningrad Affair], *Voprosy istorii KPSS* 1989, no. 3: 53-67, at pp. 62-67.

93. Aleksandr Nadzharov's interview with Rudolf Pikhoia, "Ia protivnik politicheskoi arkheologii" [I Am an Enemy of Political Archaeology], *Nezavisimaia gazeta* [Moscow], August 31, 1993; Amy Knight, *Beria: Stalin's First Lieutenant* (Princeton, N.J.: Princeton University Press, 1993), p. 151.

94. Kirk (Moscow) to Secretary of State, August 16, 1949, *FRUS*, 1949, vol. 5, pp. 646-47.

95. Reply to questions by Alexander Werth, September 17, 1946, quoted in Holloway, *Stalin and the Bomb*, p. 171; Milovan Djilas, *Conversations with Stalin* (New York: Harcourt, Brace & World, 1962), p. 153.

96. Herbert S. Dinerstein, *War and the Soviet Union: Nuclear Weapons and the Revolution in Soviet Military and Political Thinking* (New York: Praeger, 1959), pp. 3-9; Raymond L. Garthoff, *Soviet Strategy in the Nuclear Age* (New York: Praeger, 1958), p. 64.

97. Steven T. Ross, *American War Plans, 1945-1950* (New York: Garland, 1988), p. 155.

98. Cf. George H. Quester, "On the Identification of Real and Pretended Communist Military Doctrine," *Journal of Conflict Resolution* 10, no. 2 (1966): 172-79.

99. Thomas H. Hammond, "Did the United States Use Atomic Diplomacy against Russia in 1945?" in *From the Cold War to Détente*, ed. Peter J. Potichnyj and Jane P. Shapiro (New York: Praeger, 1976), pp. 26-56.

100. Russell D. Buhite and William C. Hamel, "War for Peace: The Question of an American Preventive War against the Soviet Union, 1945–1955," *Diplomatic History* 14 (1990): 367–84; Marc Trachtenberg, "A 'Wasting Asset': American Strategy and the Shifting Nuclear Balance, 1949–1954," in *History and Strategy*, ed. Marc Trachtenberg (Princeton, N.J.: Princeton University Press, 1991), pp. 128–35.

101. "Estimate of Status of Atomic Warfare in the USSR," September 20, 1949, in *CIA Cold War Records: The CIA under Harry Truman*, ed. Michael Warner (Washington, D.C.: Central Intelligence Agency, 1994), pp. 319–20.

102. Leonard Nikishin, "Stalin dormiva quando explose la bomba," *Mosca News*, November 9, 1989, pp. 12–13.

103. Pavel Sudoplatov and Anatoly Sudoplatov, *Special Tasks: The Memoirs of an Unwanted Witness—A Soviet Spymaster* (Boston: Little, Brown, 1994), p. 211.

104. TASS communiqué, September 23, quoted in Kirk (Moscow) to Secretary of State, September 25, 1949, *FRUS*, 1949, vol. 5, pp. 656–57.

105. John L. Gaddis, "The Origins of Self-Deterrence: The United States and the Non-Use of Nuclear Weapons, 1945–1958," in his *The Long Peace: Inquiries into the History of the Cold War* (New York: Oxford University Press, 1987), pp. 112–14; David A. Rosenberg, "American Atomic Strategy and the Hydrogen Bomb Decision," *Journal of American History* 66 (1979–80): 62–87, at pp. 62, 86–87.

106. Samuel R. Williamson Jr, and Steven L. Rearden, *The Origins of U.S. Nuclear Strategy, 1945–1953* (New York: St. Martin's Press, 1993), pp. 111–26.

107. "A Draft Resolution Submitted by the Soviet Union to the General Assembly on September 23, 1949," *United Nations Yearbook, 1948–1949* (New York: Columbia University Press, 1950), pp. 336–37.

108. David Holloway, "Research Note: Soviet Thermonuclear Development," *International Security* 4, no. 3 (1979–80): 192–97, at p. 193; Holloway, *Stalin and the Bomb*, pp. 295–99.

109. Matthew Evangelista, *Innovation and the Arms Race: How the United States and the Soviet Union Develop New Military Technologies* (Ithaca, N.Y.: Cornell University Press, 1988), pp. 166–75.

110. Speech by Malenkov, November 6, 1949, *For a Lasting Peace, For a People's Democracy!* November 11, 1949.

111. Memorandum by Gribanov for Vyshinskii on the state of negotiations for the Austrian State Treaty, December 17, 1949, pp. 174–78, 066/1949/30/141/AV-112, AVPRF; Cronin, *Great Power Politics and the Struggle over Austria*, pp. 89–94.

112. "The Defence of Peace and the Struggle against the Warmongers," November 16, 1949, in Procacci, *The Cominform*, pp. 676–707, at p. 699.

5 Resuming Advance

1. Marshall D. Shulman, *Stalin's Foreign Policy Reappraised* (Cambridge, Mass.: Harvard University Press, 1963), pp. 255–56.

2. "Report by the National Security Council on the position of the United States with respect to World Communism," NSC-7, March 30, 1947, *FRUS*, 1948, vol. 1, pt. 2, pp. 546–50, at p. 546.

3. "Report by the National Security Council on U.S. Objectives with Respect to the USSR to Counter Soviet Threats to U.S. Security," NSC 20/4, November 23, 1948, *FRUS*, 1948, vol. 1, pt. 2, pp. 663–69, at p. 668.

4. "United States Policy toward the Soviet Satellite States in Eastern Europe," NSC 58/2, December 8, 1949, *FRUS*, 1949, vol. 5, pp. 42–54, at p. 43.

5. Beatrice Heuser, "Covert Action within British and American Concepts of Containment, 1948–51," in *British Intelligence, Strategy and the Cold War, 1945–1951*, ed. Richard Aldrich, (London: Routledge, 1992), pp. 65–84, at p. 69.

6. Beatrice Heuser, *Western "Containment" Policies in the Cold War: The Yugoslav Case, 1948–53* (London: Routledge, 1989), p. 285.

7. Nicholas Bethell, *The Great Betrayal: The Untold Story of Kim Philby's Biggest Coup* (London: Hodder & Stoughton, 1984), pp. 111–13.

8. Minutes of the meeting of the presidium of the central committee of the Communist Party of Czechoslovakia, June 28, 1948, p. 6, 02/1/2/125, AÚV KSČ.

9. Memorandum of conversation, by the Secretary of State, September 14, 1949, *FRUS*, 1949, vol. 6, pp. 414–16, at p. 415.

10. Bethell, *The Great Betrayal*, pp. 75, 121; David Smiley, *Albanian Assignment* (London: Chatto & Windus, 1984), pp. 158–59.

11. Philby in his memoirs implies that much, although he does not explicitly admit it. Kim Philby, *My Silent War* (New York: Grove Press, 1968), pp. 194–99; Bruce Page, David Leitch, and Phillip Knightley, *Philby: The Spy Who Betrayed a Generation* (London: Deutsch, 1968), pp. 193–99.

12. "Recent Military Activity in the Balkans and Its Significance," memorandum by Joint Intelligence Committee J.I.C.522/4, undated, RG-218, CCS 092 USSR (3–27–45) S.48, NA.

13. Jon Halliday, *The Artful Albanian: The Memoirs of Enver Hoxha* (London: Chatto & Windus, 1986), pp. 134–35.

14. Peter J. Stavrakis, *Moscow and Greek Communism, 1944–1949* (Ithaca, N.Y.: Cornell University Press, 1989), p. 182.

15. Christopher Andrew and Oleg Gordievsky, *KGB: The Inside Story of Its Foreign Operations from Lenin to Gorbachev* (London: Hodder & Stoughton, 1990), pp. 317–19; John Loftus, *The Belarus Secret* (New York: Knopf, 1982), pp. 79–110.

16. Jakub Berman at the meeting of the Cominform, November 16, 1949, *The Cominform: Minutes of the Three Conferences, 1947/1948/1949*, ed. Giuliano Procacci (Milan: Feltrinelli, 1994), p. 741.

17. Bethell, *The Great Betrayal*, pp. 161, 174.

18. Speech by Slánský, November 16, 1949, "Protokol soveshchaniia Informatsionnogo Biuro kommunisticheskikh partii 16–19 noiabria 1949 g." [Minutes of the Conference of the Information Bureau of Communist Parties], p. 55, 575/1/7, RTsKhIDNI.

19. *Pravda*, January 13, 1950.

20. *Khrushchev Remembers* (Boston: Little, Brown, 1970), p. 246.

21. "Schůze předsednictva ÚV KSČ dne 22. února 1950, materiál k bodu: Příprava zasedání ÚV KSČ, referát soudruha Kopřivy" [Meeting of the Presidium of the Central Committee, February 22, 1950: Materials for the Central Committee Session, Report by Comrade Kopřiva], 02/1; 16/209, AÚV KSČ.

22. *Rudé právo* [Prague], March 3, 1950.

23. *Trybuna Ludu* [Warsaw], February 6–14, 1950.

24. Note Verbale of the Bulgarian Legation to the Department of State, Washington, January 19, 1950, *FRUS*, 1950, vol. 4, pp. 504–5.

25. The Secretary of State to the Embassy in Czechoslovakia, Washington, March 28, 1950, and "Political and Economic Developments in Czechoslovakia during May and June 1950," report prepared by the Embassy in Czechoslovakia, Prague, July 12, 1950, *FRUS*, 1950, vol. 4, pp. 540–43, 570–72.

26. U.S. Senate, Select Committee to Study Governmental Operations with Respect to Intelligence Activities, *Final Report*, book 4, April 1976, Report 94–755 (94th Cong., 2d sess.) (Washington, D.C.: U.S. Government Printing Office, 1976), pp. 31–32.

27. Situation report by Antonín Prchal, January 14, 1952, quoted in Jan Frolík, "Nástin organizačního vývoje státobezpečnostních složek Sboru národní bezpečnosti v letech 1948–1989" [An Outline of the Organizational Development of the State Security Components of the National Security Corps in 1948–1989], *Sborník archivních prací* [Prague] 41 (1991): 447–510, at p. 482.

28. Joseph E. Jacobs (Prague) to Secretary of State, May 20, 1949, RG-59, 860F.00(W)5-2049, NA; interview with Gen. Vladimír Přikryl, Petr Radosta, "Nepohodlný svědek" [An Uncomfortable Witness], *Lidové noviny* [Prague], May 14, 1990, p. 4.

29. Condoleezza Rice, *The Soviet Union and the Czechoslovak Army, 1948–1983: Uncertain Allegiance* (Princeton, N.J.: Princeton University Press, 1984), pp. 60–62.

30. Josef Slanina and Zdeněk Vališ, *Generál Karel Kutlvašr* (Prague: Naše vojsko, 1993), pp. 102–6.

31. Karel Kaplan, *Die politischen Prozesse in der Tschechoslowakei, 1948–1954* (Munich: Oldenbourg, 1986), pp. 129–35.

32. Robert T. Holt, *Radio Free Europe* (Minneapolis: University of Minnesota Press, 1958), pp. 10–11.

33. Beatrice Heuser, "NSC-68 and the Soviet Threat: A New Perspective on Western Threat Perception and Policy Making," *Review of International Studies* 17, no. 1 (January 1991): 17–40, at p. 36; Wolf Oschlies, "Die Bulgarische Volksarmee: Einführung," in *Zur Geschichte der europäischen Volksarmeen*, ed. Peter Gosztonyi (Bonn: Hochwacht, 1976), pp. 227–70, at p. 244.

34. Thomas W. Wolfe, *Soviet Power and Europe, 1945–1970* (Baltimore, Md.: Johns Hopkins Press, 1970), p. 46.

35. "The reply of the Russian Government to a Statement issued by the United States' Department of State regarding the Proposal of July 16, 1947," July 22, 1947, *Documents on International Affairs, 1947–1948*, ed. Margaret Carlyle (London: Oxford University Press, 1952), pp. 716–17; Michael Schaller, *The American Occupation of Japan: The Origins of the Cold War in Asia* (New York: Oxford University Press, 1985), pp. 98–106.

36. Minutes of Policy Planning Staff meeting, September 22, 1947, quoted in Schaller, *The American Occupation of Japan*, p. 17.

37. Quoted in Michael Schaller, "Securing the Great Crescent: Occupied Japan and the Origins of Containment in South East Asia," *Journal of American History* 69 (1982–83): 392–414, at p. 396.

38. Quoted in Schaller, *The American Occupation of Japan*, p. 106. Emphasis added.

39. Niu Jun, "The Origins of Sino-Soviet Alliance" (paper presented at the conference "New Evidence on the Cold War in Asia," Hong Kong, January 9–12, 1996), p. 19.

40. Stalin to Mao Zedong, July 14, 1948, in Andrei M. Lebovskii, "Sekretnaia missiia A.I. Mikoiana v Kitai (ianvar–fevral 1949 g.)" [Mikoian's Secret Mission to China in January–February 1949], *Problemy Dalnego Vostoka*, 1995, no. 2: 96–111, at p. 99.

41. Nakajima Mineo, "The Sino-Soviet Confrontation in Historical Perspective," in *The Origins of the Cold War in Asia*, ed. Yonosuke Nagai and Akira Iriye (New York: Columbia University Press, 1977), pp. 203–23, at p. 208.

42. Stalin to Mao Zedong, January 10–11, and Mao Zedong to Stalin, January 13, 1949, in S.L.Tikhvinskii, "Perepiska I.V. Stalina s Mao Tszedunom v ianvare 1949 g." [Stalin's Correspondence with Mao Zedong in January 1949], *Novaia i noveishaiia istoriia*, 1994, no. 4–5: 132–40, at pp. 133–36.

43. Lebovskii, "Sekretnaia missiia," pp. 99–110.

44. Nin Jun, "The Origins of the Sino-Soviet Alliance," pp. 20–23. For a different view, see Brian Murray, "Stalin, the Cold War, and the Division of China: A Multi-Archival Mystery," Working Paper no. 12, *Cold War International History Project* (Washington, D.C.: Woodrow Wilson International Center for Scholars, 1995), pp. 5–10.

45. Shu Guang Zhang, *Deterrence and Strategic Culture: Chinese-American Confrontations, 1949–1958* (Ithaca, N.Y.: Cornell University Press, 1993), pp. 14, 21–22, 24.

46. Sergei I. Tikhvinskii, "O 'sekretnom demarshe' Chzhou Enlaia i neofitsialnykh peregovorakh s amerikantsami v iune 1949 g." [Zhou Enlai's "Secret Intervention" and the Unofficial Talks between the Chinese Communists and the Americans in June 1949], *Problemy Dalnego Vostoka*, 1994, no. 3: 133–38, at pp. 134–35.

47. Gordon H. Chang, *Friends and Enemies: The United States, China, and the Soviet Union, 1948–1972* (Stanford, Calif.: Stanford University Press, 1990), pp. 21–26.

48. Niu Jun, "The Origins of the Sino-Soviet Alliance," p. 19.

49. Chen Jian, "The Ward Case and the Emergence of Sino-American Confrontation, 1948–1950," *Australian Journal of Chinese Affairs* 30 (July 1993): 149–70.

50. Chen Jian, *China's Road to the Korean War: The Making of the Sino-American Confrontation* (New York: Columbia University Press, 1994), pp. 64–70.

51. John H. Kautsky, *Moscow and the Communist Party of India: A Study in the Postwar Evolution of International Communist Strategy* (New York: Wiley, 1956), pp. 86–87.

52. Shi Zhe, "With Mao and Stalin: The Reminiscences of Mao's Interpreter—Part II: Liu Shaoqi in Moscow," *Chinese Historians* 6, no. 1 (1993): 67–90, at p. 84.

53. Haruki Wada, "Stalin and the Japanese Communist Party, 1945–1953" (paper presented at the conference "New Evidence on the Cold War in Asia," Hong Kong, January 9–12, 1996), pp. 7–11.

54. Allen S. Whiting, *China Crosses the Yalu: The Decision to Enter the Korean War* (Stanford, Calif.: Stanford University Press, 1960), pp. 35–37.

55. Cf. Joshua A. Fogel, "Mendacity and Veracity in the Recent Chinese Communist Memoir Literature," *CCP Research Newsletter*, 1 (1988): 33–37.

56. Vojtech Mastny, "From Consensus to Strains in the Sino-Soviet Alliance —A Palpable Deterioration," *Cold War International History Project Bulletin* 6–7 (1996): 22–23.

57. Note on Mao-Iudin conversation, March 31, 1956, in "Mao Tszedun o kitaiskoi politike Kominterna i Stalina" [Mao Zedong on the China Policy of the Comintern and Stalin], *Problemy Dalnego Vostoka*, 1994, no. 5: 101–10, at pp. 105–6; Mao-Gromyko conversation, November 19, 1957, cited in B. Kulik, "Kitaiskaia Narodnaia Respublika v period stanovleniia (1949–1952): Po materi-alam arkhiva vneshnei politiki RF" [The People's Republic of China in the Making: Materials from the Foreign Policy Archives of the Russian Federation], *Problemy Dalnego Vostoka*, 1994, no. 6: 73–83, at p. 77.

58. Sergei N.Goncharov, John W. Lewis, and Xue Litai, *Uncertain Partners: Stalin, Mao, and the Korean War* (Stanford, Calif.: Stanford University Press, 1993), p. 85.

59. Chen, *China's Road to the Korean War*, pp. 79–80.

60. Mao-Roshchin conversation, January 1, 1959, cited in Kulik, "Kitaiskaia Narodnaia Respublika," p. 76.

61. Nikolai T. Fedorenko, "The Stalin-Mao Summit in Moscow," *Far Eastern Affairs*, 1989, no. 2: 134–48, at p. 136.

62. Shu Guang Zhang, *Deterrence and Strategic Culture*, p. 30. The small talk is not recorded in the otherwise comprehensive Russian minutes.

63. Record of Stalin-Mao conversation, December 16, 1949, 45/1/329/9–17, at pp. 10–12, APRF.

64. Ivan V. Kovalev, "Dialog Stalina s Mao Tszedunom" [Stalin's Dialog with Mao Zedong], *Problemy Dalnego Vostoka*, 1992, no. 1–3, pp. 77–91, at pp. 82–83.

65. Mao Zedong to the Chinese party central committee, January 2, 1950, Goncharov, Lewis, and Xue Litai, *Uncertain Partners*, pp. 242–43.

66. Note on Mao-Iudin conversation, March 31, 1956, "Mao Tszedun o kitaiskoi politike," p. 106.

67. These are the fundamental flaws of the otherwise factually valuable works by Bruce Cumings, *The Origins of the Korean War*, vol. 2, *The Roaring of the Cataract, 1947–1950* (Princeton, N.J.: Princeton University Press, 1990), and Robert R. Simmons, *The Strained Alliance: Peking, P'yŏngyang, Moscow, and the Politics of the Korean Civil War* (New York: Macmillan, 1975).

68. Erik van Ree, *Socialism in One Zone: Stalin's Policy in Korea, 1945–1947* (Oxford: Berg, 1989), pp. 267–77.

69. Shtykov to Vyshinskii, January, 19, 1950, Kathryn Weathersby, "To Attack or Not to Attack? Stalin, Kim Il Sung, and the Prelude to War," *Cold War International History Project Bulletin* 5 (1995): 1–9, at p. 8.

70. Alexandre Mansourov, "Soviet-North Korean Relations and the Origins of the Korean War" (paper presented at the conference "New Evidence on the Cold War in Asia," Hong Kong, January 9–12, 1996), pp. 13–21.

71. Shtykov to Foreign Ministry, May 21, June 18, and July 27, 1949, special dossier on Korea, pp. 49–50, 64–67, 86, 87, APRF.

72. Tunkin to Vyshinskii, September 3, 1949, Weathersby, "To Attack of Not to Attack?" p. 6.

73. Politburo decision, September 24, 1949, 3/65/776/30–32, APRF.

74. Shen Zhiua, "China Had to Send Its Troops to Korea: The Policy-Making Process and Reasons" (paper presented at the conference "New Evidence on the Cold War in Asia," Hong Kong, January 9–12, 1996), pp. 2–3.

75. Shtykov to Vyshinskii, May 15, 1949, special dossier on Korea, pp. 45–48, APRF. Shtykov's rendering of Mao's phrase in Russian is *vse chernye, ne razberut.*

76. He Di, "'The Last Campaign to Unify China': The CCP's Unmaterialized Plan to Liberate Taiwan, 1949–1950," *Chinese Historians* 5, no. 1 (1992): 1–16.

77. Schaller, *The American Occupation of Japan*, pp. 213–15.

78. Odd Arne Westad, "The Sino-Soviet Alliance and the United States: Wars, Policies, and Perceptions 1950–1961" (paper presented at the conference "New Evidence on the Cold War in Asia," Hong Kong, January 9–12, 1996), p. 7.

79. *Pravda*, January 4, 1950.

80. Robert M. Slusser, "Soviet Far Eastern Policy, 1945–50: Stalin's Goals in Korea," in Nagai and Irye, *The Origins of the Cold War in Asia*, pp. 123–46, at p. 143.

81. Schaller, "Securing the Great Crescent," p. 406.

82. Simmons, *The Strained Alliance*, pp. 82–101.

83. Mao Zedong to Zhou Enlai, January 7, 1950, Goncharov, Lewis, and Xue Litai, *Uncertain Partners*, p. 246.

84. Shtykov to Vyshinskii, January, 19, 1950, in Weathersby, "To Attack or Not to Attack?" p. 8.

85. Stalin to Shtykov for Kim Il Sung, January 30, 1950, in Weathersby, "Attack or Not to Attack?" p. 9.

86. Chen, *China's Road to the Korean War*, p. 88.

87. "Recent Soviet Moves," February 8, 1950, *FRUS*, 1950, vol. 1, pp. 145–47, at p. 145.

88. John L. Gaddis and Paul H. Nitze, "NSC-68 and the Soviet Threat Reconsidered," *International Security* 4 (1980): 164–76.

89. Ernest May, ed., *American Cold War Strategy: Interpreting NSC 68* (New York: St.Martin's Press, 1993) pp. 245, 246, 238.

90. Goncharov, Lewis, and Xue Litai, *Uncertain Partners*, p. 97.

91. Record of Stalin-Mao conversation, January 22, 1950, 45/1/329/29–38, at pp. 31–34, APRF.

92. Shu Guang Zhang, *Deterrence and Strategic Culture*, p. 31.

93. Sino-Soviet treaty of alliance, February 14, 1950, in Goncharov, Lewis, and Xue Litai, *Uncertain Partners*, pp. 260–61, at p. 260.

94. Sino-Soviet agreement on the Chang Chun Railroad, Port Arthur, and Dairen, February 14, 1950, in Goncharov, Lewis, and Xue Litai, *Uncertain Partners*, pp. 261–63.

95. Wu Xiuquan, "Sino-Soviet Relations in the Early 1950s," *Beijing Review*, November 21, 1983, pp. 16–21, 30, at p. 19.

96. Sino-Soviet agreement on credit to China, February 14, 1950, in Goncharov, Lewis, and Xue Litai, *Uncertain Partners*, pp. 263–64, 95.

97. Shu Guang Zhang, "The Collapse of Sino-Soviet Economic Cooperation, 1950–1960: A Cultural Explanation" (paper presented at the conference "New Evidence on the Cold War in Asia," Hong Kong, January 9–12, 1996), pp. 14–15.

98. Peter Lowe, *The Origins of the Korean War* (London: Longman, 1986), p. 55.

99. Shtykov to Vyshinskii, March 9, and Stalin to Kim Il Sung, March 18, 1950, special dossier on Korea, pp. 131–32, 142, APRF.

100. Kautsky, *Moscow and the Communist Party of India*, p. 102.

101. Speech by Malenkov, March 9, 1950, *Pravda*, March 10, 1950.

102. Speech by Suslov, April 20, 1950, "Protokol zasedanii Sekretariata Informatsionnogo Biuro kommunisticheskikh partii" [Minutes of the Session of the Secretariat of the Information Bureau of Communist Parties], April 20–22, 1950, 575/1/8, RTsKhIDNI.

103. "N[orth] K[orea]'s Kim Initially Planned to Invade S. Korea on June 30," *The Korean Times* [Seoul], June 25, 1993. The article summarizes information from the Soviet archival documents turned over to the South Korean Government by Russian President Boris Eltsin; Ignatev to Vyshinskii, April 25, 1950, special dossier on Korea, p. 150, AVPRF.

104. Chien, *China's Road to the Korean War*, pp. 110–11.

105. "NK's Kim Initially Planned to Invade"; Kathryn Weathersby, "Soviet Aims in Korea and the Origins of the Korean War, 1945–1950: New Evidence from Russian Archives," Working Paper no. 8, *Cold War International History Project* (Washington, D.C.: Woodrow Wilson International Center for Scholars, 1993), p. 25; Goncharov, Lewis, and Xue Litai, *Uncertain Partners*, p. 150.

106. Memorandum by Jack D. Neal on conversation with a French Cominform contact, August 16, 1950, RG-59, 761.00/9-2650, NA.

107. Roshchin to Stalin, May 13, 1950, in "More Documents from the Russian Archives," *International Cold War History Project Bulletin* 4 (1994): 60–85, at p. 61.

108. Stalin to Mao, May 14, 1950, in "More Documents from the Russian Archives," p.61.

109. In a conversation with Soviet ambassador Iudin on March 31, 1956. Iudin to Foreign Ministry, April 20, 1956, special dossier on Korea, p. 157, APRF.

110. Shtykov to Vyshinskii, May 12, 1950, special dossier on Korea, pp. 151–54, APRF.

111. Christoper Andrew, *For the President's Eyes Only: Secret Intelligence and the American Presidency from Washington to Bush* (New York: Harper-Collins, 1995), p.187. The Department of the Army requested further information, but this was never supplied.

112. Shtykov to Vyshinskii, May 12, 1950, special dossier on Korea, pp. 151–54, APRF.

113. Cumings, *The Origins of the Korean War*, 2: 448–50; Richard K. Betts, *Surprise Attack: Lessons for Defense Planning* (Washington: Brookings Institution, 1982), pp. 51–56.

114. Aleksandr Paniushkin,"Politicheskii otchet Posolstva SSSR v SShA za 1950 god" [Political Report by the USSR Embassy in the U.S. for 1950], p. 5, 0129/1951/ 35/14/041/245, AVPRF.

115. Quoted in Cumings, *The Origins of the Korean War*, 2: 557–58.

116. Bruce Franklin, ed.,*The Essential Stalin: Major Theoretical Writings, 1905–1952* (New York: Doubleday, 1972), pp. 407–30, quote on p. 418.

117. Cumings, *The Origins of the Korean War*, 2: 559.

118. "NK's Kim Initially Planned to Invade."

119. Ibid.

120. Grigorii Tumanov, "Shtykovaia ataka Kim Ir Sena" [Kim Il Sung's Bayonet Attack], *Novoe Vremia*, 1993, no. 26: 32–34. In the title of the article, the word *shtyk* (bayonet) is a pun on the name of ambassador Shtykov.

121. "Information received through Russian underground sources in Western Europe, originating from a member of the dissolved Stalin secretariat," April 22, 1953, item 4050, box 9A, William J. Donovan Papers, U.S. Army Military Historical Institute, Carlisle Barracks, Pennsylvania.

122. Tumanov, "Shtykovaia ataka Kim Ir Sena," p. 33.

123. Iu.V. Votintsev, "Neizvestnye voiska izcheznuvshei sverkhderzhavy" [The-Little-Known Forces of a Vanished Superpower], *Voennoistoricheskii zhurnal*, 1993, no. 8: 54–61, at p. 57.

124. Gromyko to Stalin on message by Shtykov, June 20, 1950, in Weathersby, "The Soviet Role," p. 447.

125. Simmons, *The Strained Alliance*, pp. 114–20.

126. "On the Korean War, 1950–53, and the Armistice Negotiations," August 9, 1966, in Kathryn Weathersby, "New Findings on the Korean War," *Cold War International History Project Bulletin* 3 (1993): 15–17, at p. 16. The document is a secret summary of information compiled from Soviet archives for party General Secretary Leonid I. Brezhnev.

127. Goncharov, Lewis, and Xue Litai, *Uncertain Partners*, p. 214.

6 The Test of Strength

1. Ambassador in Korea (Muccio) to Secretary of State, June 25, 1950, *FRUS*, 1950, vol. 7, pp. 132–33.

2. Robert R. Simmons, *The Strained Alliance: Peking, P'yōngyang, Moscow, and the Politics of the Korean Civil War* (New York: Macmillan, 1975), pp. 122–23.

3. Kirk (Moscow) to Secretary of State, June 27, 1950, *FRUS*, 1950, vol. 7, p. 204.

4. Kirk (Moscow) to Secretary of State, June 29, 1950, *FRUS*, 1950, vol. 7, p. 229.

5. Stalin to Gottwald, August 27, 1950, 100/24; 99/1146, AÚV KSČ.

6. Andrei A. Gromyko, *Memories* (London: Hutchinson, 1989), p. 102.

7. John L. Gaddis, *Strategies of Containment: A Critical Appraisal of Postwar American Security Policy* (New York: Oxford University Press, 1982), p. 110.

8. Peter Lowe, *The Origins of the Korean War* (New York: Longman, 1986), p. 164.

9. Shu Guang Zhang, *Deterrence and Strategic Culture: Chinese-American Confrontations, 1949–1958* (Ithaca, N.Y.: Cornell University Press, 1993), p. 91.

10. Stalin to Shtykov, July 1, 1950, 45/1/346/104, APRF.

11. Shtykov to Stalin, July 4, 1950, 45/1/346/136–39, APRF.

12. Bruce Cumings, *The Origins of the Korean War*, vol. 2, *The Roaring of the Cataract, 1947–1950* (Princeton, N.J.: Princeton University Press, 1990), p. 644.

13. He Di, "'The Last Campaign to Unify China': The CCP's Un-materialized Plan to Liberate Taiwan, 1949–1950," *Chinese Historians* 5, no. 1 (1992): 1–16, at p. 12.

14. Iu. V. Votintsev, "Neizvestnye voiska izcheznuvshei sverkhderzhavy" [The Little-Known Forces of a Vanished Superpower], *Voennoistoricheskii zhurnal*, 1993, no. 8: 54–61, at p. 57.

15. Sergei N. Goncharov, John W. Lewis, and Xue Litai, *Uncertain Partners: Stalin, Mao, and the Korean War* (Stanford, Calif.: Stanford University Press, 1993), pp. 154–55.

16. Kim Il Sung to Shtykov, July 14, 1950, 3/65/826/109, APRF.

17. Stalin to Zhou Enlai, July 5, 1950, 45/1/331/79, APRF.

18. Hao Yufan and Zhai Zhiai, "China's Decision to Enter the Korean War: History Revisited," *China Quarterly* 121 (March 1990): 94–115, at p. 101.

19. Stalin to Roshchin for Zhou Enlai or Mao Zedong, July 13, 1950, 45/1/331/85, APRF.

20. Chen Jian, *China's Road to the Korean War: The Making of the Sino-American Confrontation* (New York: Columbia University Press, 1994), p. 137

21. He Di, "'The Last Campaign,'" pp. 14–15.

22. *Khrushchev Remembers: The Last Testament* (Boston: Little, Brown, 1974), p. 189.

23. "Political and Economic Developments in Czechoslovakia during September, October, November 1950," Tyler Thompson (Prague) to Secretary of State, December 15, 1950, 749.00/12-1250, RG-59, NA.

24. Roger Dingman, "Atomic Diplomacy during the Korean War," *International Security* 13, no. 3 (1988–89): 50–91, at pp. 55–56.

25. Speech of July 19, 1950, *Public Papers of the Presidents of the United States: Harry S. Truman, 1950* (Washington, D.C.: U.S. Government Printing Office, 1965), pp. 527–37, at p. 536.

26. Dingman, "Atomic Diplomacy," pp. 57–59.

27. Ibid., pp. 62–63.

28. Norbert Wiggershaus, "Bedrohungsvorstellungen Bundeskanzler Adenauers nach Ausbruch des Koreakrieges," *Militärgeschichtliche Mitteilungen* 25 (1979): 79–122, at p. 98.

29. Acheson to Harriman (Paris) and McCloy (Frankfurt), June 28, 1950, Moscow Embassy 320 U.S. May, RG-84, NA.

30. Steven T. Ross, *American War Plans* (New York: Garland, 1988), p. 140.

31. "The Position and Action of the United States with Respect to Possible Further Soviet Moves in the Light of the Korean Situation," NSC 73/1, August 25, 1950, *FRUS, 1950*, vol. 1, pp. 376–89.

32. Memorandum on conversation between Dean Rusk and Wellington Koo, July 3, 1950, *FRUS, 1950*, vol. 7, p. 286.

33. Cumings, *The Origins of the Korean War*, 2: 643–44.

34. Kathryn Weathersby, "The Soviet Role in the Early Phase of the Korean War: New Documentary Evidence," *Journal of American-East Asian Relations* 2 (1993): 425–58, at pp. 433–34.

35. Kirk (Moscow) to Secretary of State, July 21, 1950, Moscow Embassy, 320 U.S. May, RG-84, NA.

36. David A. Mayers, *Cracking the Monolith: U.S. Policy against the Sino-Soviet Alliance, 1949–1955* (Baton Rouge: Louisiana State University Press, 1986), p. 82; Indian Ambassador (Pandit) to Secretary of State, July 17, 1950, *FRUS, 1950*, vol. 7, pt. 1, pp. 407–8.

37. *Khrushchev Remembers* (Boston: Little, Brown, 1970), p. 369.

38. Kennan to Secretary of State, August 8, 1950, *FRUS, 1950*, vol. 1, pp. 361–67, at p. 362.

39. Vladimir Dedijer, *Novi prilozi za biografiju Josipa Broza Tita* [New Supplements to the Biography of Josip Broz Tito], vol. 3 (Belgrade: Rad, 1984), p. 443.

40. Robert McGeehan, *The German Rearmament Question: American Diplomacy and European Defense after World War II* (Urbana: University of Illinois Press, 1971), p. 24.

41. Wiggershaus, "Bedrohungsvorstellungen," pp. 103–104.

42. Quoted in Norman M. Naimark, "'To Know Everything and to Report Everything Worth Knowing': Building the East German Police State, 1945–1949," Working Paper no. 10, *Cold War International History Project* (Washington, D.C.: Woodrow Wilson International Center for Scholars, 1994), p. 25.

43. United Nations, *Security Council Official Records*, 483d meeting, August 4, 1950, pp. 2–3.

44. Allen S. Whiting, *China Crosses the Yalu: The Decision to Enter the Korean War* (Stanford, Calif.: Stanford University Press, 1960), pp. 79, 86–87.

45. United Nations, *Security Council Official Records*, 489th meeting, August 22, 1950, pp. 1–16.

46. Russell D. Buhite and William C. Hamel, "War for Peace: The Question of an American Preventive War against the Soviet Union, 1945–1955," *Diplomatic History* 14 (1990): 367–84, at p. 376.

47. Stalin to Gottwald, August 27, 1950, 100/24; 99/1146, AÚV KSČ.

48. Hao and Zhai, "China's Decision," pp. 101–2.

49. Chen, *China's Road to the Korean War*, pp 141–54.

50. Stalin to Matveev and Shtykov, September 27, 1950, 3/65/827/90–93, at p. 92, APRF.

51. Stalin to Mao Zedong or Zhou Enlai, October 1, 1950, 45/1/334/97–98, APRF.

52. Kim Il Sung to Stalin, September 29, 1950, 3/65/828/14, APRF.

53. Hao and Zhai, "China's Decision," p. 104.

54. Stalin to Mao Zedong or Zhou Enlai, October 1, 1950, 45/1/334/97–98, APRF.

55. Secretary of State to Acting Secretary of State, October 2, 1950, *FRUS*, 1950, vol. 7, pp. 838–39.

56. Mao Zedong to Stalin, October 2, 1950, in Thomas J. Christensen, "Threats, Assurances, and the Last Chance for Peace: The Lessons of Mao's Korean War Telegrams," *International Security* 17, no. 1 (1992): 122–54, at pp. 151–52.

57. Mao Zedong to Stalin, October 2, 1950, 45/1/334/105–106, APRF.

58. Hao and Zhai, "China's Decision," p. 106.

59. Shen Zhiua, "China Had to Send Its Troops to Korea: The Policy-Making Process and Reasons" (paper presented at the conference "New Evidence on the Cold War in Asia," Hong Kong, January 9–12, 1996), pp. 14–15.

60. Stalin to Mao, October 1, 1950, quoted in Stalin to Shtykov for Kim Il Sung, October 7, 1950, 45/1/347/65–67, at p. 66, APRF.

61. Chen, *China's Road to the Korean War*, p. 276.

62. Vasilevskii and Gromyko to Shtykov, October 6, 1950, 3/65/827/127, APRF.

63. Memoranda by John C. Ross of the U.S. Mission at the United Nations to Assistant Secretary of State for Eastern Affairs (Rush), October 5, 6, 7, and 10, 1950, *FRUS*, 1950, vol. 7, pp. 877–80, 897–99, 906–11, 922.

64. Alexandre Y. Mansourov, "Stalin, Mao, Kim and China's Decision to Enter the Korean War, September 16–October 15, 1950: New Evidence from the Russian Archives," *Cold War International History Project Bulletin* 6–7 (1996): 94–107, at p. 105.

65. Simmons, *The Strained Alliance*, p. 159.

66. Mao Zedong to Kim Il Sung, October 8, 1950, "Mao's Dispatch of Chinese Troops to Korea: Forty-Six Telegrams, July–October 1950," *Chinese Historians* 5, no. 1 (1992): 63–86, at p. 69.

67. Michael H. Hunt, "Beijing and the Korean Crisis, June 1950–June 1951," *Political Science Quarterly* 107 (1992–93): 453–78, at p. 460.

68. Goncharov, Lewis, and Xie Litai, *Uncertain Partners*, pp. 190–91.

69. Mansourov, "Stalin, Mao, Kim, and China's Decision," p. 103.

70. Chen, *China's Road to the Korean War*, p. 200.

71. Shu Guang Zhang, *Deterrence and Strategic Culture*, p. 98.

72. *Khrushchev Remembers: The Glasnost Tapes* (Boston: Little, Brown, 1990), p. 147.

73. Stalin to Kim Il Sung, October 13, 1950, 45/1/347/74–75, and Shtykov to Stalin, October 14, 1950, 45/1/335/3, APRF.

74. Chen, *China's Road to the Korean War*, p. 161.

75. Stalin to Kim Il Sung, October 13, 1950, 45/1/347/74–75, and October 14, 1950, 45/1/347/77, APRF.

76. Chen, *China's Road to the Korean War*, p. 196.

77. Mao Zedong to Zhou Enlai, October 13, 1950, in Christensen, "Threats, Assurances," pp. 152–53.

78. Roshchin to Stalin, October 13, 1950, 45/1/335/1–2, at p. 1, APRF.

79. Ibid.

80. Chen, *China's Road to the Korean War*, p. 204.

81. Kathryn Weathersby, "Soviet Aims in Korea and the Origins of the Korean War, 1945–1950: New Evidence from Russian Archives," Working Paper no. 8, *Cold War International History Project* (Washington, D.C.: Woodrow Wilson Inteatrnional Center for Scholars, 1993), p. 27.

82. Shtemenko to Poskrebyshev, December 9, 1951, in Weathersby, "The Soviet Role," pp. 457–58.

83. Iu.N. Semin and S.N. Ruban, "Uchastie SSSR v koreiskoi voine (novye dokumenty)" [Soviet Participation in the Korean War: New Documents], *Voprosy istorii*, 1994, no. 11: 320, no. 12: 30–45.

84. Robert F. Futrell, *The United States Air Force in Korea, 1950–1953* (Washington, D.C.: U.S. Government Printing Office, 1983), p. 692.

85. Kathryn Weathersby, "Stalin and a Negotiated Settlement in Korea, 1950–53" (paper presented at the conference "New Evidence on the Cold War in Asia," Hong Kong, January 9–12, 1996), pp. 3–4.

86. Wiggershaus, "Bedrohungsvorstellungen," p. 106.

87. Statement Issued by the Foreign Ministers of the U.S.S.R., the German Democratic Republic, Poland, Czechoslovakia, Hungary, Rumania, Bulgaria and the Albanian Minister in Moscow, at a Meeting in Prague, October 21, 1950, in *Documents on Germany under Occupation, 1945–1954*, ed. Beate Ruhm von Oppen, (London: Oxford University Press, 1955), pp. 522–27.

88. Telephone reports on the first and second sessions of the foreign ministers conference, October 20–21, 1950, 100/24; 91/1091, AÚV KSČ.

89. Note from the Soviet Government to the British Government Proposing a Meeting of the Council of Foreign Ministers to Discuss the Demilitarization of Germany, November 2, 1950, in Ruhm von Oppen, *Documents on Germany under Occupation*, p. 535.

90. Marshall D. Shulman, *Stalin's Foreign Policy Reappraised* (Cambridge, Mass.: Harvard University Press, 1963), pp. 155–56.

91. David Rees, "The Korean War and the Japanese Peace Treaty," in *The Korean War in History*, ed. James Cotton and Ian Neary (Manchester: Manchester University Press, 1989), pp. 163–74.

92. Dulles to MacArthur, November 15, 1950, *FRUS*, 1950, vol. 6, pp. 1349–52, at pp. 1350–51.

93. Memorandum on Dulles-Malik conversation, October 26–27, 1950, *FRUS*, 150, vol. 6, pp. 1332–36, at p. 1332.

94. Memorandum on Dulles-Malik conversation, November 20, 1950, *FRUS*, 1950, vol. 6, pp. 1352–54, at p. 1352.

95. Boris N. Slavinskii, "San-Frantsisskaia konferentsiia 1951 g. po mirnomu uregulirovaniiu s Iaponiei i sovetskaia diplomatiia: Novye dokumenty iz arkhiva MID Rossii" [The San Francisco Conference on the Japanese Peace Settlement and Soviet Diplomacy: New Documents from the Russian Foreign Ministry], *Problemy Dalnego Vostoka*, 1994, no. 1: 80–100, at p. 82.

96. *Pravda*, November 24, 1950; *Department of State Bulletin*, January 8, 1951.

97. "Soviet Capabilities and Intentions," NIE-3, November 15, 1950, in *CIA Cold War Records: Selected Estimates on the Soviet Union, 1950–1959*, ed. Scott A. Koch (Washington, D.C.: Central Intelligence Agency, 1993), pp. 165–78, at p. 169.

98. *Department of State Bulletin*, December 18, 1950, p. 962.

99. Editorial Note on President Truman's Statements and Answers to Questions at a News Conference of November 30, 1950, *FRUS*, 1950, vol. 7, pp. 1261–62.

100. Russell D. Buhite, *Soviet-American Relations in Asia, 1945–1954* (Norman: University of Oklahoma Press, 1981), p. 181.

101. D. Cameron Watt, "British Military Perceptions of the Soviet Union as a Strategic Threat, 1945–1950," in *Power in Europe? Great Britain, France, Italy, and Germany in a Postwar World*, ed. Josef Becker and Franz Knipping (Berlin: De Gruyter, 1986), pp. 325–36, at p. 336.

102. Roger Dingman, "Truman, Attlee, and the Korean War Crisis," *International Studies* 1 (1982): 1–42.

103. Dingman, "Atomic Diplomacy," pp. 66–69.

104. "Most Likely Period for Initiation of Hostilities between the USSR and the Western Powers," J.I.C. 530/3, August 22, 1950, CCS 092 USSR (3–27–45) S. 49, RG-218, NA.

105. Ross, *American War Plans*, pp. 142–43.

106. Christopher Andrew and Oleg Gordievsky, *KGB: The Inside Story of Its Foreign Operations from Lenin to Gorbachev* (London: Hodder & Stoughton, 1990), p. 328.

107. Wiggershaus, "Bedrohungsvorstellungen," p. 100.

108. Samuel F. Wells Jr., "The First Cold War Buildup: Europe in the United States Strategy and Policy, 1950–1953," in *Western Security, The*

Formative Years: European and Atlantic Defence, 1947–1953, ed. Olav Riste (Oslo: Universitetsforlaget, 1985), pp. 181–97, at pp. 188–90.

109. Quoted in Wells, "The First Cold War Buildup," p. 185.

110. Ibid., p. 184.

111. Editorial Note on the 521st Meeting of the United Nations' Security Council on November 10, 1950, and Ambassador in the Soviet Union (Kirk) to Secretary of State, November 14, 1950, *FRUS*, 1950, vol. 7, pp. 1126–27, 1153–55.

112. Ulbricht to Grotewohl, November 28, 1950, cited in Michael Lemke, "Eine deutsche Chance? Die innerdeutsche Diskussion um den Grotewohlbrief vom November 1950" (paper presented at the conference "The Soviet Union, Germany, and the Cold War, 1945–1962," Essen, June 28–30, 1994), note 36.

113. *DDR Werden und Wachsen: Zur Geschichte der Deutschen Demokratischen Republik* (Berlin: Dietz, 1974), pp. 191–92.

114. Exchange of Notes Between the U.S.S.R. and the Three Western Powers on a Proposed Meeting of the Council of Foreign Ministers, Russian Note, December 30, 1950, and United States' Reply, January 23, 1951, and Message from the Volkskammer of the German Democratic Republic to the Federal Bundestag Regarding All-German Elections, January 30, 1951, in *Documents on International Affairs 1951*, ed. Denise Folliot (London: Oxford University Press, 1954), pp. 248–55, 269–72.

115. Wilfried Loth, *Stalins ungeliebtes Kind: Warum Moskau die DDR nicht wollte* (Berlin: Rowohlt, 1994), pp. 173–74

116. Cited in Rosemary Foot, *The Wrong War: American Policy and the Dimensions of the Korean Conflict, 1950–1953* (Ithaca, N.Y.: Cornell University Press, 1985), p. 126.

117. Record of the Gromyko-Wang Jiaxiang conversation, December 4, 1950, 3/65/515/35–36, APRF.

118. Gromyko to Vyshinskii, December 7, 1950, 3/65/828/24, APRF.

119. Jonathan D. Pollack, "The Korean War and Sino-Soviet Relations," in *Sino-American Relations, 1945–1955: A Joint Reassessment of a Critical Decade*, ed. Yuan Ming and Harry Harding (Wilmington: Scholarly Resources, 1989), pp. 213–37, at p. 224.

120. Chen, *China's Road to the Korean War*, p. 212.

121. Hunt, "Beijing and the Korean Crisis," p. 466.

122. Stalin to Zhou Enlai, December 7, 1950, 3/1/336/20–21, APRF.

123. Memorandum on conversation by State Department Polish desk officer Vedeler with a confidential source, January 3, 1951, *FRUS*, 1951, vol. 4, pt. 2, pp. 1522–23.

124. Grant M. Adibekov, "Popytka kominternizatsii Kominforma v 1950 g.: Po novym arkhivnym materialam" [The Attempted Cominternization of the Cominform in 1950: New Archival Materials], *Novaia i noveishaia istoriia*, 1994, nos. 4–5: 51–66, at pp. 53–54.

125. "O razshirenii funktsii Informbiuro kommunisticheskikh i rabochikh partii" [The Expansion of the Functions of the Information Bureau of the Communist and Workers' Parties], November 22–24, 1950, 575/2/14/1 and 1/11, RTsKhIDNI.

126. Philippe Robrieux, *Maurice Thorez: Vie secrète et vie publique* (Paris: Fayard, 1975), pp. 393–94.

127. Irwin M. Well, *French Communism in the Era of Stalin: The Quest for Unity and Integration, 1945–1962* (Westport, Conn.: Greenwood Press, 1983), pp. 109, 145–46.

128. Helmut König, "Der Konflikt zwischen Stalin und Togliatti um die Jahreswende 1950/51," *Osteuropa* 20 (1970): 699–706 and A 705–9.

129. Giorgio Bocca, *Palmiro Togliatti* (Rome: L'Unità, 1992), pp. 491–96.

130. Togliatti to Stalin, January 4, 1951, "'Ochen tiazhelo vyrazhat mnenie ne sovpadaiushchee s Vashim': Pochemu P. Toliatti otkazalsia ot predlozheniia I. Stalina" ['It Is Very Difficult to Express an Opinion Not Coinciding with Yours': Why Togliatti Declined Stalin's Proposal], *Istochnik*, 1995, no. 3: 149–52.

131. Joan Barth Urban, *Moscow and the Italian Communist Party: From Togliatti to Berlinguer* (Ithaca, N.Y.: Cornell University Press, 1986), p. 220.

132. "Vital Tasks of the Communist Parties," *For a Lasting Peace, for a People's Democracy*, December 8, 1950.

133. The Cominform had lost any importance by the time it was formally dissolved by Khrushchev as part of his de-Stalinization drive in 1956.

134. Karel Kaplan, *Die politischen Prozesse in der Tschechoslowakei, 1948–1954* (Munich: Oldenbourg, 1986), p. 118.

135. "Soviet Capabilities and Intentions," NIE-3, November 15, 1950, in Koch, *CIA Cold War Records*, pp. 165–78, at pp. 169–70.

136. Andrew and Gordievsky, *KGB: The Inside Story*, p. 326.

137. *Khrushchev Remembers: The Last Testament*, pp. 11, 533.

138. The details appear in the account that Karel Kaplan, who as an officially appointed researcher had enjoyed unlimited access to the Czechoslovak archives before 1968, wrote after defecting to the West with his notes: *Dans les archives du comité central: Trente ans de secrets du bloc soviétique* (Paris: Michel, 1978), pp. 165–66.

139. Quoted in Walter S. Poole, *The History of the Joint Chiefs of Staff: The Joint Chiefs of Staff and National Policy*, vol. 2, 1950–1952 (Wilmington, Del.: Glazier, 1980), p. 79.

140. Memorandum by Joint Intelligence Committee, JIC 531/17, January 24, 1951, CCS 092 USSR (3-27-45) S.54, RG-218, NA.

141. "Recommended Politics and Actions in Light of the Grave World Situation," NSC-100, January 11, 1951, *FRUS*, 1951, vol. 1, pp. 7–18, at p. 18.

142. *Khrushchev Remembers: The Last Testament*, p. 11.

143. "The Soviet Union and the Atlantic Pact," September 8, 1952, in George F. Kennan, *Memoirs, 1950–1963* (Boston: Little, Brown, 1971), pp. 331–55, at pp. 345.

144. Quoted in Beatrice Heuser, *Western "Containment" Policies in the Cold War: The Yugoslav Case, 1948–53* (London: Routledge, 1989), p. 141.

7   On the Defensive

1. Answers by Marshal Stalin to Questions Put by a Correspondent of Pravda, February 16, 1951, in *Documents on International Affairs 1951*, ed. Denise Folliot (London: Oxford University Press, 1954), pp. 290–94, quote on p. 293.

2. Appeal by World Peace Council for a five-power peace pact, Berlin, February 21–25, 1951, in Folliot, *Documents on International Affairs 1951*, pp.

305–7; Jones (Berlin) to High Commissioner in Germany, March 1, 1951, *FRUS*, 1951, vol. 3, pt. 2, pp. 1997–99.

3. Arshi Pipa, "The Political Culture of Hoxha's Albania," in *The Stalinist Legacy: Its Impact on Twentieth-Century World Politics*, ed. Tariq Ali (Harmondsworth, England: Penguin, 1984), pp. 435–64, at p. 443.

4. Nicholas Bethell, *The Great Betrayal: The Untold Story of Kim Philby's Biggest Coup* (London: Hodder & Stoughton, 1984), p. 174.

5. Beatrice Heuser, *Western "Containment" Policies in the Cold War: The Yugoslav Case, 1948–53* (London: Routledge, 1989), p. 150.

6. Ibid., pp. 160–64; Memorandum of Conversation, by the Secretary of State, June, 18, 1951, the Acting Secretary of Defense (Lovett) to the Secretary of State, July 14, 1951, *FRUS*, 1951, vol. 4, pt. 2, pp. 1815–16, 1829–31.

7. Quoted in Philip Windsor, "Yugoslavia, 1951, and Czechoslovakia, 1968," in *Force without War: U.S. Armed Forces as a Political Instrument*, ed. Barry M. Blechman and Stephen S. Kaplan (Washington, D.C.: Brookings, 1978), pp. 440–512, at p. 459.

8. Excerpts from speech of July 21, 1951, in *Yugoslavia and the Soviet Union, 1939–1973*, ed. Stephen Clissold (London: Oxford University Press, 1975), p. 239.

9. Allen (Belgrade) to Secretary of State, June 1, 1951, RG-59, 768/6-151, NA; Beam (Belgrade) to Secretary of State, March 15, 1952, RG-59, 768.00/3-1552, NA.

10. James D. Marchio, "Rhetoric and Reality: The Eisenhower Administration and Unrest in Eastern Europe, 1953–1959" (Ph.D. diss., American University, 1990), p. 37.

11. Wallace Carroll and Hans Speier, "Psychological Warfare in Germany: A Report to the United States High Commissioner for Germany and the Department of State," December 1, 1950, HICOG Berlin, EAD, box 3: 352, RG-466, NA. For obtaining this document I am grateful to Christian F. Ostermann, who wrote about it in his "The United States, East Germany, and the Limits of Roll-Back in Germany, 1953" (paper presented at the conference "The Soviet Union, Germany, and the Cold War, 1945–1962," Essen, June 28–30, 1994), pp. 4–5.

12. Kai-Uwe Merz, *Kalter Krieg als antikommunistischer Widerstand: Die Kampfgruppe gegen Unmenschlichkeit, 1948–1959* (Munich: Oldenbourg, 1987), pp. 250–54.

13. *Khrushchev Remembers* (Boston: Little, Brown, 1970), p. 362.

14. Editorial Note, The Ambassador in Czechoslovakia (Briggs) to the Secretary of State, February 17, 1951, Editorial Note, The Ambassador in Czechoslovakia (Briggs) to the Secretary of State, June 16, 1951, *FRUS*, 1951, vol. 4, pt. 2, pp. 1339–40, 1345–46, 1359–61, 1370–71; *Rudé právo* [Prague], March 12, 1953.

15. *Vojenské dějiny Československa* [The Military History of Czechoslovakia], vol. 5 (1945–1955), (Prague: Naše vojsko, 1989), pp. 492, 495–96.

16. Marchio, "Rhetoric and Reality," p. 38.

17. Robert T. Holt, *Radio Free Europe* (Minneapolis: University of Minnesota Press, 1958), pp. 13–15; Allan A. Michie, *Voices through the Iron Curtain: The Radio Free Europe Story* (New York: Dodd, Mead, 1963), pp. 11–14.

18. Memorandum by the Assistant Secretary for European Affairs (Perkins) to the Secretary of State, May 31, 1951, *FRUS*, 1951, vol. 4, pt. 2, p. 1367–68.

19. Secretary of State to Legation in Hungary, April 12, 1951, Minister in Hungary (Davis) to Department of State, April 20, 1951, Minister in Austria (Donnelly) to Secretary of State, April 28, 1951, and Minister in Hungary to Department of State, May 2, 1951, *FRUS*, 1951, vol. 4, pt. 2, pp. 1452–56, 1458–64.

20. Kopřiva to Široký, April 9, 1951, box 94, Secretariat of the Minister, 1945–55, MZV.

21. Briggs (Prague) to Secretary of State, May 16, 1951, *FRUS*, 1951, vol. 4, pt. 2, pp. 1357–59, at p. 1358.

22. Briggs to Secretary of State, July 17, 1951, *FRUS*, 1951, vol. 4, pt. 2, pp. 1380–81, quote on p. 1380.

23. Note by Široký, July 1951, item 4161, box 95, Secretariat of the Minister, 1945–1955, MZV.

24. Editorial Note, *FRUS*, 1951, vol. 4, pt. 2, p. 1413.

25. "Shrnutí poznatků o špionážní službě USA," March 29, 1952, item 4161: Vyzvědačství [Intelligence], box 95, Secretariat of the Minister, MZV.

26. Karel Kaplan, *Die politischen Prozesse in der Tschechoslowakei, 1948–1954* (Munich: Oldenbourg, 1986), pp. 161, 182–83.

27. Quoted in Paul Marantz, *From Lenin to Gorbachev: Changing Soviet Perspectives on East-West Relations* (Ottawa: Canadian Institute for International Peace and Security, 1988), p. 27.

28. "Summary of the Political and Economic Developments in Hungary in 1951," February 1, 1952, *FRUS*, 1951, vol. 4, pt. 2, pp. 1482–89.

29. Bennett Kovrig, *The Myth of Liberation: East-Central Europe in U.S. Diplomacy and Politics since 1941* (Baltimore, Md.: Johns Hopkins University Press, 1973), pp. 102–3, quote on p. 103.

30. Section 101(a)(I) of the Mutual Security Act of 1951—the Kersten Amendment, October 19, 1951, in Folliot, *Documents on International Affairs 1951*, pp. 317–18, quote on p. 318.

31. Beatrice Heuser, "Subversive Operationen im Dienste der 'Roll-Back'-Politik 1948–1953," *Vierteljahrshefte für Zeitgeschichte* 37 (1989): 279–97, at pp. 291–92.

32. Note from the U.S.S.R. to the U.S.A. Protesting against the Provisions of the Kersten Amendment, November 21, 1951, in Folliot, *Documents on International Affairs 1951*, pp. 318–20.

33. *Procès contre les espions et agents de diversion du service des états imperialistes envoyés en Albanie* (Tirana: Ministère des Affaires Etrangères, 1952).

34. Bethell, *The Great Betrayal*, pp. 174–76; Beam (Belgrade) to Secretary of State, November 2, 1951, RG-59, 767.00/11-251, NA.

35. NSC 10/5, October 23, 1951, in *CIA Cold War Records: The CIA under Harry Truman*, ed. Michael Warner (Washington, D.C.: Central Intelligence Agency, 1994), pp. 437–39, quote on p. 438.

36. Quoted in Marchio, "Rhetoric and Reality," p. 43.

37. Heuser, *Western "Containment" Policies*, pp. 139–41.

38. Kirk (Moscow) to Secretary of State, November 4, 1950, *FRUS*, 1950, vol. 4, pp. 902–3; Ministry of Foreign Affairs of the Soviet Union to the Embassy of the United States, December 30, 1950, and Barbour (Moscow) to Secretary of State, January 1, 1951, *FRUS*, 1951, vol. 3, pt. 1, pp. 1051–55.

39. Secretary of State to the Embassy in France, January 8, 1951, *FRUS*, 1951, vol. 3, pt. 1, pp. 1058–60.

40. Nikolai B. Adyrkhaev, "Vstrecha Stalina s iapanskimi kommunistami" [Stalin's Meeting with Japanese Communists], *Problemy Dalnego Vostoka*, 1990, no. 2: 140–47.

41. Mao Zedong to Stalin, March 1, 1951, 45/1/337/78-82, APRF.

42. The Director of the Office of German Political Affairs (Laukhuff) to the Director of the Bureau of German Affairs (Byroade), April 6, 1951, *FRUS*, 1951, vol. 3, pt. 1, pp. 1121–22.

43. Quoted in Heuser, *Western "Containment" Policies*, p. 136.

44. Rosemary Foot, *The Wrong War: American Policy and the Dimensions of the Korean Conflict, 1950–1953* (Ithaca, N.Y.: Cornell University Press, 1985), p. 133.

45. Stalin to Mao Zedong, May 26, 1951, 45/1/338/91, APRF.

46. Nakajima Mineo, "The Sino-Soviet Confrontation in Historical Perspective," in *The Origins of the Cold War in Asia*, ed. Yonosuke Nagai and Akira Irye (New York: Columbia University Press, 1977), pp. 203–23, at p. 217.

47. See chap. 1, n. 19.

48. Roger Dingman, "Atomic Diplomacy during the Korean War," *International Security* 13, no. 3 (1988–89): 50–91, at pp. 72–74.

49. William Stueck, *The Korean War: An International History* (Princeton, N.J.: Princeton University Press, 1995), pp. 178–87.

50. Foot, *The Wrong War*, p. 134.

51. Ibid., pp. 155, 260; Dingman, "Atomic Diplomacy," pp. 75–77.

52. Xu Yan, "The Chinese Forces and Their Casualties in the Korean War: Facts and Statistics," *Chinese Historians* 6, no. 2 (1993): 45–58, at p. 57.

53. Report by Peng Dehuai, May 31, 1951, 45/1/339/10-16, APRF.

54. Mao Zedong to Stalin, June 13, 1951, 45/1/339/55-56, APRF.

55. Stalin to Mao Zedong, June 5, 1951, 45/1/339/17-18, APRF.

56. Mao Zedong to Stalin, June 5, 1951, 45/1/339/23, APRF.

57. Bernard Brodie, *War and Politics* (New York: Macmillan, 1973), pp. 91–97.

58. Mao Zedong to Gao Gang and Kim Il Sung, June 13, 1951, 45/1/339/57-60, APRF.

59. Memorandum by Kennan to the Deputy Under Secretary of State (Matthews), May 31, 1951, and Kennan to Matthews, June 5, 1951, *FRUS*, 1951, vol. 7, pp. 483–86, 507–11.

60. Mao Zedong to Stalin, June 30, 1951, 45/1/339/90-91, at p. 91, APRF.

61. Acheson to Vyshinskii, May 31, 1951, *FRUS*, 1951, vol. 3, pt. 1, pp. 1148–49.

62. U.S. Delegation at the Four-Power Exploratory Talks to the Secretary of State, June 4, 1951, *FRUS*, 1951, vol. 3, pt.1, p. 1150.

63. U.S. Representative at the Four-Power Exploratory Talks (Jessup) to Secretary of State, June 22, 1951, *FRUS*, 1951, vol. 3, pt. 1, pp. 1161–62.

64. Stalin to Mao Zedong, June 24, 1951, 45/1/339/78, APRF.

65. Extract from speech by Malik, June 23, 1951, in Folliot, *Documents on International Affairs 1951*, p. 633.

66. Memorandum on Gross-Malik conversation, June 29, 1951, *FRUS*, 1951, vol. 7, pt. 1, pp. 590–92.

67. Mao Zedong to Stalin, July 3, and Stalin to Mao Zedong, July 3, 1951, 45/1/339/8-11, APRF.

68. Record of Gromyko-Kirk conversation, June 27, 1951, 3/65/828/181-87, at p. 182, APRF; Kirk to Secretary of State, June 27, 1951, *FRUS*, 1951, vol. 7, pt. 1, pp. 560–61.

69. Boris N. Slavinskii, "San-Frantsisskaia konferentsiia 1951 g. po mirno-mu uregulirovaniiu s Iaponiei i sovetskaia diplomatiia: Novye dokumenty iz arkhiva MID Rossii" [The San Francisco Conference on the Japanese Peace Settlement and Soviet Diplomacy: New Documents from the Russian Foreign Ministry], *Problemy Dalnego Vostoka*, 1994, no. 1: 80–100, at pp. 90–93.

70. Slavinskii, "San-Frantsisskaia konferentsiia," p. 89; Memorandum by Fearey on meeting of Senate Far East Sub-Committee, March 19, 1951, *FRUS*, 1951, vol. 6, pt. 1, pp. 932–35, at p. 933.

71. "Provisional Draft of a Japanese Peace Treaty," March 23, and "Revised United States-United Kingdom Draft of a Japanese Peace Treaty," June 14, 1951, *FRUS*, 1951, vol. 6, pt. 1, pp. 944–50, at p. 945, and pp. 1119–33, at p. 1120.

72. Mao Zedong to Stalin, July 13, 1951, 45/1/340/43-45, at p. 44, APRF.

73. Mao Zedong to Stalin, July 13, Stalin to Mao Zedong, July 14, Mao Zedong to Stalin, July 20, and Stalin to Mao Zedong, July 21, 1951, 45/1/340/43-45, 48, 88–89, and 92, APRF.

74. In Zhou's conversation with Roshchin on July 24, 1951, cited in B. Kulik, "Kitaiskaia Narodnaia Respublika v period stanovleniia (1949–1952): Po materialam arkhiva vneshnei politiki RF" [The People's Republic of China at the Time of Its Foundation: Materials from the Foreign Policy Archives of the Russian Federation], *Problemy Dalnego Vostoka*, 1994, no. 6: 73–83, at p. 79.

75. Michael H. Hunt, "Beijing and the Korean Crisis, June 1950–June 1951," *Political Science Quarterly* 107 (1992–93): 453–78, at p. 468; Li Kenun to Mao Zedong, August 10, 1951, 45/1/341/37-39, at p. 38, APRF.

76. Mao Zedong to Stalin, August 27, 1951, 45/1/340/86-88, APRF; Stueck, *The Korean War*, p. 229.

77. Stalin to Mao Zedong, August 29, 1951, 45/1/340/89, APRF.

78. Slavinskii, "San-Frantsisskaia konferentsiia," pp. 97–98.

79. Michael Schaller, *The American Occupation of Japan: The Origins of the Cold War in Asia* (New York: Oxford University Press, 1985), pp. 279–94.

80. *Khrushchev Remembers: The Glasnost Tapes* (Boston: Little, Brown, 1990), p. 83.

81. Foot, *The Wrong War*, pp. 168–72.

82. Shu Guang Zhang, *Deterrence and Strategic Culture: Chinese-American Confrontations, 1949–1958* (Ithaca, N.Y.: Cornell University Press, 1993), p. 130.

83. Kirk to Secretary of State, October 5, 1951, *FRUS*, 1951, vol. 7, pt. 1, pp. 1001–1004.

84. Answers by Stalin to a *Pravda* correspondent, October 6, 1951, in Folliot, *Documents on International Affairs 1951*, pp. 341–42.

85. The Advance Headquarters of the United Nations Command, Korea, to Ridgway, October 7, 1951, *FRUS*, 1951, vol. 7, pt. 1, pp. 1004–5.

86. Cumming to Secretary of State, October 15, 1951, and Secretary of State to Embassy in the Soviet Union. October 15, 1951, *FRUS*, 1951, vol. 7, pt. 1, pp. 1042–48.

87. Barton J. Bernstein, "The Struggle over the Korean Armistice: Prisoners of Repatriation?" in *Child of Conflict: The Korean-American Relationship, 1943–1953,* ed. Bruce Cumings (Seattle: University of Washington Press, 1983), pp. 261–307.

88. Ernest R. May, John D. Steinbrunner, and Thomas Wolfe, "History of the Strategic Arms Competition 1945–1972," pt. 1 (Washington, D.C.: Office of the Secretary of Defense, 1981), p. 272, NSA.

89. Gen. Omar Bradley in United States Senate, 82nd Cong., 1st sess., *Hearings before the Committee on Armed Services and the Committee on Foreign Relations: To Conduct an Inquiry into the Military Situation in the Far East and the Facts Surrounding the Relief of General of the Army Douglas MacArthur from His Assignments in That Area,* pt. 2 (Washington, D.C.: U.S. Government Printing Office, 1951), p. 732.

90. Stalin to Mao Zedong, November 19, 1951, 3/65/828/43, APRF.

91. Gromyko to Razuvaev, November 19 and 20, 1951, 3/65/829/44-48, APRF.

92. Alex Danchev, "In the Back Room: Anglo-American Defence Cooperation, 1945–51," in *British Intelligence, Strategy and the Cold War, 1945–51,* ed. Richard J. Aldrich (London: Routledge, 1992), pp. 215–35, at pp. 224–25.

93. Statement by Vyshinsky, November 16, 1951, in Folliot, *Documents on International Affairs 1951,* pp. 348–56; Editorial Note, *FRUS,* 1951, vol. 1, pp. 582–83.

94. Paul Marantz, "Stalin and East-West Relations: A Reexamination of His Last Years" (paper presented at the meeting of the American Association for the Advancement of Slavic Studies, New York City, November 3, 1984).

95. Sheila Kerr, "NATO's First Spies: The Case of the Disappearing Diplomats—Guy Burgess and Donald Maclean," in *Securing Peace in Europe, 1945–62: Thoughts for the Post-Cold War Era,* ed. Beatrice Heuser and Robert O'Neill (London: Macmillan, 1992), pp. 293–309; Christopher Andrew and Oleg Gordievsky, *KGB: The Inside Story of Its Foreign Operations from Lenin to Gorbachev* (London: Hodder & Stoughton, 1990), p. 48.

96. Vladislav M. Zubok, "Soviet Intelligence and the Cold War: The 'Small' Committee of Information, 1952–53," Working Paper no. 4, *Cold War International History Project* (Washington, D.C.: Woodrow Wilson International Center for Scholars, 1992), pp. 3–4, 7.

97. *Khrushchev Remembers,* p. 307.

98. Svetlana Alliluyeva, *Twenty Letters to a Friend* (New York: Harper & Row, 1967), pp. 7–8.

99. Jon Halliday, *The Artful Albanian: The Memoirs of Enver Hoxha* (London: Chatto & Windus, 1986), p. 138

100. "East European Trials: II—Motives of the War against Heresy," *Times* [London], June 4, 1952, p. 7.

101. In East Germany, such a trial was planned for 1951 but then called off for unknown reasons. Hermann Weber, "Schauprozeß-Vorbereitungen in der DDR," in *Kommunisten verfolgen Kommunisten: Stalinistischer Terror und "Säuberungen" in den kommunistischen Parteien Europas seit den dreißiger Jahren,* ed. Hermann Weber and Dietrich Staritz (Berlin: Akademie, 1993), pp. 436–49, at pp. 437–44.

102. Karel Kaplan, "Zamyšlení nad politickými procesy" [Reflections on the Political Trials], *Nová mysl* [Prague] 22 (1968): 765–94, 906–40, at pp. 922–23.

103. Kaplan, *Die politischen Prozesse*, pp. 194–98.

104. Resolution of the Secretariat of the Central Committee, November 1950, RG-59, 749.00/2-851, NA.

105. Václav Brabec, "Vztah KSČ a veřejnosti k politickým procesům na počátku padesátých let" [The Attitudes of the Communist Party of Czechoslovakia and of the Public toward the Political Trials of the Early Nineteen-Fifties], *Revue dějin socialismu* [Prague], 1969, no. 3: 363–85, at p. 372.

106. "The Światło Story," *News from Behind the Iron Curtain* [Munich] 4, no. 3 (1955): 3–36, at pp. 13–19; Roman Werfel, *Trzy klęski reakcji polskiej* [Three Defeats of the Polish Reaction] (Warsaw: Książka i Wiedza, 1951).

107. Interviews with Edward Ochab and Jakub Berman, in Teresa Torańska, *"Them": Stalin's Polish Puppets* (New York: Harper & Row, 1987), pp. 47–48, 325–26.

108. Kaplan, *Die politischen Prozesse*, pp. 199–201.

109. *Khrushchev Remembers: The Last Testament* (Boston: Little, Brown, 1974), p. 182.

110. Karel Kaplan, "Zamyšlení," pp. 928–30, and his "Introduction" to Meir Cotic, *The Prague Trial: The First Anti-Zionist Show Trial in the Communist Bloc* (New York: Herzl Press, 1987), pp. 17–41, at pp. 30–31.

111. Nicholas Bethell, *Gomułka: His Poland, His Communism* (New York: Holt, Rinehart & Winston, 1969), p. 180.

112. Kaplan, "Zamyšlení," p. 930, and in Cotic, *The Prague Trial*, pp. 33–34.

113. "The Światło Story," pp. 19, 17.

114. Interview with Jakub Berman, in Torańska, *"Them,"* p. 326.

115. Interview with Edward Ochab, in Torańska, *"Them,"* pp. 47–48.

116. Alicja Zawadzka-Wetz, *Refleksje pewnego życia* [Reflections on a Lifetime] (Paris: Instytut Literacki, 1967), p. 44.

117. Ignacy Szenfeld, ed., *The Reminiscences of Władysław Gomułka* (New York: Radio Liberty Committee, 1974), p. 2; Władysław Gomułka, *Pamiętniki* [Memoirs], vol. 2 (Warsaw: BGW, 1994), pp. 305–7.

118. *Khrushchev Remembers: The Glasnost Tapes*, pp. 132–33.

119. Transcript of the meeting of the central committee, September 6, 1951, RG-59, 749.00/11-3051, NA.

120. Ibid.

121. *Rudé právo* [Prague], November 22, 1952, p. 2; Kaplan, "Zamyšlení," p. 931.

122. Marchio, "Rhetoric and Reality," p. 40.

123. Kaplan, "Zamyšlení," p. 931, and in Cotic, *The Prague Trial*, p. 39.

124. Kaplan, "Zamyšlení," p. 931.

125. Stewart Steven, *Operation Splinter Factor* (Philadelphia: Lippincott, 1974).

126. Memorandum by the Economic Section of U.S. Embassy in Moscow, October 1, 1949, *FRUS*, 1949, vol. 5, pp. 144–48, at p. 147.

127. Mordekhaï Oren, *Prisonnier politique à Prague (1951–1956)* (Paris: Julliard, 1960); Kaplan, *Die politischen Prozesse*, p. 202.

128. Amy Knight, *Beria: Stalin's First Lieutenant* (Princeton, N.J.: Princeton University Press, 1993), pp. 158–64.

129. Timothy Dunmore, *Soviet Politics 1945–53* (New York: St. Martin's Press, 1984), p. 119; Charles H. Fairbanks Jr., "National Cadres as a Force in

the Soviet System: The Evidence of Beria's Career, 1949–53," in *Soviet Nationality Policies and Practices*, ed. Jeremy R. Asrael (New York: Praeger, 1978), pp. 144–86, at pp. 149–50.

130. *Rudé právo*, February 8, 1952.

8 Fits and Starts

1. *Pravda*, April 2, 1952.

2. Manfred Kittel, "Genesis einer Legende: Die Diskussion um die Stalin-Noten in der Bundesrepublik 1952–1958," *Vierteljahrshefte für Zeitgeschichte* 41 (1993): 355–89.

3. Communiqué Issued by the Foreign Ministers of the United States, United Kingdom, and France, September 14, 1951, *FRUS*, 1951, vol. 3, pt. 1, pp. 1306–8.

4. Gromyko to Stalin, January 25, 1952, 082/1952/40/11/101a/255, p. 7, AVPRF.

5. N. Spencer Barnes (HICOG Berlin) to Secretary of State, June 20, 1952, 762B.00/6-2052, RG-59, NA.

6. In a conversation with Pietro Nenni, cited in Rolf Steininger, *Eine Chance zur Wiedervereinigung? Die Stalin-Note vom 10. März 1952* (Bonn: Neue Gesellschaft, 1985), p. 20.

7. Gromyko to Stalin, January 25, 1952, 082/1952/40/11/101a/255, p. 8, AVPRF.

8. Dietrich Staritz, "The SED, Stalin, and the German Question: Interests and Decision-Making in the Light of New Sources," *German History* 10 (1992): 274–89, at p. 285.

9. Gromyko to Stalin, January 25, 1952, 082/1952/40/11/101a/255, pp. 7–8.

10. *Beziehungen DDR-UdSSR 1949 bis 1955*, vol. 1 (Berlin: Staatsverlag der Deutschen Demokratischen Republik, 1975), pp. 338–40.

11. Draft of a peace treaty, before February 18, 1952, 082/1952/40/11/101a/255, pp. 31–46, quote on p. 35, AVPRF—a document missing in the discussion by Gerhard Wettig, "Die Deutschland-Note vom 10. März 1952 auf der Basis diplomatischer Akten des russischen Außenministeriums: Die Hypothese des Wiedervereinigungsangebots," *Deutschland-Archiv* 26 (1993): 786–805.

12. Gromyko to Stalin, February, 23, 1952, 082/1952/40/11/101a/255, pp. 14–15, AVPRF.

13. Soviet Ministry of Foreign Affairs to U.S. Embassy Moscow, March 10, 1952, *FRUS*, 1952–1954, vol. 7, pt. 1, pp. 169–72.

14. Daniil E. Melnikow, "Illusionen oder verpaßte Chance? Zur sowjetischen Deutschlandpolitik 1945–1952," *Osteuropa* 40 (1990): 593–601, at pp. 599–600.

15. "Argumentation für die Massenaufklärung zur Note der Sowjetregierung ... über den Entwurf eines Friedensvertrages mit Deutschland," March 11, 1952, ZK SED, J IV 2/2/200, SAPMO-BA.

16. Hermann-Josef Rupieper, *Der besetzte Verbündete: Die amerikanische Deutschlandpolitik 1949–1955* (Opladen: Westdeutscher Verlag, 1991), p. 254.

17. Text in Secretary of State (Acheson) to the Office of the United States High Commissioner for Germany, March 22, 1952, *FRUS*, 1952–1954, vol. 7, pt. 1, pp. 189–90. The notes were transmitted on March 25.

18. Dietrich Staritz, "Die SED, Stalin und der 'Aufbau des Sozialismus' in der DDR: Aus den Akten des Zentralen Parteiarchivs," *Deutschland-Archiv* 24 (1991): 686–70; Wilfriede Otto, "Sowjetische Deutschlandnote 1952: Stalin und die DDR: Bisher unveröffentlichte handschriftliche Notizen Wilhelm Piecks," *Beiträge zur Geschichte der Arbeiterbewegung* 33 (1991): 374–89.

19. "Plan der Besprechung" by Pieck and his notes, about April 7, 1952, Pieck 36/696, ZK SED, SAPMO-BA.

20. "Stalin and the SED Leadership, 7 April 1952: 'You Must Organize Your Own State,'" *Cold War International History Project Bulletin* 4 (1994): 34–35, 48, at p. 48.

21. Soviet Ministry for Foreign Affairs to U.S. Embassy Moscow, April 9, 1952, *FRUS*, 1952, vol. 7, pt. 1, pp. 199–202.

22. Grey to Foreign Office, April 10, 1952, cited in Steininger, *Eine Chance zur Wiedervereinigung?* p. 193.

23. Kennan to Department of State, July 25, 1952, *FRUS*, 1952–54, vol. 6, pt. 2, p. 1585.

24. Kennan to Department of State, May 25, 1952, *FRUS*, 1952–54, vol. 6, pt. 1, pp. 252–53.

25. For different opinions about the novelty of the course and East Berlin's freedom of action in pursuing it, see Wilfriede Otto, "Sowjetische Deutschland-politik 1952/1953," *Deutschland-Archiv* 26 (1993): 948–54; Elke Scherstjanoi, "Die DDR im Frühjahr 1952: Sozialismuslosung und Kollektivierungsbeschluß in sowjetischer Perspektive," *Deutschland-Archiv* 27 (1994): 354–63.

26. *SBZ von 1945 bis 1954: Die sowjetische Besatzungszone Deutschlands in den Jahren 1945–1954* (Bonn: Bundesministerium für Gesamtdeutsche Fragen, 1961), pp. 144–45, 188, 190; Martin Jänicke, *Der dritte Weg: Die antistalinis-tische Opposition gegen Ulbricht seit 1953* (Cologne: Neuer Deutscher Verlag, 1964), pp. 23–24.

27. Wilhelm Fricke, *Warten auf Gerechtigkeit: Kommunistische Säuber-ungen und Rehabilitierungen: Bericht und Dokumentation* (Cologne: Wissen-schaft und Politik, 1971), pp. 90–91.

28. Wilfried Loth, *Stalin's ungeliebtes Kind: Warum Moskau die DDR nicht wollte* (Berlin: Rowohlt, 1994), p. 189.

29. Kennan to Department of State, June 20, 1952, *FRUS*, 1952–54, vol. 8, pp. 1014–15.

30. Shtemenko to Khrushchev, July 21, 1953, 5/30/4, TsKhSD.

31. Entry for January 13, 1984, in Feliks Chuev, *Sto sorok besed s Molotovym* [Hundred-and-Forty Conversations with Molotov] (Moscow: Terra, 1991), p. 324.

32. *Khrushchev Remembers* (Boston: Little, Brown, 1970), pp. 307–8.

33. Excerpts from Miasnikov's unpublished memoirs, *Literaturnaia gazeta*, March 1, 1989.

34. Iu. S. Aksenov, "Apogei Stalinizma: Poslevoennaia piramida vlasti" [The Apogee of Stalinism: The Postwar Pyramid of Power], *Voprosy istorii KPSS*, 1990, no. 11: 90–104, at p. 100.

35. Niels E. Rosenfeldt, *Knowledge and Power: The Role of Stalin's Secret Chancellery in the Soviet System of Government* (Copenhagen: Rosenkilde & Begger, 1978), p. 191.

36. *Khrushchev Remembers*, p. 297.

37. Feodosii Vidrashku, *Rumynskaia rapsodiia* [The Romanian Rhapsody] (Moscow: Sovetskii pisatel, 1986), p. 322.

38. Vladimir Tismaneanu, "Miron Constantinescu or the Impossible Heresy," *Survey* 28, no. 4 (1984): 175–87, at pp. 181–83.

39. *Documents Concerning Right Deviation in Romanian Workers' Party* (Bucharest: Rumanian Workers' Party Publishing House, 1952); Gautenbein (Bucharest) to Secretary of State, June 1, 1952, RG-59, 766.00/6-152, NA.

40. Ghita Ionescu, *Communism in Rumania, 1944–1962* (London: Oxford University Press, 1964), pp. 213–14. See also chap. 4, n. 26.

41. François Fejtö, *A History of the People's Democracies: Eastern Europe since Stalin* (New York: Praeger, 1971), pp. 276–77.

42. Kenneth Jowitt, *Revolutionary Breakthroughs and National Development: The Case of Romania, 1944–1965* (Berkeley: University of California Press, 1971), pp. 141–42; Alexander Sulla, "General Epishev: Na partiino-politicheskom Olimpe" [General Epishev: On the Party's Political Olympus], *Voennoistoricheskii zhurnal*, 1993, no. 3: 62–70, at p. 65.

43. Pavel Sudoplatov and Anatoly Sudoplatov, *Special Tasks: The Memoirs of an Unwanted Witness—A Soviet Spymaster* (Boston: Little, Brown, 1994), p. 352.

44. Charles Tillon, *Un "procès de Moscou" à Paris* (Paris: Éditions du Seuil, 1971), pp. 76–77.

45. Philippe Robrieux, *Histoire intérieure du Parti Communiste*, vol. 2 (Paris: Fayard, 1981), pp. 309–30; Irwin M. Wall, *French Communism in the Era of Stalin: The Quest for Unity and Integration, 1945–1962* (Westport, Conn.: Greenwood Press, 1983), pp. 145–47.

46. André Marty, *L'affaire Marty* (Paris: Deux Rives, 1955), pp. 68–75, 55–59, 91; Yves Le Braz (pseud.), *Les Rejetés: L'affaire Marty-Tillon: Pour une histoire différente du PCF* (Paris: La Table Ronde, 1974).

47. Philippe Robrieux, *Maurice Thorez: Vie secrète et vie publique* (Paris: Fayard, 1975), p. 399.

48. Wall, *French Communism*, pp. 148–53; Robrieux, *Maurice Thorez*, pp. 400–407.

49. Wall, *French Communism*, p. 142.

50. Ibid., pp. 148–53.

51. Ibid., p. 155.

52. Ibid., p. 142.

53. George F. Kennan, *Memoirs, 1950–1963* (Boston: Little, Brown, 1972), pp. 147–49.

54. Georgii Arbatov, *The System: An Insider's Life in Soviet Politics* (New York: Times Books, 1992), p. 44.

55. Hermann-Josef Rupieper, "Zu den sowjetischen Deutschlandnoten 1952: Das Gespräch Stalin-Nenni," *Vierteljahrshefte für Zeitgeschichte* 33 (1985): 547–57, at pp. 548–49, 553–55.

56. Pietro Nenni, *Tempo di Guerra Fredda: Diari, 1943–1956* (Milan: Sugar, 1981), pp. 534–38. Presumably because of its lack of substance, the interview has not been found listed in the diary of Stalin's appointments.

57. Georges Bortoli, *The Death of Stalin* (New York: Praeger, 1975), p. 48.

58. Kennan to Deputy Undersecretary of State, August 25, 1952, *FRUS*, 1952–54, vol. 8, pp. 1042–45.

59. Report by NATO Council of Deputies for Lisbon ministerial meeting on February 21–22, 1952, quoted in Robert Spencer, "Alliance Perceptions of the Soviet Threat, 1950–1988," in *The Changing Western Analysis of the Soviet Threat,* ed. Carl-Christoph Schweitzer (London: Pinter, 1990), pp. 9–48, at pp. 15–16.

60. See chap. 4, n. 85.

61. "The Soviet Union and the Atlantic Pact," September 8, 1952, in Kennan, *Memoirs, 1950–1963,* pp. 332–55, at pp. 350–52.

62. Matthew Evangelista, *Innovation and the Arms Race: How the United States and the Soviet Union Develop New Military Technologies* (Ithaca, N.Y.: Cornell University Press, 1988), p. 152.

63. Memorandum for own use, May 18, 1952, in *Off the Record: The Private Papers of Harry S. Truman,* ed. Robert H. Ferrell (New York: Harper & Row, 1980), p. 251; Barton J. Bernstein, "Truman's Secret Thoughts on Ending the Korean War," *Foreign Service Journal,* November 1980.

64. Briggs (Prague) to Secretary of State, May 26, 1952, RG-59, 749.00/5-2652, NA.

65. Thompson (Prague) to Secretary of State, May 26, 1952, RG-59, 749.00/5-2652, NA.

66. Scherstjanoi, "Die DDR im Frühjahr 1952," pp. 360–61.

67. Kennan to Secretary of State, May 22, 1952, *FRUS,* 1952–54, vol. 8, pp. 971–77, at p. 972.

68. Robert Harris and Jeremy Paxman, *A Higher Form of Killing: The Secret Story of Gas and Germ Warfare* (London: Chatto & Windus, 1982), pp. 162–63.

69. *Pravda,* August 29, 1952.

70. James D. Marchio, "Rhetoric and Reality: The Eisenhower Administration and Unrest in Eastern Europe, 1953–1959" (Ph.D. diss., American University, 1990), pp. 70–72.

71. John Foster Dulles, "A Policy of Boldness," *Life,* May 19, 1952.

72. Quoted in Robert A. Divine, *Foreign Policy and U.S. Presidential Election: 1940–1960,* vol. 2 (New York: New Viewpoints, 1974), p. 51.

73. Marchio, "Rhetoric and Reality," pp. 57–64, 73.

74. Rosemary Foot, *The Wrong War: American Policy and the Dimensions of the Korean Conflict, 1950–1953* (Ithaca, N.Y.: Cornell University Press, 1985), p. 178.

75. Harold C. Hinton, *China's Turbulent Quest* (Bloomington: Indiana University Press, 1973), p. 54; John M. Mackintosh, *Strategy and Tactics of Soviet Foreign Policy* (New York: Oxford University Press, 1962), p. 30.

76. Kim Il Sung to Stalin, July 16, 1952, 45/1/348/66–68, Kim Il Sung to Mao Zedong, August 24, and Zhou Enlai to Stalin, August 25, 1952, 45/1/343/89–90, APRF.

77. Record of Stalin-Zhou Enlai conversations, August 20, September 3, and September 19, 1952, 45/1/329/54–101, APRF.

78. Mao Zedong to Stalin, March 28, 1952, 45/1/342/126–30, at p. 127, APRF.

79. Stalin to Mao Zedong, September 17, 1952, 45/1/343/97, APRF.

80. Robert R. Simmons, *The Strained Alliance: Peking, P'yŏngyang, Moscow, and the Politics of the Korean Civil War* (New York: Macmillan, 1975), pp. 217–24.

81. Kathryn Weathersby, "Stalin and a Negotiated Settlement in Korea, 1950–53" (paper presented at the conference "New Evidence on the Cold War in Asia," Hong Kong, January 9–12, 1996), pp. 34–40, comes closer to this view than Chen Jian, "China's Strategies to End the Korean War" (paper presented at the same conference), pp. 28–30.

82. Mackintosh, *Strategy and Tactics of Soviet Foreign Policy*, p. 57.

83. *Izvestiia*, November 4, 1952.

84. William Stueck, *The Korean War: An International History* (Princeton, N.J.: Princeton University Press, 1995), pp. 303–6.

85. Simmons, *The Strained Alliance*, p. 228.

86. Geoge D. Embree, *The Soviet Union between the 19th and 20th Party Congresses, 1952–1956* (The Hague: Nijhoff, 1952), p. 5.

87. Wladimir S. Semjonow, *Von Stalin bis Gorbatschow: Ein halbes Jahrhundert in diplomatischer Mission, 1939–1991* (Berlin: Nikolai, 1995), p. 282.

88. Speech of October 14, 1952, in *The Essential Stalin: Major Theoretical Writings, 1905–1952*, ed. Bruce Franklin (Garden City: Doubleday, 1972), pp. 508–11.

89. "Economic Problems of Socialism in the U.S.S.R.," in Franklin, *The Essential Stalin*, pp. 445–81, at pp. 467–73.

90. Ibid., pp. 445–51, 473–77.

91. Werner G. Hahn, *Postwar Soviet Politics: The Fall of Zhdanov and the Defeat of Moderation, 1946–1953* (Ithaca, N.Y.: Cornell University Press, 1982), pp. 149–56.

92. *Khrushchev Remembers*, pp. 277–78.

93. Leo Gruliow, ed., *Current Soviet Policies: The Documentary Record of the 19th Communist Party Congress and Reorganization after Stalin's Death* (New York: Praeger, 1953).

94. Charles H. Fairbanks Jr., "National Cadres as a Force in the Soviet System: The Evidence of Beria's Career, 1949–53," in *Soviet Nationality Policies and Practices*, ed. Jeremy R. Azrael (New York: Praeger, 1978), pp. 146–86, at pp. 144–49.

95. Robert Conquest, *Power and Policy in the U.S.S.R.: The Study of Soviet Dynastics* (New York: St. Martin's Press, 1961), pp. 158–63.

96. Cf. Yoram Gorlitzki, "Party Revivalism and the Death of Stalin," *Slavic Review* 54 (1995): 1–22.

97. Editor's note, "Plenum TsK KPSS: Iiul 1953 goda [The CPSU Central Committee Plenum of July 1953], *Izvestiia TsK KPSS*, 1991, no. 1: 145; Konstantin Simonov, *Glazami cheloveka moego pokoleniia* [Through the Eyes of a Man of My Generation] (Moscow: Pravda, 1991), pp. 236–38.

98. R. W. Davies, *Soviet History in the Gorbachev Revolution* (London: Macmillan, 1989), p. 64.

99. Entry for January 13, 1984, in Chuev, *Sto sorok besed*, p. 325.

100. *Khrushchev Remembers*, pp. 601, 392.

101. Vladislav M. Zubok, "Soviet Intelligence and the Cold War: The 'Small' Committee of Information, 1952–53," Working Paper no. 4, *Cold War International History Project* (Washington, D.C.: Woodrow Wilson International Center for Scholars, 1992), p. 11.

102. *Khrushchev Remembers: The Glasnost Tapes* (Boston: Little, Brown, 1990), pp. 100–101.

103. Quoted in Martin Gilbert, *"Never Despair": Winston S. Churchill, 1945–1965* (London: Heinemann, 1988), p. 773.

104. Barton J. Bernstein, "Crossing the Rubicon: A Missed Opportunity to Stop the H-Bomb?" *International Security* 14, no. 2 (fall 1989): 132–60, at p. 149.

105. Stephen E. Ambrose, *Eisenhower*, vol. 2, *The President* (New York: Simon & Schuster, 1984), pp. 31–35.

106. Cited in J. Ronald Caridi, *The Korean War and American Politics: The Republican Party as a Case Study* (Philadelphia: University of Pennsylvania Press, 1968), p. 253.

107. Foot, *The Wrong War*, p. 205.

108. Anatolii Dobrynin, "Politicheskii otchet Posolstva SSSR v SShA za 1953 god" [The Political Report of the USSR Embassy in the USA for 1953], February 12, 1954, 0129/1953/37/14/265, pp. 50–80, AVPRF.

109. Kennan, "The Soviet Union and the Atlantic Pact," in *Memoirs 1950–1963*, p. 348.

9   The Dead End

1. "Estimate Prepared by the Board of National Estimates," November 21, 1951, *FRUS*, 1952–1954, vol. 2, pt. 1, pp. 186–96, at p. 187.

2. George D. Embree, *The Soviet Union between the 19th and 20th Party Congresses, 1952–1956* (The Hague: Nijhoff, 1952), p. 16.

3. Václav Brabec, "Vztah KSČ a veřejnosti k politickým procesům na počátku padesátých let" [The Attitudes of the Communist Party of Czechoslovakia and the Public toward the Political Trials of the Early Fifties], *Revue dějin socialismu* [Prague], 1969, no. 3: 363–85.

4. Letters by Lise Margolius and Tomáš Frejka, cited in Meir Cotic, *The Prague Trial: The First Anti-Zionist Show Trial in the Communist Bloc* (New York: Herzl Press, 1987), pp. 155–56. The son later committed suicide.

5. "Žaloba proti vedení protistátního spikleneckého centra v čele s Rudolfem Slánským" [Indictment of the Antistate Conspiratorial Center Headed by Rudolf Slánský], *Rudé právo* [Prague], November 20, 1952.

6. Yaacov R'oi, *Soviet Decision-Making in Practice: The U.S.S.R. and Israel, 1947–1954* (New Brunswick, N.J.: Transaction Books, 1982), p. 366; Peter Brod, *Die Antizionismus- und Israelpolitik der UdSSR: Voraussetzungen und Entwicklung bis 1956* (Baden-Baden: Nomos, 1980), pp. 90–91.

7. The transcript is translated in *Procès des dirigeants du centre de conspiration contre l'Etat dirigé par Rudolf Slánský* (Prague: Ministère de la Justice, 1953).

8. The aspect of the proceedings best understood by its protagonist, Konni Zilliacus, in his *A New Birth of Freedom? World Communism after Stalin* (London: Secker & Warburg, 1957), pp. 143–49.

9. Vojtech Mastny, *Russia's Road to the Cold War: Diplomacy, Warfare, and the Politics of Communism, 1941–1945* (New York: Columbia University Press, 1979), pp. 190–95, 274–79.

10. Ivo Banac, "Yugoslav Cominformist Organizations and Insurgent Activity: 1948–1954," in *At the Brink of War and Peace*, ed. Wayne S. Vucinich (New York: Brooklyn College Press, 1982), pp. 239–52, at p. 241.

11. They are described, but inflated, in Stephen Koch, *Double Lives: Spies and Writers in the Secret Soviet War of Ideas against the West* (New York: Free Press, 1994), pp. 75–95, 321–32.

12. King (Prague) to Secretary of State, November 20, 1952, 749.00/11–2052, RG-59, NA.

13. Karel Kaplan, *Die politischen Prozesse in der Tschechoslowakei, 1948–1954* (Munich: Oldenbourg, 1986), p. 205.

14. Hence the assertion by former Polish politburo member Edward Ochab that Beriia had initiated the Komar case is open to doubt. Teresa Torańska, *"Them:" Stalin's Polish Puppets* (New York: Harper & Row, 1987), p.50.

15. "Lehren aus dem Prozeß gegen das Verschwörerzentrum Slansky," in *Dokumente der Sozialistischen Einheitspartei Deutschlands*, vol. 4 (Berlin: Dietz, 1954), pp. 199–219.

16. Jeffrey Hurf, "East German Communists and the Jewish Question: The Case of Paul Merker" (paper presented at the conference "Archive und jüngere Forschungen zur Frühgeschichte von SBZ und DDR," Potsdam, July 1–2, 1994).

17. "O tak nazyvaemom 'dele Evreiskogo Antifashistskogo Komiteta,'" *Izvestiia TsK KPSS*, 1989, no. 12: 35–40, at p. 38.

18. Pavel Sudoplatov and Anatoly Sudoplatov, *Special Tasks: The Memoirs of an Unwanted Witness—A Soviet Spymaster* (Boston: Little, Brown, 1994), p. 301.

19. Amy Knight, *Beria: Stalin's First Lieutenant* (Princeton, N.J.: Princeton University Press, 1993), pp. 147–48.

20. Robert Conquest, *Power and Policy in the U.S.S.R.: The Study of Soviet Dynastics* (New York: St. Martin's Press, 1961), pp. 178–80. See chap. 4, n. 90.

21. Stalin's responses, sent on December 24, 1952, to Reston's questions sumitted on December 18, 1952, in Walt W. Rostow, *Europe after Stalin: Eisenhower's Three Decisions of March 11, 1953* (Austin: University of Texas Press, 1982), pp. 136–37.

22. Benjamin Pinkus, *The Soviet Government and the Jews, 1948–1967: A Documented Study* (Cambridge: Cambridge University Press, 1984), pp. 219–20.

23. Louis Rapoport, *Stalin's War against the Jews: The Doctors' Plot and the Soviet Solution* (New York: Free Press, 1990), pp. 50, 147.

24. Entry for February 3, 1972, in Feliks Chuev, *Sto sorok besed s Molotovym* [Hundred-and-Forty Conversations with Molotov] (Moscow: Terra, 1991), p. 477; Miasnikov in *Literaturnaia gazeta*, March 1, 1989.

25. Sudoplatov and Sudoplatov, *Special Tasks*, p. 298.

26. "Konets 'dela vrachei'" [The End of the "Doctors' Affair"], *Moskovskie Novosti*, February 7, 1988.

27. *Khrushchev Remembers* (Boston: Little, Brown, 1970), p. 286.

28. *Rudé právo* [Prague], November 20 and 21, 1952.

29. "Delo Beriia (Plenum TsK KPSS, Iiul 1953 goda: Stenograficheskii otchet)" [The Beriia Affair: Stenographic Record of the Plenary Session of the Central Committee of the Communist Party of the Soviet Union in July 1953], *Izvestiia TsK KPSS*, January 1991, pp. 139–214, at p. 142.

30. *Khrushchev Remembers*, p. 171.

31. Rapoport, *Stalin's War against the Jews*, pp. 145–46.

32. Myron Rush, *Political Succession in the USSR* (New York: Columbia University Press, 1965), pp. 52–54.

254 NOTES TO PAGES 160-164

33. *Khrushchev Remembers*, p. 311.

34. Harrison E. Salisbury, *Moscow Journal: The End of Stalin* (Chicago: University of Chicago Press, 1961), pp. 312–22.

35. Alexander Solzhenitsyn, *The Gulag Archipelago, 1918–1956* (London: Collins/Fontana, 1974), pp. 157–58.

36. Boris I. Nicolaevsky, *Power and the Soviet Elite: "The Letter of an Old Bolshevik" and Other Essays* (New York: Praeger, 1965), pp. 148–57; Sudoplatov and Sudoplatov, *Special Tasks*, p. 242.

37. Konstantin Simonow, "Ein schrecklicher Mensch," in *Berija, Henker in Stalins Diensten: Ende einer Karriere*, ed. Wladimir F. Nekrassow (Berlin: Edition q, 1992), pp. 228–39, at pp. 232–33.

38. Sudoplatov and Sudoplatov, *Special Tasks*, p. 351.

39. Speech of February 25, 1956, in *Khrushchev Remembers*, pp. 559–618, at p. 601.

40. Raymond L. Garthoff, *Soviet Strategy in the Nuclear Age* (New York: Praeger, 1958), pp. 20–21.

41. Conquest, *Power and Policy in the U.S.S.R.*, p. 167.

42. John A. Armstrong, *The Politics of Totalitarianism: The Communist Party of the Soviet Union from 1934 to the Present* (New York: Random House, 1961), p. 236.

43. Kaplan, *Die politischen Prozesse*, p. 204.

44. Salisbury, *Moscow Journal*, pp. 318–26.

45. Mario Keßler, "Zwischen Repression und Toleranz: Die SED-Politik und die Juden (1949 bis 1967)," in *Historische DDR-Forschung: Aufsätze und Studien*, ed. Jürgen Kocka (Berlin: Akademie, 1993), pp. 149–67, at pp. 152–61; Wilhelm Fricke, *Warten auf Gerechtigkeit: Kommunistische Säuberungen und Rehabilitierungen: Bericht und Dokumentation* (Cologne: Wissenschaft und Politik, 1971), pp. 87–88.

46. Bennett Kovrig, *Communism in Hungary: From Kun to Kádár* (Stanford, Calif.: Hoover Institution Press, 1979), p. 247.

47. George H. Hodos, *Show Trials: Stalinist Purges in Eastern Europe, 1948–1954* (New York: Praeger, 1987), pp. 66–67.

48. Arnold Krammer, *The Forgotten Friendship: Israel and the Soviet Bloc, 1947–53* (Urbana: University of Illinois Press, 1974), p. 192.

49. Rapoport, *Stalin's War against the Jews*, pp. 200–201.

50. Quoted in Torańska, *"Them,"* p. 171.

51. Iakov Etinger, "'Delo vrachei' sorok let spustia" [The "Doctors' Plot" Forty Years After], *Novoe vremia* 1993, no. 2–3: 47–49, at p. 49.

52. Rapoport, *Stalin's War against the Jews*, pp. 176–79.

53. Anatol Goldberg, *Ilya Erenburg: Writing, Politics and the Art of Survival* (London: Weidenfeld & Nicolson, 1984), pp. 6–7, 281–82.

54. Sudoplatov and Sudoplatov, *Special Tasks*, p. 308.

55. Conquest, *Power and Policy in the U.S.S.R.*, p. 178.

56. "Delo Beriia," January 1991, p. 155.

57. Niels E. Rosenfeldt, *Knowledge and Power: The Role of Stalin's Secret Chancellery in the Soviet System of Government* (Copenhagen: Rosenkilde & Begger, 1978), pp. 194–96.

58. Entry for January 13, 1984, in Chuev, *Sto sorok besed*, p. 325.

59. Peter Deriabin, *Watchdogs of Terror* (Frederick, Md.: University Publications, 1984), p. 239.

60. Dmitrii Volkogonov, *Triumf i tragediia: Politicheskii portret I. V. Stalina* [Triumph and Tragedy: The Political Portrait of J. V. Stalin], vol. 2, pt. 2 (Moscow: Novosti, 1989), p. 189.

61. Embree, *The Soviet Union between the 19th and 20th Party Congresses*, pp. 22–24, 27–28; Salisbury, *Moscow Journal*, p. 328; I.S. Glebov, "Intrigi v generalnom shtabe" [Intrigues in the General Staff], *Voennoistoricheskii zhurnal*, 1993, no. 11: 37–43, at pp. 41–42.

62. Wolfgang Leonhard, *The Kremlin since Stalin* (New York: Praeger, 1962), pp. 49–50.

63. "Soviet Bloc Capabilities, through mid-1953," NIE-64, pt. 1, November 12, 1952, p. 2, Central Intelligence Agency.

64. Minutes of National Security Council meeting, June 4, 1953, *FRUS*, 1952–54, vol. 2, p. 369.

65. "Annual Message to the Congress on the State of the Union," February 2, 1953, *Public Papers of the Presidents: Dwight D. Eisenhower, 1953* (Washington, D.C.: Government Printing Office, 1960), p. 17.

66. Rosemary Foot, "Nuclear Coercion and the Ending of the Korean Conflict," *International Security* 13, no. 3 (1988–89): 92–110, at pp. 96–97.

67. Vladislav M. Zubok, "Soviet Intelligence and the Cold War: The 'Small' Committee of Information, 1952–53," Working Paper no. 4, *Cold War International History Project*, (Washington, D.C.: Woodrow Wilson International Center for Scholars, 1992), p. 12.

68. Minutes of National Security Council meeting, February 11, 1953, *FRUS*, 1952–54, vol. 15, p. 770.

69. Cited in John L. Gaddis, "The Origins of Self-Deterrence: The United States and the Non-Use of Nuclear Weapons, 1945–1958," in his *The Long Peace: Inquiries into the History of the Cold War* (New York: Oxford University Press, 1987), pp. 105–46, at p. 123.

70. Quote from his *The Art of War*, as translated in Xue Litai, "China's Nuclear Strategy and Capabilities" (paper presented at the conference "New Evidence on the Cold War in Asia," Hong Kong, January 9–12), p. 7.

71. He Di, "Paper or Real Tiger: America's Nuclear Deterrence and Mao Zedong's Response" (paper presented at the conference "New Evidence on the Cold War in Asia," Hong Kong, January 9–12, 1996), pp. 6–10.

72. Stalin to Mao Zedong, December 27, 1952, 45/1/343/115–16, APRF.

73. He Di, "Paper or Real Tiger," p. 8.

74. U.S. Embassy Moscow, 320 USSR January–June, RG-84, NA.

75. Harold C. Hinton, *China's Turbulent Quest* (Bloomington: Indiana University Press, 1970), p. 56.

76. Krishna P. S. Menon, *The Flying Troika* (London: Oxford University Press, 1963), pp. 26–32.

77. Ernest R. May, John D. Steinbrunner, and Thomas Wolfe, "History of the Strategic Arms Competition 1945–1972," pt. 1 (Washington, D.C.: Office of the Secretary of Defense, 1981), pp. 278–83, NSA.

78. Salisbury, *Moscow Journal*, p. 331.

79. Beam to Secretary of State, February 19, 1953, *FRUS*, 1952–54, vol. 8, pp. 1078–79.

80. The assassination plan is reported by Gen. Volkogonov, the amateur historian whose privileged access to otherwise inaccessible Soviet documents has been no guarantee or reliability, as having been devised by Beriia on

Stalin's orders already in the late fall of 1952. Dmitrii A. Volkogonov, "Nesostoiavsheesia pokushenie: Kak sovetskii agent Maks gotovilsia k teroristicheskomu aktu protiv Tito" [The Assassination Attempt That Never Took Place: How Soviet Agent Max Was Preparing Terrorist Action against Tito], *Izvestiia*, June 11, 1993.

81. Sudoplatov and Sudoplatov, *Special Tasks*, pp. 334–39.

82. Embree, *The Soviet Union between the 19th and 20th Party Congresses*, p. 25.

83. Ibid.

84. *Khrushchev Remembers*, pp. 316–20.

85. A. T. Rybin, "Riadom s I.V. Stalinym" [Side by Side with Stalin], *Sotsiologicheskie issledovaniia*, 1988, no. 3: 84–94.

86. Entries for January 13, 1984, and March 11, 1983, in Chuev, *Sto sorok besed*, pp. 323–24, 476–77. Volkogonov deducts from his unverifiable sources that Beriia helped Stalin to die by advising him to take a steam bath which was certain to kill him. Enrico Franceschini, "Ucciderò Stalin," *La Repubblica* [Rome], February 26, 1993, pp. 38–39.

87. Entries for August 24, 1971, and June 9, 1976, in Chuev, *Sto sorok besed*, p. 328.

88. Svetlana Alliluyeva, *Twenty Letters to a Friend* (New York: Harper & Row, 1967), pp. 215–16.

89. Ibid., p. 23.

90. The best is the study by the émigré Soviet author, translated into German, Abdurachman Awtorchanow, *Das Rätsel um Stalins Tod* (Frankfurt: Ullstein, 1984); see also Anton Antonow-Owsejenko, "Der Weg nach oben: Skizzen zu einem Berija-Porträt," in Nekrassow, *Berija*, pp. 11–172, at p. 159.

91. Antonow-Owsejenko, "Der Weg nach oben," p. 160.

92. James Richter, "Reexamining Soviet Policy towards Germany during the Beria Interregnum," Working Paper no. 3, *Cold War International History Project* (Washington, D.C.: Woodrow Wilson International Center for Scholars, 1992), p. 6.

93. For the best account of Stalin's last days, see Robert H. McNeal, *Stalin: Man and Ruler* (London: Macmillan, 1988), pp. 300–309.

94. Excerpts from Miasnikov's unpublished memoirs, *Literaturnaia gazeta*, March 1, 1989.

95. "Protokol sovmestnogo zasedaniia plenuma TsK KPSS, Soveta Ministrov SSSR i Prezidiuma Verkhovnogo Soveta SSSR" [Minutes of the Joint Session of the CPSU Central Committee, the USSR Council of Ministers, and the Presidium of the USSR Supreme Soviet], March 5, 1953, *Istochnik*, 1994, no. 4: 107–11.

96. "On the Korean War, 1950–53, and the Armistice Negotiations," August 9, 1966, *Cold War International History Project Bulletin* 3 (1993): 15–17, at p. 17.

97. Iu. A. Poliakov, "Pokhorony Stalina: Vzgliad istorika-ochevidtsa" [The Stalin Burial: The Perspective of an Eyewitness Historian], *Novaia i noveishaia istoriia*, 1994, nos. 4–5: 195–207.

98. Report by Captain Lang, March 6, 1953, Moscow Embassy, 361.1 Stalin, RG-83, NA.

99. Beam to Secretary of State, March 6, *FRUS*, 1952–54, vol. 8, p. 1099.

100. Merkulov to Khrushchev, July 23, 1953, Party Secretariat, 5/30/4, TsKhSD.

101. Juri Krotkow, "Ich führte Berijas Befehl aus," in Nekrassow, *Berija*, p. 315; *Khrushchev Remembers*, pp. 340–41.

10 Coping with the Stalin Legacy

1. "Delo Beriia (Plenum TsK KPSS, Iiul 1953 goda: Stenograficheskii otchet)" [The Beriia Affair: Stenographic Record of the Plenary Session of the Central Committee of the Communist Party of the Soviet Union in July 1953], *Izvestiia TsK KPSS*, January 1991, pp. 139–214, at pp. 160–61.

2. The picture is reproduced in Sergei N. Goncharov, John W. Lewis, and Xue Litai, *Uncertain Partners: Stalin, Mao, and the Korean War* (Stanford, Calif.: Stanford University Press, 1993), as illustration no. 21.

3. N. A. Barsukov et al., *XX sezd KPSS i ego istoricheskie realnosti* [The 20th Congress of the CPSU and Its Historical Realities] (Moscow: Politizdat, 1991), pp. 12–13.

4. Decision by the Council of Ministers, March 19, 1953, 3/65/830/60-71, APRF.

5. Speech by Malenkov, March 15, 1953, in *Documents on International Affairs 1953*, ed. Denise Folliot (London: Oxford University Press, 1956), pp. 12–13.

6. Decision by the Council of Ministers, March 19, 1953, 3/65/830/60-71, at p. 61, APRF.

7. Chen Jian, "China's Strategies to End the Korean War" (paper presented at the conference "New Evidence on the Cold War in Asia," Hong Kong, January 9–12, 1996), p. 28.

8. See chap. 6, n. 68 and 69.

9. See chap. 8, n. 77.

10. Memorandum by Carlton Savage of the Policy Planning Staff, April 1, 1953, *FRUS*, 1952–1954, vol. 8, p. 1138.

11. Draft of statement by Molotov, March 31, 1953, 3/65/830/106-12, APRF.

12. *Khrushchev Remembers: The Last Testament* (Boston: Little, Brown, 1974), p. 220.

13. By decision of the party presidium, referred to in P2/6/22/5, TsKhSD.

14. "Delo Beriia," January 1991, pp. 153, 161–62.

15. *Khrushchev Remembers: The Glasnost Tapes* (Boston: Little, Brown, 1990), p. 73.

16. John A. Armstrong, *The Politics of Totalitarianism: The Communist Party of the Soviet Union from 1934 to the Present* (New York: Random House, 1961), p. 241; Robert Conquest, *Power and Policy in the U.S.S.R.: The Study of Soviet Dynastics* (New York: St. Martin's Press, 1961), p. 221.

17. "Delo Beriia," January 1991, p. 144.

18. Ibid., p. 151.

19. Ibid., pp. 204–8, 182.

20. Printed in Benjamin Pinkus, *The Soviet Government and the Jews, 1948–1967: A Documented Study* (Cambridge: Cambridge University Press, 1984), pp. 220–21.

21. *Pravda* editorial cited in Beam (Moscow) to Secretary of State, April 6, 1953, *FRUS*, 1952–1954, vol. 8, p. 1142.

22. "Delo Beriia," January 1991, pp. 174, 184.

23. Andrei G. Malenkov, *O moem ottse Georgii Malenkove* [My Father, Georgii Malenkov] (Moscow: Tekhnoekos, 1992), p. 70.

24. James Richter, "Reexamining Soviet Policy toward Germany during the Beria Interregnum," *Cold War International History Project*, Working Paper no. 3 (Washington, D.C.: Woodrow Wilson International Center for Scholars, 1992), p. 5.

25. Sergo Mikoian, "Pokaianie i iskuplenie" [Repentance and Redemption], *Sovetskaia kultura*, August 13, 1988.

26. Richter, "Reexamining Soviet Policy," pp. 6–7.

27. John W. Young, "Cold War and Détente with Moscow," in *The Foreign Policy of the Churchill Peacetime Administration, 1951–1955*, ed. John W. Young (Leicester, England: Leicester University Press, 1988), pp. 55–80.

28. Quoted in John Yurechko, "The Day Stalin Died: American Plans for Exploiting the Soviet Succession Crisis of 1953," *Journal of Strategic Studies* 3, no. 1 (May 1980): 44–73, at p. 67.

29. "Delo Beriia," January 1991, p. 171.

30. *Khrushchev Remembers: The Last Testament*, p. 362.

31. John O. Iatrides, *Balkan Triangle: Birth and Decline of an Alliance across Ideological Boundaries* (The Hague: Mouton, 1968), pp. 114–15.

32. "Delo Beriia," January 1991, p. 164.

33. Ferenc Vali, *Bridge across the Bosporus: The Foreign Policy of Turkey* (Baltimore, Md.: Johns Hopkins Press, 1971), pp. 174–75.

34. Arnold Krammer, *The Forgotten Friendship: Israel and the Soviet Bloc, 1947–53* (Urbana: University of Illinois Press, 1974), p. 196.

35. Yurechko, "The Day Stalin Died," pp. 44–73; Steven Fish, "After Stalin's Death: The Anglo-American Debate over a New Cold War," *Diplomatic History* 10 (1986): 333–55.

36. Churchill to Eisenhower, March 11, 1953, in *The Churchill-Eisenhower Correspondence, 1953–1955*, ed. Peter G. Boyle (Chapel Hill: University of North Carolina Press, 1990), p. 31.

37. Proposal of March 11, 1953, quoted in Yurechko, "The Day Stalin Died," p. 65.

38. Walt W. Rostow, "Notes on the Origin of the President's Speech of April 16, 1953," May 11, 1953, *FRUS*, 1952–1954, vol. 8, pp. 1173–83.

39. Walt W. Rostow, *Europe after Stalin: Eisenhower's Three Decisions of March 11, 1953* (Austin: University of Texas Press, 1982), pp. 113–22.

40. Bohlen to Department of State, April 20, 1953, *FRUS*, 1952–54, vol. 8, pp. 1155–56.

41. Cf. James Richter, "Action and Reaction in Soviet Foreign Policy" (Ph.D. diss. University of California at Berkeley, 1989), pp. 53–54.

42. Vladislav M. Zubok, "Soviet Intelligence and the Cold War: The 'Small' Committee of Information, 1952–53," *Diplomatic History* 19 (1995): 453–72, at pp. 460–61.

43. Speech by Dulles, April 18, 1953, in Rostow, *Europe after Stalin*, pp. 122–31.

44. Note by Arkadev on conversation with Beam, April 16, 1953, 0129/1953/37/6/264, pp. 9–11, AVPRF.

45. "Predlozheniia po germanskomu voprosu" [Proposals on the German Question], by Malik and Pushkin, April 24, 1953, 082/1953/41/19/112 vol.2/271, pp. 2–12, quote on p. 5, AVPRF.

46. "Zapiska po germanskomu voprosu" [Memorandum on the German Question], by Semenov, May 2, 1953, 082/1953/41/18/112/271, pp. 52–59, AVPRF.

47. Norman Naimark, "Was the SBZ/GDR in the Soviet Bloc?" *Potsdamer Bulletin für Zeithistorische Studien* 4 (1995): 6–18.

48. *SBZ von 1945 bis 1954: Die sowjetische Besatzungszone Deutschlands in den Jahren 1945–1954* (Bonn: Bundesministerium für gesamtdeutsche Fragen, 1961), p. 241.

49. Aleksei M. Filitov and Tamara V. Domracheva, "Soviet Policy on Germany, 1953–1955" (paper presented at the conference "New Evidence on Cold War History," Moscow, January 12–15, 1993), p. 3.

50. "Über die Auswertung des Beschlußes des Zentralkomitees zu den 'Lehren aus dem Prozeß gegen das Verschwörerzentrum Slansky,'" *Dokumente der Sozialistischen Einheitspartei Deutschlands*, vol. 4 (Berlin: Dietz, 1954), pp. 394–409.

51. *Tägliche Rundschau*, May 20, 1953.

52. Martin Jänicke, *Der dritte Weg: Die antistalinistische Opposition gegen Ulbricht seit 1953* (Cologne: Neuer Deutscher Verlag, 1964), pp. 28–32.

53. Gerhard Wettig, "Sowjetische Wiedervereinigungsbemühungen im ausgehenden Frühjahr 1953? Neue Aufschlüsse über ein altes Problem," *Deutschland-Archiv* 25 (1992): 943–58, at pp. 945–46; cf. Valur Ingimundarson, "Cold War Misperceptions: The Communist and Western Responses to the East German Refugee Crisis in 1953," *Journal of Contemporary History* 29 (1994): 463–81.

54. Amy Knight, *Beria: Stalin's First Lieutenant* (Princeton, N.J.: Princeton University Press, 1993), pp. 226, 183–91.

55. George Daniel Embree, *The Soviet Union between the 19th and 20th Party Congresses, 1952–1956* (The Hague: Nijhoff, 1952), p. 48.

56. "Delo Beriia," January 1991, pp. 153–54.

57. Gerhard Simon, *Nationalismus und Nationalitätenpolitik in der Sowjetunion: Von der totalitären Diktatur zur nachstalinistischen Gesellschaft* (Baden-Baden: Nomos, 1986), pp. 256–63; Boris Starkov, "Sto dnei 'Lubianskogo Marshala'" [The Hundred Days of the "Marshal of Lubianka"], *Istochnik*, 1993, no. 4: 82–90, at pp. 87–88.

58. Referred to in 6/30/5, TsKhSD.

59. "Delo Beriia," January 1991, p. 142; O. A. Gorchakov, "Dokumenty General-Leitenanta T. A. Strokacha o podgotovke Beriei zagovora v 1953 g." [Documents of Lt.-Gen. T. A. Strokach on Beriia's Preparation of a Conspiracy in 1953], *Novaia i noveishaia istoriia*, 1989, no. 3: 166–76.

60. "Delo Beriia," January 1991, p. 142.

61. Ibid., p. 143; Pavel Sudoplatov and Anatoly Sudoplatov, *Special Tasks: The Memoirs of an Unwanted Witness—A Soviet Spymaster* (Boston: Little, Brown, 1994), p. 367.

62. Cf. Gerhard Wettig, "Zum Stand der Forschung über Berijas Deutschlandpolitik im Frühjahr 1953," *Deutschland-Archiv* 26 (1993): 674–82.

63. Zubok, "Soviet Intelligence and the Cold War," p. 463.

64. Sudoplatov and Sudoplatov, *Special Tasks*, pp. 363–64.

65. Starkov, "Sto dnei," p. 85; "Delo Beriia," January 1991, p. 213.

66. Ibid., pp. 143–44, 162–63.

67. Feliks Chuev, *Sto sorok besed s Molotovym* [Hundred and Forty Conversations with Molotov] (Moscow: Terra, 1991), pp. 334–35; Richter, "Reexamining Soviet Policy," p. 21.

68. "Delo Beriia," January 1991, p. 173.

69. Chuev, *Sto sorok besed*, pp. 334–35.

70. It would be sent in a revised form in a different situation on August 23, without bringing the desired solution of the German question any closer.

71. Karel Kaplan, "La crisi cecoslovacca, 1953–56," in *Annali della Fondazione Giangiacomo Feltrinelli 1982* (Milan: Feltrinelli, 1983), pp. 267–75.

72. Karel Kaplan, *The Overcoming of the Regime Crisis after Stalin's Death in Czechoslovakia, Poland, and Hungary* (Cologne: Index, 1986), p. 12

73. "Delo Beriia," January 1991, p. 157.

74. Andrei Gromyko, *Memories* (London: Hutchinson, 1989), p. 316.

75. Rudolf Herrnstadt, *Das Herrnstadt-Dokument: Das Politbüro der SED und die Geschichte des 17. Juni 1953* (Reinbeck bei Hamburg: Rowohlt, 1990), pp. 58–59.

76. Cited in Wilfried Loth, *Stalins ungeliebtes Kind: Warum Moskau die DDR nicht wollte* (Berlin: Rowohlt, 1994), p. 258.

77. "Über die Maßnahmen zur Gesundung der politischen Lage in der Deutschen Demokratischen Republik," *Beiträge zur Geschichte der Arbeiterbewegung* 32 (1990): 651–54.

78. Herrnstadt, *Das Herrnstadt-Dokument*, p. 59.

79. John W. Young, "Churchill, the Russians and the Western Alliance: The Three-Power Conference in Bermuda, December 1953," *English Historical Review* 101 (1986): 889–912, at p. 893.

80. Frank Roberts, "Stalin, Khrushchev i Berlinskie krizisy" [Stalin, Khrushchev, and the Berlin Crises], *Mezhdunarodnaia zhizn*, 1991, no. 10: 130–43, at p. 138; Zubok, "Soviet Intelligence and the Cold War," p. 469.

81. Charles Gati, *Hungary and the Soviet Bloc* (Durham, N.C.: Duke University Press, 1986), pp. 130–31.

82. Elke Scherstjanoi, "Wollen wir den Sozialismus? Dokumente aus der Sitzung des Polibüros der SED am 6. Juni 1953," *Beiträge zur Geschichte der Arbeiterbewegung* 33 (1991): 658–80.

83. Jochen Hansen in Nordwestdeutscher Rundfunk, 762 B.00/6-1653, RG-59, NA.

84. Herrnstadt, *Das Herrnstadt-Dokument*, pp. 62–65, 72–74.

85. Thomas Powers, *The Man Who Kept the Secrets: Richard Helms and the CIA* (New York: Knopf, 1979), p. 51.

86. Quoted in Christian F. Ostermann, "The United States, the East German Uprising of 1953, and the Limits of Rollback," Working Paper No. 11, *Cold War International History Project* (Washington, D.C.: Woodrow Wilson International Center for Scholars, 1994), p. 20.

87. The corps never attained any importance and eventually faded away. James D. Marchio, "Rhetoric and Reality: The Eisenhower Administration and Unrest in Eastern Europe, 1953–1959" (Ph.D. diss. American University, 1990), pp. 119–20; Henry W. Brands, Jr., "A Cold War Foreign Legion? The Eisenhower Administration and the Volunteer Freedom Corps," *Military Affairs* 52 (January 1988): 7–11.

88. Statement by Dulles at press conference, June 30, 1953, *Department of State Bulletin*, July 13, 1953, p. 40.

89. Wladimir S. Semjonow, *Von Stalin bis Gorbatschow: Ein halbes Jahrhundert in diplomatischer Mission, 1939–1991* (Berlin: Nicolai, 1995), p. 295.

90. "Probable Soviet Bloc Courses of Action through mid-1953," NIE-64, pt. 2, December 11, 1952, Central Intelligence Agency.

91. "Der neue Kurs und die Erneuerung der Partei," in Wilfriede Otto, "Dokumente zur Auseinandersetzung in der SED 1953," *Beiträge zur Geschichte der Arbeiterbewegung* 32 (1990): 655–72, at p. 659.

92. James Richter, "Reexamining Soviet Policy toward Germany during the Beria Interregnum" (paper presented at the conference "New Evidence on Cold War History," Moscow, January 12–15, 1993), p. 17.

93. Cf. Victor Baras, "Beria's Fall and Ulbricht's Survival," *Soviet Studies* 27 (1975): 381–95.

94. Kiril Moskalenko, "Wie Berija verhaftet wurde," in *Berija, Henker in Stalins Diensten: Ende einer Karriere,* ed. Wladimir F. Nekrassow (Berlin: Edition q, 1922), pp. 345–51; *Khrushchev Remembers* (Boston: Little, Brown, 1970), pp. 333–38; Chuev, *Sto sorok besed,* pp. 343–45; Fedor Burlatskii, "Posle Stalina: Zametki o politicheskoi ottepeli" [After Stalin: Notes on the Political Thaw], *Novyi mir,* 1988, no. 10: 153–97, at p. 164.

95. The persons include Anton Kolendić (*Posljednji dani kulta ličnosti: Od Staljinove do Berijne smrti* [The Last Days of the Cult of Personality: From Stalin's to Beriia's Deaths] [Rijeka: Keršovani, 1980]), Boris Popov, and Vitalii Oppokov ("Die Berija-Zeit [Nach Unterlagen der Beweisaufnahme], in Nekrasow, *Berija,* pp. 369–454).

96. Knight, *Beria,* pp. 218–22.

97. Giulio Seniga, *Togliatti e Stalin* (Milan: Sugar, 1978), pp. 39–43.

98. *Pravda,* July 10, 1953.

99. *Khrushchev Remembers,* pp. 319–20.

100. "Delo Beriia," January 1991, p. 146.

101. The thirty letters known to have been written, which can be seen in the archives, were received between July 1 and 13. Memorandum by Kabashkin, July 31, 1953, 5/30/6, TsKhSD.

102. Cited in Loth, *Stalins ungeliebtes Kind,* p. 204.

103. Imre Nagy, *On Communism,* quoted in Gati, *Hungary and the Soviet Bloc,* p. 132.

104. "Delo Beriia," January 1991, pp. 174, 213.

105. Sudoplatov and Sudoplatov, *Special Tasks,* pp. 360–63.

106. "Delo Beriia," January 1991, pp. 213, 145.

107. Ibid., p. 157.

108. See chap. 5, n. 121.

109. Sudoplatov and Sudoplatov, *Special Tasks,* pp. 342–46.

110. Alexander Nekrich, "The Arrest and Trial of I. M. Maisky," *Survey* 118 (1976): 313–20.

111. Conquest, *Power and Policy in the U.S.S.R.,* pp. 252–53.

112. R. J. Service, "The Road to the Twentieth Party Congress: An Analysis of the Events Surrounding the Central Committee Plenum of July 1953," *Soviet Studies* 33 (1981): 232–45.

113. Juri Krotkow, "Ich führte Berijas Befehl aus," in Nekrassow, *Berija,* p. 319.

114. "Analyse über die Vorbereitung, den Ausbruch und die Niederschlagung des faschistischen Abenteuers vom 16.-22.6.1953," July 20, 1953, J IV 2/202/15, and "Disposition für eine Entschließung des Zentralkomitees der SED," about July 26, 1953, NL 90/290, SAPMO-BA; cf. Gerhard Wettig,

"Sowjetische Wiedervereinigungsbemühungen im ausgehenden Frühjahr 1953? Neue Aufschlüsse über ein altes Problem," *Deutschland-Archiv* 25 (1992): 943–58, at p. 950.

115. Gati, *Hungary and the Soviet Bloc*, p. 132.

116. "Review of Basic National Security Policy," September 30, 1953, *FRUS*, 1952–54, vol. 2, pt. 1, pp. 491–514, at p. 493.

117. Jim Marchio, "Resistance Potential and Rollback: U.S. Intelligence and the Eisenhower Administration's Policies toward Eastern Europe, 1953–56," *Intelligence and National Security* 10 (1995): 219–41.

118. Marchio, "Rhetoric and Reality," pp. 137–50; Henry W. Brands Jr., "The Age of Vulnerability: Eisenhower and the National Insecurity State," *American Historical Review* 94 (1989): 963–89.

119. Conant to Dulles, August 8, 1953, *FRUS*, 1952–1954, vol. 7, p. 1640; James G. Hershberg, *James B. Conant: Harvard to Hiroshima and the Making of the Nuclear Age* (New York: Knopf, 1993), pp. 660–63.

120. Speech by Malenkov, August 8, 1953, in *The International Situation and Soviet Foreign Policy: Key Reports by Soviet Leaders from the Revolution to the Present*, ed. Myron Rush (Columbus, Ohio: Merrill, 1970), pp. 156–65, at p. 162.

121. Raymond L. Garthoff, *Deterrence and the Revolution in Soviet Military Doctrine* (Washington, D.C.: Brookings, 1990), pp. 42–46.

Conclusion

1. *Khrushchev Remembers: The Glasnost Tapes* (Boston: Little, Brown, 1990), p. 72.

2. Odd Arne Westad, *Cold War and Revolution: Soviet-American Rivalry and the Origins of the Chinese Civil War, 1944–1946* (New York: Columbia University Press, 1993), p. 55.

3. Dean Acheson, *Present at the Creation* (New York: Norton, 1969), p. 430.

4. Henry Kissinger, "Reflections on Containment," *Foreign Affairs* 73, no. 3 (1994): 113–30, at p. 124.

5. Frank Umbach, "Die Evolution des Warschauer Paktes als außen- und militärpolitisches Instrument sowjetischer Sicherheitspolitik 1955–1991" (Ph.D. diss., University of Bonn, 1995), pp. 222–30.

6. Edward N. Luttwak, "Where Are the Great Powers? At home with the Kids," *Foreign Affairs* 73, no. 4 (1994): 23–28.

Archival Sources

Moscow

(The citations of documents in Russian archives consist, when available, of numbers indicating collection *[fond]*/inventory *[opis]*/file *[delo]*/ page *[list]*.)

Archives of the President of the Russian Federation (APRF)

Collections 3 and 45

Special dossier on Korea (cited according to page numbers)

Foreign Policy Archives of the Russian Federation, Ministry of Foreign Affairs of the Russian Federation (AVPRF)

06	- Molotov's Secretariat [Sekretariat Molotova]
07	- Vyshinskii's Secretariat [Sekretariat Vyshinskogo]
066	- Austrian Desk [Referentura po Avstrii]
082	- German Desk [Referentura po Germanii]
0129	- U.S. Desk [Referentura po SShA]

Russian Center for the Preservation and Study of Documents of Recent History (RTsKhIDNI)

575   Records of the Cominform [Arkhiv Kominforma] (cited according to original filing, prior to archival rearrangement)

Storage Center for Contemporary Documentation (TsKhSD)

5   Party Secretariat

Berlin

Stiftung Archiv der Parteien und Massenorganisationen der DDR im Bundesarchiv,

Archives of the Central Committee of the Socialist Unity Party of Germany (SAPMO-BA)

J IV   - Ulbricht Office [Büro Ulbricht]

NL 36   - Pieck Papers [Nachlaß Wilhelm Pieck]

NL 90   - Grotewohl Papers [Nachlaß Otto Grotewohl]

Prague

Archives of the Central Committee of the Communist Party of Czechoslovakia (AÚV KSČ)

02   Records of the Presidium

Archives of the Ministry of Foreign Affairs of the Czech Republic (MZV)

Secretariat of the Minister [Sekretariát ministra]

Reports of Foreign Posts, Moscow [Zprávy zastupitelských úřadů, Moskva]

Rome

The Gramsci Institute, Archives of the Italian Communist Party (IG)

Minutes of the Meetings of the Party Directorate [Verbali Direzione PCI]

Amsterdam

Archives of the Communist Party of the Netherlands (CPN)

III. After 1945, 1. Organization

Washington, D.C.

Central Intelligence Agency

National Intelligence Estimates

National Archives (NA)

RG-59   General Records of the Department of State

RG-84   Records of the Foreign Service Posts of the Department of State: United States Embassy Moscow

RG-218   Records of the United States Chiefs of Staff

RG-466   Records of the U.S. High Commissioner in Germany

National Security Archive

Ernest R. May, John D. Steinbrunner, and Thomas Wolfe, "History of the Strategic Arms Competition 1945–1972" (Washington, D.C.: Office of the Secretary of Defense, 1981).

Index

Abakumov, Viktor S., 76, 132
Acheson, Dean: and Japanese peace treaty, 126; and Korean War, 101, 109; and McCarthyism, 146; miscalculations of, 194; and "Morgenthau plan," 154; prediction about Asian chaos by, 92; "Strategy for Freedom" speech of, 109; and subversive activities in Eastern Europe, 81; and U.S. defense perimeter in Asia, 91
Adenauer, Konrad, 110
"Agreement on Berlin Security" (NATO), 110
Albania: and Britain, 81–82; and Cominform, 32, 81; and Greece, 35, 36, 82; purges in, 68, 70; repression in, 120; Soviet embassy bombed in, 116–17; sovietization of, 44; Stalin's views about, 35–36, 38, 39; subversive activities in, 81–82, 116–17, 120; and Yugoslavia, 35–36, 37, 38, 44, 81
Allied Control Council, 43
Amery, Julian, 81
Anders, Władysław, 120

Anti-Fascist Jewish Committee, 157
Anti-Semitism, 154, 157–58, 162, 167, 173, 175, 179
Antonov-Ovseenko, Anton, 168
Arbatov, Georgii, 144
Arms race, 78, 122, 176, 196
Asia: atomic/nuclear weapons deployed in, 145; colonialism in, 55, 85; Mao's views about U.S. military action in, 90; nationalism in, 85; Soviet policies in, 85–97; U.S. strategy in, 91, 92, 93, 97. See also specific nation
Association of the Victims of the Nazi Regime, 162
Atomic/nuclear weapons: banning of, 78; and Berlin blockade, 49, 58; as deterrence to Soviet aggression, 49, 74, 146; Dulles's views about, 164; European deployment of, 145; and Korean War, 109–10, 123, 165; MacArthur's views about, 152; and massive retaliation strategy, 189; and militarization of Cold War, 58, 60; and "mutual assured destruction," 197;

Atomic/nuclear weapons
   (*continued*)
   and post-Stalin government, 197;
   role in Cold War of, 196, 197;
   Soviet development of, 76–78,
   79, 126, 187, 190; Soviet
   knowledge about U.S., 74, 76;
   and specter of global war, 74, 75,
   76–79, 109–10, 114; Stalin's
   views about, 75, 76–77; U.S.
   monopoly of, 58; and U.S.
   preparations for strike at Soviet
   Union, 101, 109–10, 114; U.S.
   reliance on, 78; U.S. response to
   Soviet testing of, 77–78; U.S.
   stockpile of, 58, 125, 192; U.S.
   warnings to China about, 114,
   122–23, 165, 166. *See also*
   Hydrogen bombs
Attlee, Clement, 109
Austria: communism in, 25, 57;
   elections in, 25; Stalin's views
   about, 19, 23; state treaty with,
   65, 67, 78, 79, 176; and unopened
   window of opportunity, 165; and
   Yugoslavia, 23
Azerbaijan, 23

"Babice case," 119
Balance of power, 17, 85, 92–93
Balkan confederation, 20, 21, 26,
   33, 37, 39, 73, 174
Balkans, 20, 22, 25–26, 35–40, 44,
   46. *See also* Balkan
   confederation; *specific country*
Baltic states, 16, 82, 179
Beam, Jacob, 167, 176
Bear Island, 19
Begin, Menachem, 162
Belgium, 27
Belkin, Fedor, 70, 162
Ben Gurion, David, 154
Beneš, Edvard, 22, 42
Beriia, Lavrentii P.: and Anti-
   Fascist Jewish Committee, 56;

and atomic/nuclear weapons, 76;
   blame on, 185–88, 194;
   conspiracy against, 186, 187;
   decline of, 132, 142, 143, 151,
   157, 160; and "doctors' plot,"
   159–60, 161; domestic intrigues
   of, 179–80, 188; and foreign
   affairs, 23, 179–85; and Germany,
   178, 180–85; and Kennan
   intruder, 144; and Khrushchev,
   160, 168, 174, 179, 180, 185,
   186–87; legacy of, 188–90; as
   mediator with West, 187–88;
   Molotov's views about, 180; and
   nationalism, 150, 179; ouster of,
   185–88, 190; power of, 179, 188;
   and prison system, 178–79; and
   purges, 76, 141, 155, 156, 157,
   158, 186; and reform in Soviet
   Union, 171, 173, 174, 179, 188;
   and Soviet imperialism, 23; on
   Stalin, 170; and Stalin's
   illness/death, 163, 164, 167,
   168–69, 170, 186; Stalin's
   relationship with, 23, 128, 157,
   159, 160, 163, 167, 168
Berlin, Germany: airlift to, 49, 50,
   52; and beginnings of Cold War,
   24; blockade of, 43, 47–53, 58,
   61, 62, 63–65, 111, 193; and
   creation of West German state,
   48; and currency issue, 47–48,
   50–51, 52, 63–64; international
   peace rally in, 116; and
   militarization of Cold War, 60;
   People's Congress in, 43; political
   parties in, 24; pro-Western
   sentiments in, 24, 50; Soviet
   disentanglement from, 63–67,
   111; as Soviet fiasco, 47–53; and
   Stalin's meeting with Western
   ambassadors, 50–51; and U.N.,
   52–53; and U.S. elections of
   1948, 61; West's ultimatums
   about, 52

Berman, Jakub, 130, 156
Bevin, Ernest, 37, 40–41, 43, 81
Bidault, Georges, 43
Bieńkowski, Władysław, 68
Bierut, Bolesław, 29, 68, 130, 156
Billoux, François, 142, 143
Biological warfare, 146
Bled agreement (1947), 33
Bohlen, Charles, 122, 176
Bolshevik (journal), 75, 158
Bolshevik Old Guard, 11, 14
Bolshevik party, 11–12, 13–14, 15
Bornholm Island, 19
Bosnia, 102
Bourgeois nationalism, 54, 55, 56,
   87. *See also* "National roads to
   socialism"
Bradley, Omar, 110
Brankov, Lazar, 53, 73
Brezhnev, Leonid, 190, 192
Britain. *See* Great Britain
Brussels pact (1948), 43, 44, 51
Budennyi, Semen, 23
Bulganin, Nikolai A., 162–63, 166,
   167, 168, 180
Bulgaria: communism-socialism in,
   54; and Greek civil war, 35;
   human rights in, 122; political
   parties in, 21; and post-Stalin
   government, 175, 181; purges in,
   69, 70, 72; riots in, 181; and
   Romania, 37, 38, 39; Soviet
   relations with, 37, 38, 39;
   sovietization of, 34; and Stalin's
   "invasion" of Yugoslavia, 71–72;
   subversive activities in, 82, 83;
   and Yugoslavia, 33, 38, 39, 73,
   117
Burgess, Guy, 128
Burma, 55, 85

Čepička, Alexej, 113
Caffery, Jefferson, 67
Capitalism: and benefits of Cold
   War, 196; and creation of

Cominform, 31–32; and Great
   Depression, 196; impending
   breakdown of, 14, 28, 60, 75, 78,
   149; imperialism as product of,
   15; Lenin's views about, 149; and
   militarization of Cold War, 60;
   Varga's views about, 28, 60, 75;
   Zhdanov divides world into
   camps of socialism and, 31
Catholic church, 33, 84, 119
Central Committee (Soviet), 62,
   169, 174, 179, 185, 186, 187, 189
Central Intelligence Agency (CIA),
   80, 84, 109, 131–32, 183
Chataigneau, Yves, 122
Chişnevschi, Iosif, 141
Chiang Kai-shek, 86, 91, 92, 164
Chiaureli, Mikhail, 170
China: communists gain control of,
   78, 85–88, 97; Great Britain
   recognizes, 90; India recognizes,
   90; and international
   communism, 55, 87, 97; and
   Japanese peace settlement, 109;
   and North Korea's design for
   aggression, 90–91, 92, 94, 95, 96;
   repatriation of prisoners of war
   to, 127, 148; revolution in, 12;
   Soviet aid/assistance to, 94, 104,
   105–7, 122, 126, 127, 147–48;
   Soviet loan to, 94; Soviet plans
   to isolate, 92; Soviet relations
   with, 12, 87–91, 93–94, 95, 96,
   104, 105–7, 113, 122, 124, 127,
   132, 147–49, 166, 193; Soviet
   treaties with, 86, 89–90, 91,
   93–94; and specter of global war,
   114; Stalin's views about, 12, 13,
   55, 85–90; and U.N., 92, 99, 102,
   111, 124; U.S. relations with,
   86–87, 105, 165, 166; U.S.
   warnings about atomic/nuclear
   weapons aimed at, 114, 122–23,
   165, 166. *See also* Korean War;
   Mao Zedong; Taiwan; Zhou Enlai

Chuev, Feliks, 167

Churchill, Winston, 20, 151, 174, 175, 182, 183, 184, 194

Clark, Mark, 152

Class struggle, 139

Clausewitz, Carl von, 165

Clay, Lucius D., 43, 49

Coalition governments: communist participation in. *See* "National road to socialism"

Cold War: and atomic/nuclear weapons, 196, 197; benefits of, 195–96; continuation of, 74–79; division of Europe during, 16; end of, 190, 196; inevitability of, 27; Korean War as distraction in, 164; lessons of, 197–98; militarization of, 58–62; new era in, 190; as predetermined, 23; security issues as basis of, 23–29, 191; and unopened window of opportunity, 164–66, 170; as unwanted, 23–29; U.S. strategy in, 194–95; as war, 164–65. *See also specific person or event*

"Collective" security policy, 14–15, 29

Colonialism, 31, 55, 85

Comecon (Council for Mutual Economic Assistance), 57–58

Cominform: Albania requests membership in, 81; Belgrade headquarters of, 33; birth of, 30–35; conferences of, 69, 78–79, 82; creation of, 46; decline of, 113; and Greek problems, 57; and Korea, 94; strengthening of, 112, 142; and subversive activities in Eastern Europe, 82; and Tito/Yugoslavia, 39, 53

Comintern, 12, 13, 14, 16, 17–18, 26, 30, 31, 36

Committee of Information (Soviet foreign ministry), 59

Communist Party: congresses of, 144–45, 149–50, 159, 161; "foreign policy" department of, 31; international department of, 18, 31; purges as means to destroy, 151; and reform in Soviet Union, 173, 188; Stalin's planned reorganization of, 163–64

"The Conditions in the Ministry of State Security and Medical Sabotage" resolution, 159

Congress, U.S., 46, 91, 96, 120, 123

Containment policy, 26, 41, 85, 146, 194

Contingency plans, U.S., 74, 77–78

Correlation of forces, 13, 25, 29, 33, 92–93, 113, 133, 191

Cosmopolitanism, 54, 56–57, 75, 158

Council of Foreign Ministers: and Allied-Soviet relations, 22; and Austria, 65, 78; and Germany, 26, 40–41, 48, 49, 51, 64–65, 124; inauguration of, 22; and Japan's future, 85; Soviet attempts to reconvene, 108, 110, 121, 135; Soviet views about, 22

Council of ministers, Soviet, 169, 173

Council for Mutual Economic Assistance (Comecon), 57–58

Currency issue: and Berlin blockade, 47–48, 50–51, 52, 63–64

Czechoslovakia: anticommunism in, 33; "Babice case" in, 119; and beginnings of Cold War, 25, 28, 29; Catholic church in, 119; defections from, 83, 119; and Germany, 41; government-in-exile of, 21–22; heresy charges against leaders in, 45; intelligence department in, 83; Israeli diplomats expelled from, 162; Jews receive assistance from, 56;

Czechoslovakia (*continued*)
and Korean War, 100, 103; and
Marshall Plan, 28, 29; and
"national roads to socialism," 25,
53; and Poland, 39; and post-
Stalin government, 181; purges
in, 68–79, 128–29, 130–32,
153–56; rehabilitation of
communists in, 112–13;
resignation of non-communist
ministers in, 42; Soviet fears of
U.S. invasion of, 100, 103; Soviet
"friendship" treaty with, 21–22;
Soviet intelligence in, 17;
sovietization of, 33, 41–42, 43,
44, 46; and specter of global war,
145; subversive activities in, 82,
83–84, 117, 118–19, 120, 129,
131, 155, 181; uprisings in, 155,
181; U.S. views about, 46;
Western reaction to Soviet
"clampdown" on, 43; and
Western-Soviet relations, 44, 46;
and Yugoslavia, 35, 117

Dahlem, Franz, 139, 178
Dairen, 86, 93–94
Death penalty, Soviet:
reintroduction of, 82
"Declaration of Liberated Europe,"
22
Defections: of Czechoslovaks, 83,
119
Defense budget, U.S., 58, 127
Defense perimeter, U.S., 91
Defense spending, Soviet, 74,
166–67
Deng Hua, 103
Denmark, 19
Derevianko, Kuzma N., 96
Dertinger, Georg, 108, 137
Despotism, 11–16, 29, 191, 192–93
"Dictatorship of the proletariat,"
54
Đilas, Milovan, 32, 37, 38, 71

Dimitrov, Georgii, 18, 31, 33, 37,
38, 39, 54, 69, 73
Djurić, Dragoje, 174
"Doctors' plot," 158–64, 169, 173
Duclos, Jacques, 30, 32, 33, 143
Dulles, Allen, 120, 183
Dulles, John Foster: on
atomic/nuclear weapons, 164;
and Germany, 183–84; and
Japanese peace treaty, 108, 109,
121; and post-Stalin government,
174, 176; and Soviet domination
of Eastern Europe, 176; and
specter of global war, 146–47,
164, 165
Dunkirk treaty (1946), 37, 43

East Germany: and aftermath of
Beriia's demise, 188–89; and
Beriia's intrigues, 180–85; call for
all-German unification
conference by, 135; and
Cominform, 32; economy of,
138, 178; and German peace
treaty, 135–36, 137–38; Jews in,
162; as "model democratic
state," 184; "New Course" in,
182–83, 184, 188–89; and post-
Stalin government, 177–78,
180–85, 187, 188–89; purges in,
128, 135, 139, 156–57, 178;
repression in, 139; socialism in,
138, 145, 177–78; Soviet aid to,
102; Soviet troops in, 74, 167;
and specter of global war, 102,
110; Stalin's reversal of policy
concerning, 63; subversive
activities in, 82, 117; uprisings
in, 183–84, 187, 189, 190. *See
also* German Democratic
Republic; Germany
Eastern Europe: and benefits of Cold
War, 195; Cominform mission
in, 33; economy of, 27–29;
expansion of militaries of, 84–85;

Eastern Europe (*continued*)
and Korean War, 116, 132; and
Marshall Plan, 27–29, 30; and
"Morgenthau plan," 154; purges
in, 67–74, 75, 79, 128–32, 133,
141–42, 161, 162, 179, 193, 195;
Soviet security advisers in,
68–69, 128, 129; as Soviet sphere
of influence, 16, 20, 37, 45, 134,
189; and specter of global war,
113–15, 145; and Stalin's decline,
141–42; Stalin's views about, 21;
subversive activities in, 73,
80–85, 97, 116–21, 131–32, 133,
146–47, 154, 189, 195; Western
views about, 20, 22. *See also*
specific person or country
"Economic Problems of Socialism
in the USSR" (Stalin), 149–50,
158
Economy: of East Germany, 178; of
Eastern Europe, 27–29; of
Germany, 25, 26, 40, 41, 48, 50,
52; of Soviet Union, 75–76; of
Western Europe, 196; of
Yugoslavia, 39. *See also*
Capitalism; Council for Mutual
Economic Assistance (Comecon);
Currency issue
Eden, Anthony, 16, 41
Egypt, 31
Eisenhower, Dwight D.: and arms
race, 176; and Austria, 176;
"Chance for Peace" speech of,
175–76; and Churchill's proposed
summit, 175; and Korean War,
152, 176; and post-Stalin Soviet
government, 175–76; reluctance
to challenge Soviets by, 189;
Soviet perception of militancy of,
146, 151–52; Stalin's views
about, 167; as supreme allied
commander in Europe, 110; and
unopened window of
opportunity, 164, 165

Elections (U.S.), 60–61, 145, 146,
151
Epishev, Aleksei A., 141
Erenburg, Ilia, 56, 163
Europe: atomic/nuclear weapons
deployed in, 145; and beginnings
of Cold War, 27–29; division
during Cold War of, 16; and
Marshall Plan, 27–29; self-
determination in, 22; Soviet
Union as arbiter of, 20; and
specter of global war, 101, 102,
113–15; unification of, 196, 197;
U.S. ascendancy in, 46, 124, 133;
U.S. military bases in, 124; U.S.
role in, 27–29. *See also* Eastern
Europe; Western Europe; *specific
country*
European Coal and Steel
Community, 136
European Defense Community,
110, 134–35, 137, 138, 139, 143,
175, 176

Far Eastern Commission, 85
Farge, Yves, 143
Farkas, Mihály, 70–71
Fascism, 13, 14–15, 17, 196, 198
Federal Republic of Germany, 65,
66. *See also* West Germany
Fedorenko, Nikolai T., 89
Fedoseev, Petr, 157–58
Field, Noel, 70–71
Finder, Paweł, 17, 19
Finland, 21, 31, 33, 42–43, 44–45,
46
Five-Year Plan, 150
Foreign ministers: Allied, 134–35;
satellite, 41, 48, 108, 110. *See
also* Council of Foreign
Ministers; *specific person*
France: and beginning of Cold
War, 26–27, 28; and Brussels
pact, 43; and colonialism,
85; and Cominform, 32;

France (*continued*)
  communists in, 26–27, 32, 33, 112, 142–43; demise as great power of, 20; and German questions, 26, 50, 66, 183; and Great Britain, 37; international brigades interned in, 16; and international communism, 57; and Marshall Plan, 28; and NATO, 61; "peace" movement in, 142, 143; and "popular fronts," 14; purges in, 112, 142–43; riots/strikes in, 142; Soviet relations with, 15; Soviet views about, 22, 57; subversive activities in, 121, 155
François-Poncet, André, 101
Free Europe Committee, 84, 147
Freedom and Independence, 156
"Friendship" treaties, 21–22, 42, 44–45
Frunze, Mikhail, 159

Gao Gang, 103, 123, 124
Geminder, Bedřich, 155
Georgescu, Teohari, 141
Georgia, 132, 150, 174, 179
Geró, Ernö, 71
German Democratic Republic, 66, 177–78. *See also* East Germany
Germany: administrative agencies of, 26; and all-German State Council, 65, 110; anticommunist protest in, 24; and beginning of Cold War, 24–25, 26; British views about, 41, 50, 66; communist exiles of, 19; Council of Foreign Ministers meetings about, 40–41, 64–65, 124; debts of, 136; democracy in, 196; economy of, 25, 26, 40, 41, 48, 50, 52; four-power commission for, 134, 136, 139; French views about, 50, 66, 142; indoctrination of prisoners of war from, 18;
industry in, 25, 50, 136; and international communism, 66–67; Kennan's memorandum about, 65; Korean War's effects on, 107, 108, 110, 121; Malik-Jessup talks about, 64; military of, 136; and NATO, 124, 137; and Nazi-Soviet pact (1939), 15–16; as neutral versus communist, 180; political structure of, 40, 43, 108, 110–11; and post-Stalin government, 176–77, 180–85; postwar government of, 24–25; and Potsdam conference, 22; recovery of, 149; satellite foreign ministers' recommendation about, 41, 48, 108, 110; socialism in, 25; and Soviet postwar planning, 18, 19; Soviet proposals concerning, 50, 135–40, 176–77, 182; Soviet rapprochement with, 15–16; Soviet views about unification of, 19, 24–25, 41, 46, 50, 51, 63, 65, 66–67, 134, 135–38, 193; Stalin's miscalculations about, 13–14, 16, 24, 134–40, 177, 193; Stalin's reversal of policy concerning, 63–67, 79; and Tito-Stalin relationship, 40; and U.N., 110; and unopened window of opportunity, 165; U.S. views about, 41, 50, 66; Weimar Republic of, 137; Western Europe's fear of, 28; and Western-Soviet relations, 41–45. *See also* Berlin, Germany; East Germany; West Germany
Gheorghiu-Dej, Gheorghe, 68, 141
Global war: and atomic/nuclear weapons, 76–79; and collapse of Soviet system, 194; and Eastern Europe, 145; Hitler's views about, 114; inevitability of, 127–28, 149,

Global war: inevitability of
  (*continued*)
  158, 190; and Korean War, 101,
  109–10, 113–15; and militarization of Cold War, 60–61; and
  post-Stalin leadership, 189–90;
  probability of, 151–52; Stalin's
  views about, 13, 114, 134, 147,
  149, 151, 158; and subversive
  activities, 195, 196; as tool of
  U.S. policy, 164; and U.S.
  contingency plans, 74, 77–78;
  U.S. views about, 61; and
  Western rearmament, 126
Goli Otok concentration camp, 71
Gomułka, Władysław, 20, 32, 33,
  68, 129, 130, 131
Gorbachev, Mikhail, 190, 192
Gorkii, Maxim, 159
Gottwald, Klement, 42, 60, 68–69,
  129, 130, 156, 159
Govorov, Leonid A., 161
Great Britain: and Albania, 81–82;
  and Brussels pact, 43; and
  France, 37; and Germany, 40–41,
  48, 50, 66, 183; and Greece, 23,
  81; and "popular fronts," 14;
  recognizes China, 90; Soviet
  relations with, 15, 16, 17, 25,
  40–41; and subversive activities,
  121; unpreparedness for Stalin's
  death of, 175; Yugoslavia receives
  armaments from, 117. *See also*
  Intelligence, British
Great Purge trials (1930s), 13, 14,
  16, 62, 132, 157, 159
Greece: and Balkan confederation,
  37; and beginnings of Cold War,
  23, 26; British influence in, 23;
  and British-Soviet relations, 81;
  civil war in, 23, 26, 35–36, 38,
  44, 78; and Cominform, 32, 57;
  communists in, 32, 33, 35, 78;
  and Greek exiles in Albania, 82;
  and international communism,

57; and post-Stalin government,
  174, 175; and purges, 69; Soviet
  relations with, 44; Stalin's views
  about, 20, 23, 26, 35, 38, 81;
  subversive activities in, 81; and
  Tito, 57, 69; U.S. assistance to,
  26, 35; and Voice of America,
  118; and Yugoslavia, 33, 35, 39
Grigorian, Vagan, 62, 112
Gromyko, Andrei A.: and German
  peace treaty, 135, 136; and
  Korean War, 98, 99, 111, 121–22,
  125; and Soviet boycott of U.N.,
  99; and U.S.-Soviet relations, 21
Gross, Ernest A., 125
Grotewohl, Otto, 110, 135

Herrnstadt, Rudolf, 183, 184, 188,
  189
Himmler, Heinrich, 187
Hirohito (emperor of Japan), 88
Hiroshima, Japan, 76
Hitler, Adolf, 13, 14, 20, 64, 76,
  114, 194
Horák, Jiři, 40
Horáková, Milada, 84
"Hoßbach minutes" (Hitler, 1937),
  114
House of Representatives, U.S., 120
Hoxha, Enver, 35–36, 37, 38, 39,
  44, 68, 70, 81–82, 128
Hungary: and aftermath of Beriia's
  demise, 189; and beginnings of
  Cold War, 25; communism in,
  25; heresy charges against leaders
  in, 45; human rights in, 122;
  indoctrination of prisoners of war
  from, 18; Jews in, 71, 162, 182;
  "New Course" in, 182, 189;
  political parties in, 21; and post-
  Stalin government, 182, 187,
  189; purges in, 69–71; repression
  in, 119; and Romania, 38; Soviet
  treaty with, 38; Soviet use of
  force in, 21; sovietization of, 33;

Hungary (*continued*)
and Stalin's "invasion" of
Yugoslavia, 71–72; subversive
activities in, 118, 119; and Yugoslavia, 33, 38, 69–70, 73, 117
Huntington, Samuel, 197
Hydrogen bomb, 77, 78, 151, 187,
189

Iakovlev, N. N., 144
Ignatev, Semen D., 132, 161, 163,
164, 169, 173
Imperialism, 15–16, 17–22, 23–24,
29, 31
India, 31, 55, 90, 94, 102, 148, 166,
167
Indochina, 31, 55, 85
Indonesia, 31, 55, 85
Intelligence, British, 48, 81, 109,
114, 162
Intelligence, Czechoslovakia, 83
Intelligence, Israeli, 162
Intelligence, Soviet: in
Czechoslovakia, 17; and Field
case, 70; and Korean War, 99,
164; and Marshall Plan, 28; and
militarization of Cold War, 59;
and Nazis, 16; and recruitment
of spies, 14; reorganization of,
128, 167; and specter of global
war, 113; weakness of, 17. *See
also specific person*
Intelligence, U.S.: and Albania,
81; and Germany, 184, 189; and
Korean War, 109, 114, 188; and
North Korea's design for
aggression, 95, 96; and purges
in Eastern Europe, 131–32;
reorganization of, 59; on
Soviet views of West, 184;
and specter of global war, 109,
114
Intelligentsia, Soviet, 31
International brigades, 14–15, 16,
67, 72

International communism: and
China, 87, 97; and colonialism,
55; and German unification,
66–67; and imperialism, 17–22;
and nationalism, 15; as policy
tool, 45; and purges, 72; splits
within, 53–58, 62; and Stalin's
decline, 143; and Stalin's
delusions, 53–58; Stalin's
miscalculation about using, 46;
and Stalin's prestige, 30; Stalin's
reliance on, 29; and subversive
activities in Eastern Europe,
131–32. *See also* Cominform;
"National roads to socialism";
Zhdanov, Andrei; *specific
country*
International peace rally (Berlin,
Germany, 1951), 116
International politics: as zero-sum
game, 13
International youth conference
(Calcutta, 1948), 55
Internationalism, 196
Iran, 23
Israel, 55–56, 132, 154, 155, 157,
162, 175
Italy: African colonies of, 24; and
beginning of Cold War, 23, 24,
27; and Cominform, 32;
communists in, 27, 41, 45,
112–13; indoctrination of
prisoners of war from, 18; and
international communism, 57;
purge in, 112; Soviet views
about, 22, 23, 57; subversive
activities in, 80; and Trieste
dispute, 122; and Yugoslavia, 23.
See also specific person
Iudin, Pavel, 53–54, 89, 178, 181,
184

Jackson, C. D., 175
Japan: as anti-Soviet buffer state, 85;
communists in, 91–92, 121, 142;

Japan (continued)
  and Khabarovsk trials, 88; and
    Korean unification, 94; peace
    treaty with, 91–92, 93, 94, 108–9,
    121, 125; recommendations
    (1947) on future of, 85; recovery
    of, 149; Soviet diplomatic
    mission withdraws from, 96;
    Soviet entry into war against, 22;
    U.S. ascendancy in, 88, 133; and
    U.S. defense perimeter, 91
Jessup, Philip C., 64
Jewish Anti-Fascist Committee, 56
Jews: assimilation of, 71; as
    communists, 71; and "doctors'
    plot," 158–64; in Hungary, 182;
    pogrom against, 157–58, 167; and
    purges, 71, 154, 179; in Soviet
    Union, 157–58, 163, 167; Stalin's
    views about, 55–57, 71
Joint Chiefs of Staff, U.S., 49, 58,
    74, 101, 114
Joxe, Louis, 144

Kaganovich, Lazar M., 157, 163,
    181
Kardelj, Edvard, 32, 45, 117
Katz, Otto, 155
Kelly, David, 74, 99
Kennan, George F.: containment
    policy of, 41, 194; and
    Czechoslovakia, 41–42; and
    Germany, 65, 139; mentioned,
    64, 147; as persona non grata,
    143–44; and Radio Free Europe,
    118; and sovietization of Eastern
    Europe, 41–42, 62; on Stalin's
    personality, 140; on Stalin's
    views about capitalism, 145; and
    subversive activities, 72
Kersten amendment, 120, 183
Khabarovsk trials, 88
Khrushchev, Nikita S.: ascendancy
    of, 132, 150; and Beriia, 160, 168,
    174, 179, 180, 185, 186–87; and

Churchill's proposal for summit,
    174, 175; and "doctors' plot,"
    159–60, 161; and economic
    superiority of Soviet-style
    socialism, 192; and Germany,
    180; and Japanese peace treaty,
    126; and Korean War, 100, 102,
    106, 176; mentioned, 178; as
    party secretary, 171; on pogrom,
    162; on purges in Eastern Europe,
    131; and reform in Soviet Union,
    173, 174; rise of, 141–42; on
    Shcherbakov, 159; on specter of
    global war, 114; on Stalin, 128,
    140, 151, 163, 191; and Stalin's
    illness/death, 167, 168, 169, 170,
    186; Stalin's relationship with,
    83, 168; and subversive activities
    in Eastern Europe, 118
Kim Il, 91
Kim Il Sung: and Korean War
    negotiations, 123, 124, 125, 147;
    and plans to attack South Korea,
    90, 92, 94, 95, 96, 97, 193; and
    strategy/tactics during Korean
    War, 99, 100, 101, 103, 104, 105,
    106
Kirk, Alan G., 76, 101–2, 126, 127
Kirov, Sergei M., 67
Kitchlew, Saffrudin, 166, 167
Kobulov, Amaiak Z., 179
Kolman, Arnošt, 68
Komar, Wacław, 156
Konev, Ivan S., 161
Kopřiva, Ladislav, 83, 118
Korean War: as adventure gone
    wrong, 98–103; and
    atomic/nuclear weapons, 109–10,
    123, 145, 165; beginning of, 97,
    98; casualties in, 107, 109; cease-
    fire in, 104–5, 122, 124, 148, 165;
    Chinese military in, 100, 104–10,
    111, 115, 116, 121, 122, 123;
    Chinese negotiations in, 102,
    111, 123, 124, 125–26, 127,

Korean War: Chinese negotiations in (*continued*) 147–48, 165, 172–73; Chinese strategy/tactics in, 102, 103, 104, 106, 111; as distraction in Cold War, 164; and Eastern Europe, 116, 132; and Eisenhower's visit to Korea, 152; end of, 164, 165, 172, 176; expansion of, 104–10, 115, 152; and Germany, 107, 108, 110, 121; negotiations concerning, 101–3, 104–5, 106, 111, 121–28, 132, 145, 147–48, 165, 172–73, 174, 176; and North Korean plans for aggression, 91–97, 188, 193; prisoners of war in, 127, 148, 165, 172, 176; and Sino-Soviet relations, 104, 105–7, 113, 124, 147–49, 193, 105–7, 113, 124, 125–26, 132; Soviet aid/assistance in, 99, 100, 101, 103, 104, 105–7, 122, 126, 127; Soviet response to beginning of, 98; Soviet role in negotiations about, 103, 104–5, 106, 111, 123, 124–26, 127, 132, 172–73, 174, 176; Soviet role in strategy/tactics in, 99, 102, 103, 104–6, 107, 111, 115, 116, 123, 127; and specter of global war, 101, 109–10, 113–15; stalemate in, 115, 116, 123, 127, 193; and U.N., 98–99, 102–3, 104, 105, 110, 111, 127, 148; and unopened window of opportunity, 165, 166; and U.S. bombing of North Korea, 147, 164; U.S. public opinion about, 123; U.S. role in, 99, 100, 101, 103, 109–10, 111. *See also specific person*
Kostov, Traicho, 21, 69, 70, 72, 73
Kosynkin, Petr, 163
Kovalev, Ivan V., 75, 89, 93–94
Kozlov, Frol R., 161, 162

Kuprešanin, M., 38
Kurile Islands, 22, 89, 108, 125
Kuusinen, Hertta, 42
Kuznetsov, Vasilii V., 169

Labor party (Britain), 25
Laski, Harold, 25
Lavrentev, Anatolii I., 36, 38, 39, 141
Leino, Yrjö, 42–43
Lenin, V. I., 12, 13, 15, 132, 149
Leningrad: purges in, 75–76, 132, 158, 186
Levchenko, G. I., 161
Li Lisan, 96
"Limnes" plan, 35
Linguistics, theory of, 96
Lippmann, Walter, 194
Litvinov, Maksim, 15, 18, 23, 24, 64
Liu Shaoqi, 87, 91, 148
Locarno treaty (1925), 182
Longo, Luigi, 32
Lozovskii, Solomon A., 56, 157
Luca, Vasile, 141

MacArthur, Douglas, 95, 103, 105, 109, 123, 151–52
McCarthy, Joseph, 146, 165
Maclean, Donald, 43, 44, 110, 128
Maidun rebellion, 55
Maiskii, Ivan M., 18, 19, 188
Malaya, 55, 85
Malenkov, Andrei, 168, 174
Malenkov, Georgii M.: and atomic/nuclear weapons, 78; and "doctors' plot," 162–63; as heir apparent, 150, 151; and international communism, 31–32, 79, 94; and Korean War, 172; and Leningrad purge, 76, 158; mentioned, 179; nominated for premiership, 171; and reform in Soviet Union, 174;

Malenkov, Georgii M. (*continued*) and retouched photograph, 171–72; rise of, 54; and Stalin's illness/death, 163–64, 167, 168, 169; on Stalin's offer to resign, 151; Stalin's relationship with, 151, 168; Stassen's views about, 175; Zhdanov's rivalry with, 54

Malik, Iakov, 64, 102–3, 108–9, 124–25, 176

Manchuria, 85–86, 87, 93, 94, 107, 123

Mao Zedong: mentioned, 172, 175; Moscow visits of, 88–91; Stalin's relationship with, 86, 87, 88–89, 92, 95, 96, 97, 113, 123, 124, 125–26, 166; Tito compared with, 87; and U.S. designs in Asia, 99; and U.S.-China relations, 86–87, 165. *See also* China; Korean War

Marshall, George C., 27, 40, 49, 52, 85

Marshall Plan, 27–29, 30, 40, 57–58, 80, 110, 196

Marty, André, 142, 143

Marx, Karl, 12, 30

Marxism, 12

Massive retaliation strategy, 189

Matthews, Francis, 103

Mauriac, François, 66

Mekhlis, Lev Z., 162

Menon, Krishna, 166, 167

Menshikov, Mikhail A., 62

Merker, Paul, 139, 157

Merkulov, Vsevolod N., 170

Mexico, 157

Meyersohn, Golda, 56

Miasnikov, Ia. L., 140, 169

Mikhoels, Shlomo, 56, 60

Mikoian, Anastas A., 62, 86, 131, 166

Mikoian, Sergo, 174

Mikunis, Shmuel, 56

Militarization: of Cold War, 58–62

Military, Eastern European, 84–85

Military, German, 136

Military Governors of Germany, 49, 51–52

Military power: as policy instrument, 58

Military, Soviet: as basis of Soviet system, 192; and Beriia, 164, 185; buildup of, 13, 84–85; defensive nature of planning by, 74–75; expenditures for, 127; mentioned, 187; modernization of, 74; and purges, 74, 157, 161; role in Cold War of, 196; and Soviet postwar planning, 18; and Stalin's death, 171; and subversive activities in Eastern Europe, 84–85; U.S. perceptions of, 175

Military, U.S., 74–75, 110, 127, 196

Military, Western, 127. *See also* North Atlantic Treaty Organization (NATO)

"Mingrelian" purge, 132, 174

Molotov, Viacheslav M.: and Allied-Soviet relations, 21; and atomic/nuclear weapons, 77; and Balkans, 38, 39, 40; and Beriia, 180, 188; and Berlin blockade, 43, 48, 51, 53; on Bevin, 41; and British-Soviet relations, 40–41; and Bulgarian-Romanian treaty, 39; and Cold War, 23; and Council of Foreign Ministers, 22; and economy of Eastern Europe, 27–28; on Eden, 41; and German questions, 40, 48, 51, 53, 108, 177, 178, 180; and Korean War, 106, 172–73, 174; and Mao, 88; and militarization of Cold War, 59, 60; and Nazi-Soviet relations, 15, 16; and reform in Soviet Union, 174; and reparations, 26;

Molotov, Viacheslav M. (*continued*)
and self-determination in Europe,
22; and Sino-Soviet relations, 93,
106, 166; Smith's talks with, 46;
on Stalin's decline, 140; and
Stalin's illness/death, 167–68,
169, 170; on Stalin's offer to
resign, 151; Stalin's relationship
with, 151, 168; and U.S. defense
perimeter in Asia, 91; and U.S.-
Soviet relations, 46; Vyshinskii
replaces, 62, 64; and Yugoslavia,
38, 39, 40, 44, 117, 174
Morgenthau, Henry, 154
"Morgenthau plan," 154
Murphy, Robert, 41, 49
"Mutual assured destruction," 197
Mutual Security Act, 120

Nagy, Imre, 182, 189
"National front" strategy, 17, 25
"National road to socialism": and
purges, 68
"National roads to socialism," 25,
30–35, 53, 54, 149–50, 155. *See
also* Bourgeois nationalism
National Security Council, 58–59,
80, 101, 114, 120, 164, 183. *See
also* NSC–68
Nationalism, 11, 15, 68, 82, 85,
141, 179, 196. *See also* Bourgeois
nationalism; "National roads to
socialism"
Navy, Soviet, 85
Nazi-Soviet pact (1939), 15–16
Nenni, Pietro, 135, 139, 144
Nitze, Paul H., 92
North Atlantic Treaty Organization
(NATO): "Agreement on Berlin
Security" of, 110; atomic/nuclear
weapons as main deterrent used
by, 74; and Balkan Pact, 174;
Beriia's probe of defenses of, 187;
Congress approves U.S.
participation in, 46; creation of,
74, 80; Eisenhower as supreme
commander of, 110; formal
inauguration of, 61; France as
member of, 61; and Germany,
124, 137; growth of, 125;
mentioned, 145; military
capabilities of, 74; and Norway,
61; role in Cold War of, 196;
Soviet attempts to derail, 61;
Soviet response to creation of,
74; Stalin's views about, 74–75;
and U.N., 61; and U.S.
contingency plans, 74; and West
Germany, 124
North Korea: Chinese aid to, 102,
103; and Chinese support for
plan to attack South Korea,
90–91, 92, 94, 95, 96; design for
aggression by, 90, 91–97; and end
of Korean War, 172; repatriation
of prisoners of war to, 127, 148;
Soviet aid/assistance to, 94–95,
97, 99, 100, 103, 104; Soviet
approval for invasion of South
Korea by, 95, 96–97; Soviet treaty
with, 94; Stalin's views about,
90, 92, 95, 96; U.S. bombing of,
147, 164; U.S. prisoners in, 176;
U.S. warnings about invasion
plans of, 95, 96. *See also*
Korean War
Norway, 19, 42–43, 61
Nosek, Václav, 42
Nowotko, Marceli, 17
NSC–68, 93, 96
Nuclear weapons. *See* Arms race;
Atomic/nuclear weapons

Oatis, William, 118–19
Ochab, Edward, 130, 156
Office of Policy Coordination, U.S.,
58, 80, 83
Office of Special Projects, U.S., 58
Oren, Mordechai, 132
Outer Mongolia, 86, 94

Paasikivi, Juho, 42–43
Pajetta, Gian Carlo, 61
Paniushkin, Aleksandr, 31, 59, 148–49, 165, 169
Pătrășcanu, Lucrețiu, 68, 141
Pauker, Ana, 32, 141
Pavlík, Gejza, 70
"Peace" congress (Warsaw, Poland, 1950), 108
Peace Council of France, 61
"Peace" movement, 142, 143
Peasantry, 12
Peng Dehuai, 123
Personnel changes, Soviet: and policy changes, 62
Péter, Gábor, 70, 71, 162
Philby, Kim, 82
Philippines, 55, 85
Pieck, Wilhelm, 135, 138
Poland: anti-Soviet attitudes in, 20; and beginning of Cold War, 28, 29; Catholics in, 33; Cominform mission in, 33; communist divisiveness in, 17; and Czechoslovakia, 39; dissolution of communist party in, 15; and Germany, 15, 41; government-in-exile of, 20; heresy charges against leaders in, 45; Israeli diplomats expelled from, 162; as key to Soviet security, 19–20; and Marshall Plan, 28, 29; and Nazi-Soviet pact (1939), 15, 16; purges in, 68, 128, 129–30, 156; Red Army crusade against (1920), 11–12; Rokossovskii appointed defense minister of, 85; Soviet annexation of eastern, 16; and Soviet postwar planning, 19–20; sovietization of, 33; Stalin's views about, 19–20, 39; subversive activities in, 82, 83, 156; Western views about, 22; during World War II, 15, 17; and Yugoslavia, 35

Policy, Soviet: and personnel changes, 62; and purges as means of reassessing, 67–74, 79, 154–55; and reasons for turning out differently than intended, 72; subordinates' initiation of, 192. See also specific country, topic, or event
Politburo, 83, 140, 150, 151, 158, 183
Political prisoners, 173
Ponomarenko, Panteleimon, 167
Popivoda, Pero, 53
Popović, Koča, 117
Popović, Vladimir, 36
"Popular fronts," 14, 17
Port Arthur, 86, 89, 93, 107, 148
Poskrebyshev, Aleksandr N., 140, 163
Postwar planning, Soviet, 17–22
Potsdam Conference (1945), 22, 137
Presidium, 150, 151, 167, 169, 173
Prison system, Soviet, 173, 178–79
Prisoner-of-war committees, 18
Prisoners of war, 33, 127, 148, 165, 172, 176
Proletarian internationalism, 54
Propaganda campaign (1946), Soviet, 24
Psychological Strategy Board, 118, 131, 183
Purges: of 1930s, 13, 14, 16, 132, 157, 159; of 1950s, 128, 157–58, 161; in East Germany, 135; in Eastern Europe, 67–74, 75, 79, 128–32, 133, 141–42, 161, 162, 179, 193, 195; in France, 112, 142–43; of Hitler, 14, 76; in Italy, 112; and Jews, 154, 179; in Leningrad, 75–76, 132, 158, 186; as means of reassessing Soviet policies, 67–74, 79, 154–55; as means to destroy Communist party, 151; as panacea, 152;

Purges (*continued*)
in Romania, 141–42; and
socialism-communism merger,
54; and Soviet military, 74, 157,
161; in Soviet Union, 157–58,
161, 193; and Stalin's paranoia,
128; Stalin's views about, 150;
and subversive activities in
Eastern Europe, 131–32, 133
Pushkin, Georgii M., 70–71, 137,
176

Radio Free Europe, 118, 131
Rajk, László, 70–71, 72, 73, 162
Rákosi, Mátyás, 70, 71, 182, 189
Ranković, Aleksandar, 73, 179
"Reaper" plan, U.S., 109–10
Reicin, Bedřich, 84, 155
Reid, Escott, 44
Reparations, 18, 24, 25, 26, 40, 67,
136
Reston, James, 158
Revolution: and Soviet security,
11–15, 29
Rhee Syngmann, 90
RIAS (U.S. radio station), 117
Ridgway, Matthew, 123, 143
Riumin, Mikhail, 160–61, 173
Robertson, Brian, 49, 50
Röhm, Erich, 76
Rokossovskii, Konstantin, 85
"Rollback" policy, 146
Romania: and Bulgaria, 37, 38, 39;
human rights in, 122; and
Hungary, 38; indoctrination of
prisoners of war from, 18;
nationalism in, 141; purges in,
68, 141–42; Soviet treaty with,
38; sovietization of, 33; and
Stalin's "invasion" of Yugoslavia,
71–72; Stalin's views about, 39;
and Yugoslavia, 33, 38
Roosevelt-Litvinov agreement
(1933), 120
Roshchin, Nikolai V., 104, 147

Ruhr region: industrial resources
of, 50, 136
Russia Committee, British, 81

Sakhalin Islands, 89, 108, 123,
125
Satellites, Soviet: foreign ministers
of, 41, 48, 108, 110; and Soviet
postwar planning, 17–22. *See
also specific country*
Security, Soviet: as basis of Cold
War, 23–29, 191; and Beriia
legacy, 188–90; internal versus
external, 12, 14, 16, 192–93; and
intrinsic Soviet insecurity, 11;
and preemption, 189–90; by
revolution, 11–15, 29. *See also*
Imperialism
Self-criticism, ritual of, 32, 131
Self-determination, 22
Semenov, Vladimir, 19, 122, 150,
177, 178, 181, 183, 184
Sharet, Moshe, 154
Shcherbakov, Aleksandr, 159, 161
Shi Zhe, 105
Shtemenko, Sergei M., 140, 161,
164
Shtykov, Terentii F., 92, 95, 96, 97,
111
Sidorovich, G. S., 36, 53
Simone, André, 155
Simonov, Konstantin, 160
Široký, Viliam, 118, 119
Slánský, Rudolf: as Czechoslovak
general secretary, 33; and
"national road to socialism,"
33, 53–54, 56; purge of, 129,
130, 131–32, 153–56, 157, 158,
159, 160, 162, 178; and
subversive activities, 82, 84,
141
Šling, Ota, 129
Smirnov, A. A., 19, 43
Smirnov, Efim, 159
Smith, Kingsbury, 61, 63

Smith, Walter B., 29, 46, 50, 51, 52

Socialism: and benefits of Cold War, 196; and Dimitrov, 54; Stalin's prediction about consolidation of, 103; and Tito, 54; Zhdanov divides world into camps of capitalism and, 31. *See also* "National roads to socialism"; *specific country*

Sokolovskii, Vasilii D., 43, 48, 49, 52, 184

"Solarium" plan, 189

Solzhenitsyn, Aleksander I., 160

South East Asia, 142

South Korea, 90, 91. *See also* Korean War

Soviet Union: collapse of, 192, 194; domestic relations in, 24, 191; economic basis to system in, 192; economy of, 75–76; and evil of Stalinism, 194; as great power, 31; and internal versus external security, 12, 14, 16; intrinsic insecurity of, 11; military as basis of system in, 192; nationalism in, 82; Nazi invasion of, 16; paradox about, 192–93; politics as basis of system in, 192; purge in (1950s), 157–58, 161, 193; reform in, 171–78, 188, 190; and saving the Soviet system, 185–90; as "second front" in World War II, 18; as second-rate power, 13; subversive activities in, 82; and U.S. views about post-Stalin government, 189; and Western unpreparedness for Stalin's death, 175

Soviet-American nonaggression pact, 61

Spaatz, Carl, 59

Spain: civil war in, 14–15, 16, 67, 72, 155, 156

Spheres of influence, 16, 20, 22, 27

Spiru, Nako, 35

Spychalski, Marian, 129

Stalin, Josef: administrative style of, 12, 192; delusions of, 53–58; deposing of, 169; and doctors, 159; as Europeanist, 12; as evil, 194; funeral of, 170; illness/death of, 166–70, 175, 193–94; illusion of, 193; indispensability of, 151; legacy of, 194; as Marxist, 12; offers to resign as general secretary, 151; paranoia of, 24, 128–32; physical decline of, 140–45, 147, 152, 163; power of, 12, 54–55, 169, 191; prestige of, 30; rise to power of, 11; and scapegoats, 13, 103; seventieth birthday celebration of, 89. *See also specific person or topic*

Stalin, Svetlana, 128, 163, 168

Stalin, Vasilii, 168

Stalingrad, battle at, 17

Stalinism: evil of, 194; nadir of, 153–58; repudiation of, 194

Stassen, Harold, 175

State Department, U.S., 61, 105, 124, 175. *See also specific person*

Stöckler, Lájos, 162

Strang, William, 115

Strategic missiles, 78

"Strategy for Freedom" speech (Acheson), 109

Strokach, Timofei, 179

Submarine fleet, Soviet, 60

Subversive activities: and British, 121; in Eastern Europe, 80–85, 97, 116–21, 131, 132, 133, 146–47, 154, 189, 195; and France, 121; by Soviets, 187; and specter of global war, 195, 196. *See also specific country*

Sudoplatov, Pavel and Antoly, 159, 163, 167, 180, 187, 188

"Summary of Our Knowledge about the Espionage Services of the U.S." (Czech foreign office), 119

Sun Tzu, 165

Supreme Soviet, 171

Suslov, Mikhail, 56, 78–79, 157–58, 162–63

Šváb, Karel, 155

Svalbard (Spitsbergen) archipelago, 19

Światło, Józef, 130, 131

Syria, 31

Szklarska Poręba, Poland: communist meeting (1947) at, 30–33

Taiwan, 91, 99, 100, 103, 104, 108, 111, 124, 164

Teheran conference (1944), 18

Thorez, Maurice, 17, 61, 112, 143

Tillon, Charles, 142, 143

Timashuk, Lidia, 158, 159, 160, 170

Tito, Josip Broz: and Albania, 35, 36; attempts to avoid break with Stalin by, 53; and Balkan confederation, 20, 21, 26, 33; Beriia proposes meeting with, 179; Cominform pronounces against, 53; declares Yugoslavia a "people's democracy," 33; and Greece, 36, 57, 69; Mao compared with, 87; mentioned, 63; and purges, 69, 70–71, 72–73; and socialism, 54; Stalin plans murder of, 167; Stalin's miscalculations about, 155, 193; Stalin's relationship with, 20, 26, 36–37, 43–44, 45–46, 53–54; 57, 62, 70–71, 193; Stalin's views about, 35; Western views of, 34–35. *See also* Yugoslavia

Titoism: and purges, 68, 70, 71, 79; U.S. support for, 72

Tiulpanov, Sergei I, 41, 52

Togliatti, Palmiro, 33, 112, 113

Trieste dispute, 122

"Trojan" plan, 74

Trotskii, Leon, 12, 167, 185

Trotskyism, 12, 13, 15, 45, 67, 68, 71, 84

Truman Doctrine (1947), 26, 27,30

Truman, Harry S.: and atomic/nuclear weapons, 77, 101, 123; and elections of 1948, 60–61; hardening of U.S. policies by, 151; and Korean War, 145; and militarization of Cold War, 58; and military power as policy instrument, 58; and "Morgenthau plan," 154; Soviet views about, 60–61, 167

Trygve Lie, 96, 122

Tsarapkin, Semen K., 97

Tumanov, Grigorii, 96, 97

Turkey, 23, 26, 174, 175

Ukraine, 82, 141, 179, 186

Ulbricht, Walter, 52, 102, 139, 177–78, 181, 182, 183, 184–85, 188, 189

United Nations (U.N.): and banning of atomic/nuclear weapons, 78; and Berlin blockade, 52–53; and China, 92, 99, 102, 111, 124; and disarmament proposals, 127; and Germany, 110; and Korean War, 98–99, 102–3, 104, 105, 110, 111, 127, 148; and NATO, 61; Soviet boycott of, 98–99, 102

United States: benefits of Cold War to, 195–96; Soviet misconceptions about, 28, 111, 125, 126; Stalin's exaggerations of strength of, 59–60; unpreparedness for Stalin's death of, 175. *See also specific person or country*

Unopened window of opportunity, 164–66, 170
U.S. Board of National Estimates, 153

Vafiades, Markos, 33, 57, 69
Vandenberg resolution (1948), 46
Varga, Evgenii, 28, 60, 75
Vasilevskii, Aleksandr M., 161
Vermeersch, Jeannette, 143
Versailles system, 13–14
Vinson, Fred, 61
Vladivostok, 123
Vlasik, N. S., 163
Voice of America, 118
Voitinskii, Georgii, 96–97
Volunteer Freedom Corps, 183
Voroshilov, Klimenti Ye., 18, 151, 167, 168, 176
Voznesenskii, Nikolai A., 62, 75–76, 150, 158, 160
Vyshinskii, Andrei Ia.: appointed foreign minister, 62; and Austrian state treaty, 65, 67; disarmament proposals of, 127; and German questions, 64, 65, 139; and Korean War, 98, 111–12, 126, 127, 148; as prosecutor in Great Purge trials, 62; replaces Molotov, 64; and U.S. as desperate, 111–12

Wallace, Henry, 60
Wang Jiaxiang, 111
War: Cold War as, 164–65; obsolescence of, 197; "permanently operating factors" in, 76, 190; Stalin's views about, 76–77; and "war psychosis" of U.S., 59. See also Specter of global war
Weimar Republic, 137
Welfare state, 197
Werth, Alexander, 35

West German Fighting Group Against Inhumanity, 117–18
West Germany: Allied foreign ministers meeting (1951) about, 134–35; Allied troops in, 107, 110; constitution for, 51; end of occupation of, 137; and European Defense Community, 137, 139, 143; General Treaty with, 137, 139, 143; and NATO, 124; plans to create state of, 46, 48, 61, 64–65, 165; rearmament of, 121; threats of uprising in, 102. See also Federal Republic of Germany
Western Europe: American ascendancy in, 63; and beginnings of Cold War, 27; benefits of Cold War to, 195, 196; Bevin's plea for unification of, 37; change in Soviet tactics in, 32–33; Cominform as basis of Soviet policy for, 32–33; economy of, 196; socialism in, 25; Soviet attempts to destabilize, 80; and specter of global war, 114; Stalin's views about, 27. See also specific country
Western military pact. See Brussels treaty; European Defense Community; North Atlantic Treaty Organization (NATO)
Wiley, Alexander, 120
Working class, 31

Xinjiang, 86, 94
Xoxe, Koçi, 35–36, 37, 39, 44, 68, 69, 70
Xu Xiangqian, 126

Yalta Conference (1945), 22, 89, 146
Yugoslavia: and Albania, 35–36, 37, 39, 44, 81; and Austria, 23, 65;

Yugoslavia (*continued*)

and beginnings of Cold War, 23, 28, 29; British armaments shipped to, 117; and British-Soviet relations, 20; and Bulgaria, 33, 38, 39, 117; and Cominform, 32, 33; and Czechoslovakia, 35, 117; declared "people's democracy," 33; economy of, 39, 72, 73; and Germany, 41; and Greek civil war, 33, 35, 39; heresy charges against leaders in, 45, 87; and Hungary, 33, 38, 69–70, 117; and Italy, 23; Mao condemns heresy in, 87; and Marshall Plan, 28, 29; and "national roads to socialism," 53; and Poland, 35; and post-Stalin government, 174; and purges, 67–68, 69–70, 72–73; and Romania, 33, 38; Soviet army withdraws from, 73; Soviet relations with, 33, 35–40, 43–44, 45–46, 71–74, 117, 174, 179; Soviets sever relations with, 45–46, 71–74; Stalin's decides to "invade," 71–72, 102; Stalin's misjudgments about, 72–74; Stalin's views about, 20, 23, 33, 35, 38, 39; and Trieste dispute, 122; U.S. relations with, 72, 73, 117, 120; during World War II, 17. *See also* Tito, Josip Broz

Žujović, Sreten, 39, 45–46
Zaisser, Wilhelm, 188, 189
Zakhariades, Nikos, 82
Zero-sum game, 13
Zhdanov, Andrei A.: and coalition-building, 142; and Cominform, 32; death of, 54, 69; and "doctors' plot," 158, 159, 161; fall of, 54; and French communists, 27; and international communism, 27, 31–32, 33; and Italian communists, 41; and Leningrad purge, 158; Malenkov's rivalry with, 54; mentioned, 56; as politician, 31; repudiation of, 158; and Yugoslav communists, 36
Zhdanov, Iurii, 158
Zhou Enlai, 90, 93, 105, 106–7, 147, 165, 172
Zhukov, Georgii A., 102
Zilliacus, Konni, 33, 53
Zionism, 56, 132, 154, 157, 162
Zorin, Valerian, 42